A RADICAL APPROACH TO LEBESGUE'S THEORY OF INTEGRATION

Meant for advanced undergraduate and graduate students in mathematics, this lively introduction to measure theory and Lebesgue integration is rooted in and motivated by the historical questions that led to its development. The author stresses the original purpose of the definitions and theorems and highlights some of the difficulties that were encountered as these ideas were refined.

The story begins with Riemann's definition of the integral, a definition created so that he could understand how broadly one could define a function and yet have it be integrable. The reader then follows the efforts of many mathematicians who wrestled with the difficulties inherent in the Riemann integral, leading to the work in the late nineteenth and early twentieth centuries of Jordan, Borel, and Lebesgue, who finally broke with Riemann's definition. Ushering in a new way of understanding integration, they opened the door to fresh and productive approaches to many of the previously intractable problems of analysis.

David M. Bressoud is the DeWitt Wallace Professor of Mathematics at Macalester College. He was a Peace Corps Volunteer in Antigua, West Indies, received his PhD from Temple University, and taught at The Pennsylvania State University before moving to Macalester. He has held visiting positions at the Institute for Advanced Study, the University of Wisconsin, the University of Minnesota, and the University of Strasbourg. He has received a Sloan Fellowship, a Fulbright Fellowship, and the MAA Distinguished Teaching Award. He has published more than 50 research articles in number theory, partition theory, combinatorics, and the theory of special functions. His other books include *Factorization and Primality Testing, Second Year Calculus from Celestial Mechanics to Special Relativity, A Radical Approach to Real Analysis*, and *Proofs and Confirmations*, for which he won the MAA Beckenbach Book Prize.

A Radical Approach to Real Analysis, 2nd edition, David M. Bressoud
Real Infinite Series, Daniel D. Bonar and Michael Khoury, Jr.
Topology Now!, Robert Messer and Philip Straffin
Understanding Our Quantitative World, Janet Andersen and Todd Swanson

MAA Service Center
P.O. Box 91112
Washington, DC 20090-1112
1-800-331-1MAA   FAX: 1-301-206-9789

A RADICAL APPROACH TO LEBESGUE'S THEORY OF INTEGRATION

DAVID M. BRESSOUD

Macalester College

CAMBRIDGE
UNIVERSITY PRESS

CAMBRIDGE UNIVERSITY PRESS
Cambridge, New York, Melbourne, Madrid, Cape Town, Singapore,
São Paulo, Delhi, Dubai, Tokyo, Mexico City

Cambridge University Press
32 Avenue of the Americas, New York, NY 10013-2473, USA

www.cambridge.org
Information on this title: www.cambridge.org/9780521711838

The Mathematical Association of America
1529 Eighteenth Street, NW, Washington, DC 20036

© David M. Bressoud 2008

First published 2008
Reprinted 2008, 2009

A catalog record for this publication is available from the British Library.

Library of Congress Cataloging in Publication Data

Bressoud, David M., 1950–
A Radical approach to Lebesgue's theory of integration / David M. Bressoud.
p.   cm. – (Classroom resource materials)
Includes bibliographical references and index.
ISBN-13: 978-0-521-88474-7 (hardback)
ISBN-10: 0-521-88474-8 (hardback)
ISBN-13: 978-0-521-71183-8 (pbk.)
ISBN-10: 0-521-71183-5 (pbk.)
1. Integrals, Generalized. I. Title. II. Series.
QA312.B67   2008
515'.42–dc22   2007035326

ISBN 978-0-521-88474-7 Hardback
ISBN 978-0-521-71183-8 Paperback

Dedicated to Herodotus,
the little lion of Cambridge Street,
and to the woman who loves him

Contents

Preface

I look at the burning question of the foundations of infinitesimal analysis without sorrow, anger, or irritation. What Weierstrass – Cantor – did was very good. That's the way it had to be done. But whether this corresponds to what is in the depths of our consciousness is a very different question. I cannot but see a stark contradiction between the intuitively clear fundamental formulas of the integral calculus and the incomparably artificial and complex work of the "justification" and their "proofs." One must be quite stupid not to see this at once, and quite careless if, after having seen this, one can get used to this artificial, logical atmosphere, and can later on forget this stark contradiction.

— *Nikolaĭ Nikolaevich Luzin*

Nikolaĭ Luzin reminds us of a truth too often forgotten in the teaching of analysis; the ideas, methods, definitions, and theorems of this study are neither natural nor intuitive. It is all too common for students to emerge from this study with little sense of how the concepts and results that constitute modern analysis hang together. Here more than anywhere else in the advanced undergraduate/beginning graduate curriculum, the historical context is critical to developing an understanding of the mathematics.

This historical context is both interesting and pedagogically informative. From transfinite numbers to the Heine–Borel theorem to Lebesgue measure, these ideas arose from practical problems but were greeted with a skepticism that betrayed confusion. Understanding what they mean and how they can be used was an uncertain process. We should expect our students to encounter difficulties at precisely those points at which the contemporaries of Weierstrass, Cantor, and Lebesgue had balked.

Throughout this text I have tried to emphasize that no one set out to invent measure theory or functional analysis. I find it both surprising and immensely satisfying that the search for understanding of Fourier series continued to be one of the principal driving forces behind the development of analysis well into the twentieth century. The tools that these mathematicians had at hand were not adequate to the task. In particular, the Riemann integral was poorly adapted to their needs.

It took several decades of wrestling with frustrating difficulties before mathematicians were willing to abandon the Riemann integral. The route to its eventual replacement, the Lebesgue integral, led through a sequence of remarkable insights into the complexities of the real number line. By the end of 1890s, it was recognized that analysis and the study of sets were inextricably linked. From this rich interplay, measure theory would emerge. With it came what today we call Lebesgue's dominated convergence theorem, the holy grail of nineteenth-century analysis. What so many had struggled so hard to discover now appeared as a gift that was almost free.

This text is an introduction to measure theory and Lebesgue integration, though anyone using it to support such a course must be forewarned that I have intentionally avoided stating results in their greatest possible generality. Almost all results are given only for the real number line. Theorems that are true over any compact set are often stated only for closed, bounded intervals. I want students to get a feel for these results, what they say, and why they are important. Close examination of the most general conditions under which conclusions will hold is something that can come later, if and when it is needed.

The title of this book was chosen to communicate two important points. First, this is a sequel to *A Radical Approach to Real Analysis* (ARATRA). That book ended with Riemann's definition of the integral. That is where this text begins. All of the topics that one might expect to find in an undergraduate analysis book that were not in ARATRA are contained here, including the topology of the real number line, fundamentals of set theory, transfinite cardinals, the Bolzano–Weierstrass theorem, and the Heine–Borel theorem. I did not include them in the first volume because I felt I could not do them justice there and because, historically, they are quite sophisticated insights that did not arise until the second half of the nineteenth century.

Second, this book owes a tremendous debt to Thomas Hawkins' *Lebesgue's Theory of Integration: Its Origins and Development*. Like ARATRA, this book is not intended to be read as a history of the development of analysis. Rather, this is a textbook informed by history, attempting to communicate the motivations, uncertainties, and difficulties surrounding the key concepts. This task would have been far more difficult without Hawkins as a guide. Those who are intrigued by the historical details encountered in this book are encouraged to turn to Hawkins and other historians of this period for fuller explanation.

Even more than ARATRA, this is the story of many contributions by many members of a large community of mathematicians working on different pieces of the puzzle. I hope that I have succeeded in opening a small window into the workings of this community. One of the most intriguing of these mathematicians is Axel

Harnack, who keeps reappearing in our story because he kept making mistakes, but they were *good* mistakes. Harnack's errors condensed and made explicit many of the misconceptions of his time, and so helped others to find the correct path. For ARATRA, it was easy to select the four mathematicians who should grace the cover: Fourier, Cauchy, Abel, and Dirichlet stand out as those who shaped the origins of modern analysis. For this book, the choice is far less clear. Certainly I need to include Riemann and Lebesgue, for they initiate and bring to conclusion the principal elements of this story. Weierstrass? He trained and inspired the generation that would grapple with Riemann's work, but his contributions are less direct. Heine, du Bois-Reymond, Jordan, Hankel, Darboux, or Dini? They all made substantial progress toward the ultimate solution, but none of them stands out sufficiently. Cantor? Certainly yes. It was his recognition that set theory lies at the heart of analysis that would enable the progress of the next generation. Who should we select from that next generation: Peano, Volterra, Borel, Baire? Maybe Riesz or one of the others who built on Lebesgue's insights, bringing them to fruition? Now the choice is even less clear. I have settled on Borel for his impact as a young mathematician and to honor him as the true source of the Heine–Borel theorem, a result that I have been very tempted to refer to as he did: the first fundamental theorem of measure theory.

I have drawn freely on the scholarship of others. I must pay special tribute to Soo Bong Chae's *Lebesgue Integration*. When I first saw this book, my reaction was that I did not need to write my own on Lebesgue integration. Here was someone who had already put the subject into historical context, writing in an elegant yet accessible style. However, as I have used his book over the years, I have found that there is much that he leaves unsaid, and I disagree with his choice to use Riesz's approach to the Lebesgue integral, building it via an analysis of step functions. Riesz found an elegant route to Lebesgue integration, but in defining the integral first and using it to define Lebesgue measure, the motivation for developing these concepts is lost. Despite such fundamental divergences, the attentive reader will discover many close parallels between Chae's treatment and mine.

I am indebted to many people who read and commented on early drafts of this book. I especially thank Dave Renfro who gave generously of his time to correct many of my historical and mathematical errors. Steve Greenfield had the temerity to be the very first reader of my very first draft, and I appreciate his many helpful suggestions on the organization and presentation of this book. I also want to single out my students who, during the spring semester of 2007, struggled through a preliminary draft of this book and helped me in many ways to correct errors and improve the presentation of this material. They are Jacob Bond, Kyle Braam, Pawan Dhir, Elizabeth Gillaspy, Dan Gusset, Sam Handler, Kassa Haileyesus, Xi Luo, Jake Norton, Stella Stamenova, and Linh To.

I am also grateful to the mathematicians and historians of mathematics who suggested corrections and changes or helped me find information. These include Roger Cooke, Larry D'Antonio, Ivor Grattan-Guinness, Daesuk Han, Tom Hawkins, Mark Huibregtse, Nicholas Rose, Peter Ross, Jim Smoak, John Stillwell, and Sergio B. Volchan. I am also indebted to Don Albers of the MAA and Lauren Cowles of Cambridge who so enthusiastically embraced this project, and to the reviewers for both MAA and Cambridge whose names are unknown to me but who gave much good advice.

Corrections, commentary, and additional material for this book can be found at www.macalester.edu/aratra.

David M. Bressoud
bressoud@macalester.edu
June 19, 2007

1

Introduction

By 1850, most mathematicians thought they understood calculus. Real progress was being made in extending the tools of calculus to complex numbers and spaces of higher dimensions. Equipped with appropriate generalizations of Fourier series, solutions to partial differential equations were being found. Cauchy's insights had been assimilated, and the concepts that had been unclear during his pioneering work of the 1820s, concepts such as uniform convergence and uniform continuity, were coming to be understood. There was reason to feel confident.

One of the small, nagging problems that remained was the question of the convergence of the Fourier series expansion. When does it converge? When it does, can we be certain that it converges to the original function from which the Fourier coefficients were derived? In 1829, Peter Gustav Lejeune Dirichlet had proven that as long as a function is piecewise monotonic on a closed and bounded interval, the Fourier series converges to the original function. Dirichlet believed that functions did not have to be piecewise monotonic in order for the Fourier series to converge to the original function, but neither he nor anyone else had been able to weaken this assumption.

In the early 1850s, Bernard Riemann, a young protegé of Dirichlet and a student of Gauss, would make substantial progress in extending our understanding of trigonometric series. In so doing, the certainties of calculus would come into question. Over the next 60 years, five big questions would emerge and be answered. The answers would be totally unexpected. They would forever change the nature of analysis.

1. **When does a function have a Fourier series expansion that converges to that function?**
2. **What is integration?**
3. **What is the relationship between integration and differentiation?**

1

4.   What is the relationship between continuity and differentiability?
5.   When can an infinite series be integrated by integrating each term?

   This book is devoted to explaining the answers to these five questions – answers
that are very much intertwined. Before we tackle what happened after 1850, we
need to understand what was known or believed in that year.

1.1 The Five Big Questions

Fourier Series

Fourier's method for expanding an arbitrary function F defined on $[-\pi, \pi]$ into a
trigonometric series is to use integration to calculate coefficients:

$$a_k = \frac{1}{\pi} \int_{-\pi}^{\pi} F(x) \cos(kx) \, dx \quad (k \geq 0), \tag{1.1}$$

$$b_k = \frac{1}{\pi} \int_{-\pi}^{\pi} F(x) \sin(kx) \, dx \quad (k \geq 1). \tag{1.2}$$

The Fourier expansion is then given by

$$F(x) = \frac{a_0}{2} + \sum_{k=1}^{\infty} \left[a_k \cos(kx) + b_k \sin(kx) \right]. \tag{1.3}$$

   The heuristic argument for the validity of this procedure is that if F really can
be expanded in a series of the form given in Equation (1.3), then

$$\int_{-\pi}^{\pi} F(x) \cos(nx) \, dx$$

$$= \int_{-\pi}^{\pi} \left(\frac{a_0}{2} + \sum_{k=1}^{\infty} \left[a_k \cos(kx) + b_k \sin(kx) \right] \right) \cos(nx) \, dx$$

$$= \int_{-\pi}^{\pi} \frac{a_0}{2} \cos(nx) \, dx + \sum_{k=1}^{\infty} \int_{-\pi}^{\pi} a_k \cos(kx) \cos(nx) \, dx$$

$$+ \sum_{k=1}^{\infty} \int_{-\pi}^{\pi} b_k \sin(kx) \cos(nx) \, dx. \tag{1.4}$$

Since n and k are integers, all of the integrals are zero except for the one involving
a_n. These integrals are easily evaluated:

$$\int_{-\pi}^{\pi} F(x) \cos(nx) \, dx = \pi a_n. \tag{1.5}$$

Similarly,

$$\int_{-\pi}^{\pi} F(x) \sin(nx)\, dx = \pi b_n. \tag{1.6}$$

This is a convincing heuristic, but it ignores the problem of interchanging integration and summation, and it sidesteps two crucial questions:

1. Are the integrals that produce the Fourier coefficients well-defined?
2. If these integrals can be evaluated, does the resulting Fourier series actually converge to the original function?

Not all functions are integrable. In the 1820s, Dirichlet proposed the following example.

Example 1.1. The **characteristic function of the rationals** is defined as

$$f(x) = \begin{cases} 1, & x \text{ is rational}, \\ 0, & x \text{ is not rational}. \end{cases}$$

This example demonstrates how very strange functions can be if we take seriously the definition of a function as a well-defined rule that assigns a value to each number in the domain. Dirichlet's example represents an important step in the evolution of the concept of function. To the early explorers of calculus, a function was an algebraic rule such as $\sin x$ or $x^2 - 3$, an expression that could be computed to whatever accuracy one might desire.

When Augustin-Louis Cauchy showed that any piecewise continuous function is integrable, he cemented the realization that functions could also be purely geometric, representable only as curves. Even in a situation in which a function has no explicit algebraic formulation, it is possible to make sense of its integral, provided the function is continuous.

Dirichlet stretched the concept of function to that of a rule that can be individually defined for each value of the domain. Once this conception of function is accepted, the gates are opened to very strange functions. At the very least, integrability can no longer be assumed.

The next problem is to show that our trigonometric series converges. In his 1829 paper, Dirichlet accomplished this, but he needed the hypothesis that the original function F is piecewise monotonic, that is the domain can be partitioned into a finite number of subintervals so that F is either monotonically increasing or monotonically decreasing on each subinterval.

The final question is whether the function to which it converges is the function F with which we started. Under the same assumptions, Dirichlet was able to show that this is the case, provided that at any points of discontinuity of F, the value taken by the function is the average of the limit from the left and the limit from the right.

Dirichlet's result implies that the functions one is likely to encounter in physical situations present no problems for conversion into Fourier series. Riemann recognized that it was important to be able to extend this technique to more complicated functions now arising in questions in number theory and geometry. The first step was to get a better handle on what we mean by integration.

Integration

It is ironic that integration took so long to get right because it is so much older than any other piece of calculus. Its roots lie in methods of calculating areas, volumes, and moments that were undertaken by such scientists as Archimedes (287–212 BC), Liu Hui (late third century AD), ibn al-Haytham (965–1039), and Johannes Kepler (1571–1630). The basic idea was always the same. To evaluate an area, one divided it into rectangles or triangles or other shapes of known area that together approximated the desired region. As more and smaller figures were used, the region would be matched more precisely. Some sort of limiting argument would then be invoked, some means of finding the actual area based on an analysis of the areas of the approximating regions.

Into the eighteenth century, integration was identified with the problem of "quadrature," literally the process of finding a square equal in area to a given area and thus, in practice, the problem of computing areas. In section 1 of Book I of his *Mathematical Principles of Natural Philosophy*, Newton explains how to calculate areas under curves. He gives a procedure that looks very much like the definition of the Riemann integral, and he justifies it by an argument that would be appropriate for any modern textbook.

Specifically, Newton begins by approximating the area under a decreasing curve by subdividing the domain into equal subintervals (see Figure 1.1). Above each subinterval, he constructs two rectangles: one whose height is the maximum value of the function on that interval (the circumscribed rectangle) and the other whose height is the minimum value of the function (the inscribed rectangle). The true area lies between the sum of the areas of the circumscribed rectangles and the sum of the areas of the inscribed rectangles.

The difference between these areas is the sum of the areas of the rectangles $aKbl$, $bLcm$, $cMdn$, $dDEo$. If we slide all of these rectangles to line up under $aKbl$, we see that the sum of their areas is just the change in height of the function multiplied by the length of any one subinterval. As we take narrower subintervals, the difference in the areas approaches zero. As Newton asserts: "The ultimate ratios which the inscribed figure, the circumscribed figure, and the curvilinear figure have

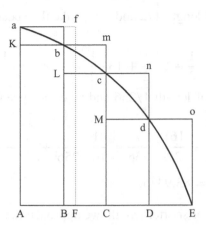

Figure 1.1. Newton's illustration from *Mathematical Principles of Natural Philosophy*. (Newton, 1999, p. 433)

to one another are ratios of equality," which is his way of saying that the ratio of any two of these areas approaches 1. Therefore, the areas are all approaching the same value as the length of the subinterval approaches 0.

In Lemma 3 of his book, Newton considers the case where the subintervals are not of equal length (using the dotted line fF in Figure 1.1 in place of lB). He observes that the sum of the differences of the areas is still less than the change in height multiplied by the length of the longest subinterval. We therefore get the same limit for the ratio so long as the length of the longest subinterval is approaching zero.

This method of finding areas is paradigmatic for an entire class of problems in which one is multiplying two quantities such as

- area = height × width,
- volume = cross-sectional area × width,
- moment = mass × distance,
- work = force × distance,
- distance = speed × time, or
- velocity = acceleration × time,

where the value of the first quantity can vary as the second quantity increases. For example, knowing that "distance = speed × time," we can find the distance traveled by a particle whose speed is a function of time, say $v(t) = 8t + 5, 0 \leq t \leq 4$. If we split the time into four intervals and use the velocity at the start of each interval, we get an approximation to the total distance:

$$\text{distance} \approx 5 \cdot 1 + 13 \cdot 1 + 21 \cdot 1 + 29 \cdot 1 = 68.$$

If we use eight intervals of length 1/2 and again take the speed at the start of each interval, we get

$$\text{distance} \approx 5 \cdot \frac{1}{2} + 9 \cdot \frac{1}{2} + 13 \cdot \frac{1}{2} + \cdots + 33 \cdot \frac{1}{2} = 76.$$

If we use 1,024 intervals of length 1/256 and take the speed at the start of each interval, we get

$$\text{distance} \approx 5 \cdot \frac{1}{256} + \frac{161}{32} \cdot \frac{1}{256} + \frac{81}{16} \cdot \frac{1}{256} + \cdots + \frac{1,183}{32} \cdot \frac{1}{256}$$

$$= \frac{1,343}{16} = 83.9375.$$

As we take more intervals of shorter length, we approach the true distance, which is 84. How do you actually *get* 84? We can think of this as taking infinitely many intervals of infinitely short length.

Leibniz's notation is a brilliant encapsulation of this process:

$$\int f(x)\, dx.$$

The product is $f(x)\, dx$, the value of the first quantity times the infinitesimal increment. The elongated S, $\int$, represents the summation.

This is all precalculus. The insight at the heart of calculus is that if $f(x)$ represents the slope of the tangent to the graph of a function F at x, then this provides an easy method for computing limits of sums of products: If x ranges over the interval $[a, b]$, then the value of this integral is $F(b) - F(a)$. Thus, to find the area under the curve $v = 8t + 5$ from $t = 0$ to $t = 4$, we can observe that $f(t) = 8t + 5$ is the derivative of $F(t) = 4t^2 + 5t$. The desired area is equal to

$$(4 \cdot 4^2 + 5 \cdot 4) - (4 \cdot 0^2 + 5 \cdot 0) = 84.$$

The calculating power of calculus comes from this dual nature of the integral. It can be viewed as a limit of sums of products or as the inverse process of differentiation.

It is hard to find a precise definition of the integral from the eighteenth century. The scientists of this century understood and exploited the dual nature of the integral, but most were reluctant to define it as the sum of products of $f(x)$ times the infinitesimal dx, for that inevitably led to the problem of what exactly is meant by an "infinitesimal." It is a useful concept, but one that is hard to pin down. George Berkeley aptly described infinitesimals as "ghosts of departed quantities." He would object, "Now to conceive a quantity infinitely small, that is, infinitely less than any sensible or imaginable quantity or than any the least finite magnitude is, I confess, above my capacity."[1]

---

[1] George Berkeley, *The Analyst*, as quoted in Struik (1986, pp. 335, 338).

The result was that when a definition of $\int f(x)\,dx$ was needed, the integral was simply defined as the operator that returns you to the function (or, in modern use, the class of functions) whose derivative is f. One of the early calculus textbooks written for an undergraduate audience was S. F. Lacroix's *Traité élémentaire de Calcul Différentiel et de Calcul Intégral* of 1802 (*Elementary Treatise of Differential Calculus and Integral Calculus*). Translated into many languages, it would serve as the standard text of the first half of the nineteenth century. It provides no explicit definition of the integral, but does state that

> Integral calculus is the inverse of differential calculus. Its goal is to restore the functions from their differential coefficients.

After this clarification of what is meant by integration, Lacroix then proceeds to deal with the definite integral which "is found by successively calculating the value of the integral when $x = a$, then when $x = b$, and subtracting the first result from the second."

This would continue to be the standard definition of integration in calculus texts until the 1950s and 1960s. There is no loss in the power of calculus. The many textbook writers who took this approach then went on to explain how the definite integral can be used to evaluate limits of sums of products. Pedagogically, this approach has merit. It starts with the more intuitively accessible definition. Mathematically, this definition of integration is totally inadequate.

Cauchy and Riemann Integrals

Fourier and Cauchy were among the first to fully realize the inadequacy of defining integration as the inverse process of differentiation. It is too restrictive. Fourier wanted to apply his methods to arbitrary functions. Not all functions have antiderivatives that can be expressed in terms of standard functions. Fourier tried defining the definite integral of a nonnegative function as the area between the graph of the function and the x-axis, but that begs the question of what we mean by area. Cauchy embraced Leibniz's understanding as a limit of products, and he found a way to avoid infinitesimals.

To define $\int_a^b f(x)\,dx$, Cauchy worked with finite approximating sums. Given a partition of $[a, b]$: $(a = x_0 < x_1 < \cdots < x_n = b)$, we consider

$$\sum_{k=1}^{n} f(x_{k-1})(x_k - x_{k-1}).$$

If we can force all of these approximating sums to be as close to each as other as we wish simply by limiting the size of the difference between consecutive values in the partition, then these summations have a limiting value that is designated as the value of the definite integral, and the function f is said to be integrable over $[a, b]$.

Equipped with this definition, Cauchy succeeded in proving that *any* continuous or piecewise continuous function is integrable. The class of functions to which Fourier's analysis could be applied was suddenly greatly expanded.

When Riemann turned to the study of trigonometric series, he wanted to know the limits of Cauchy's approach to integration. Was there an easy test that could be used to determine whether or not a function could be integrated? Cauchy had chosen to evaluate the function at the left-hand endpoint of the interval simply for convenience. As Riemann thought about how far this definition could be pushed, he realized that his analysis would be simpler if the definition were stated in a slightly more complicated but essentially equivalent manner. Given a partition of $[a, b]$: $(a = x_0 < x_1 < \cdots < x_n = b)$, we assign a **tag** to each interval, a number x_j^* contained in that interval, and consider all sums of the form

$$\sum_{k=1}^{n} f(x_k^*)(x_k - x_{k-1}).$$

A partition together with such a collection of tags, $x_j^* \in [x_{j-1}, x_j]$, is called a **tagged partition**. If we can force all of these approximating sums to be as close to each other as we wish simply by limiting the size of the difference between consecutive values in the partition, then these summations have a limiting value. We call this limiting value the definite integral, and the function f is said to be integrable over $[a, b]$. In the next chapter, we shall see why this seemingly more complicated definition of the integral simplifies the process of determining when a function is integrable.

Riemann succeeded in clarifying what is meant by integration. In the process, he was able to clearly identify and delimit the set of functions that are integrable and to make it possible for others to realize that this limit definition introduces serious difficulties, difficulties that eventually would lead to the rejection of Riemann's definition in favor of a radically different approach to integration proposed by Henri Lebesgue. In particular, Riemann's definition greatly complicates the relationship between integration and differentiation.

The Fundamental Theorem of Calculus

The fundamental theorem of calculus is, in essence, simply a statement of the equivalence of the two means of understanding integration, as the inverse process of differentiation and as a limit of sums of products. The precise theorems to which this designation refers today arise from the assumption that integration is defined as a limiting process. They then clarify the precise relationship between integration and differentiation. The actual statements that we shall use are given by the following theorems.

Theorem 1.1 (FTC, evaluation). *If f is the derivative of F at every point on $[a, b]$, then under suitable hypotheses we have that*

$$\int_a^b f(t)\, dt = F(b) - F(a). \tag{1.7}$$

Theorem 1.2 (FTC, antiderivative). *If f is integrable on the interval $[a, b]$, then under suitable hypotheses we have that*

$$\frac{d}{dx} \int_a^x f(t)\, dt = f(x). \tag{1.8}$$

The first of these theorems tells us how we can use any antiderivative to obtain a simple evaluation of a definite integral. The second shows that the definite integral can be used to create an antiderivative, the definite integral of f from a to x is a function of x whose derivative is f. Both of these statements would be meaningless if we had defined the integral as the antiderivative. Their meaning and importance comes from the assumption that $\int_a^b f(t)\, dt$ is defined as a limit of summations.

In both cases, I have not specified the hypotheses under which these theorems hold. There are two reasons for this. One is that much of the interesting story that is to be told about the creation of analysis in the late nineteenth century revolves around finding necessary and sufficient conditions under which the conclusions hold. When working with Riemann's definition of the integal, the answer is complicated. The second reason is that the hypotheses that are needed depend on the way we choose to define the integral. For Lebesgue's definition, the hypotheses are quite different.

A Brief History of Theorems 1.1 and 1.2[2]

The earliest reference to Theorem 1.1 of which I am aware is Siméon Denis Poisson's 1820 *Suite du Mémoire sur les Intégrales Définies*. There he refers to it as "the fundamental proposition of the theory of definite integrals." Poisson's work is worth some digression because it illustrates the importance of how we define the definite integral and the difficulties encountered when it is defined as the difference of the values of an antiderivative at the endpoints.

---

[2] With thanks to Larry D'Antonio and Ivor Grattan-Guinness for uncovering many of these references.

Siméon Denis Poisson (1781–1840) studied and then taught at the École Polytechnique. He succeeded to Fourier's professorship in mathematics when Fourier
departed for Grenoble to become prefect of the department of Isère. It was Poisson
who wrote up the rejection of Fourier's *Theory of the Propoagation of Heat in
Solid Bodies* in 1808. When, in 1815, Poisson published his own article on the flow
of heat, Fourier pointed out its many flaws and the extent to which Poisson had
rediscovered Fourier's own work.

Poisson, as a colleague of Cauchy at the École Polytechnique, almost certainly
was aware of Cauchy's definition of the definite integral even though Cauchy had
not yet published it. But the relationship between Poisson and Cauchy was far from
amicable, and it would have been surprising had Poisson chosen to embrace his
colleague's approach. Poisson defines the definite integral as the difference of the
values of the antiderivative. It would seem there is nothing to prove. What Poisson
does prove is that if F has a Taylor series expansion and $F' = f$, then

$$F(b) - F(a) = \lim_{n \to \infty} \sum_{j=1}^{n} t\, f\left(a + (j-1)t\right), \quad \text{where } t = \frac{b-a}{n}.$$

Poisson begins with the observation that for $1 \le j \le n$ and $t = (b-a)/n$, there
is a $k \ge 1$ and a collection of functions R_j such that

$$F(a + jt) = F(a + (j-1)t) + tf(a + (j-1)t) + t^{1+k} R_j(t),$$

and therefore

$$F(b) - F(a) = \sum_{j=1}^{n} \left[F(a + jt) - F(a + (j-1)t) \right]$$

$$= \sum_{j=1}^{n} tf(a + (j-1)t) + t^{1+k} \sum_{j=1}^{n} R_j(t).$$

Poisson now asserts that the functions $R_j(t)$ stay bounded. In fact, we know
by the Lagrange remainder theorem that we can take $k = 1$ and these functions
are bounded by the supremum of $|f'(x)|/2$ over all x in $[a, b]$. It follows that
$t^{1+k} \sum_{j=1}^{n} R_j(t)$ approaches 0 as n approaches infinity.

The confusion over the meaning of the definite integral is revealed in Poisson's
attempt to complete the proof by connecting this limit back to the definite integral.
He appeals to the Leibniz conception of the integral as a sum of products:

> Using the language of infinitesimals, we shall say what we needed to show, that $F(b) - F(a)$
> is the sum of the values of $f(x)\,dx$ as x increases by infinitesimal amounts from $x = a$ to
> $x = b$, dx being the difference between two consecutive values of this variable.[3]

---

[3] Poisson (1820, pp. 323–324).

The statement and proof of Theorem 1.2 can be found in Cauchy's *Résumé des Leçons Données a L'École Polytechnique* of 1823, the same place where he first defines the definite integral. It is not stated as a fundamental theorem. In fact, it is not identified as a theorem or proposition, simply a result mentioned in the text en route to the real problem which is to define the indefinite integral, the general class of functions that have f as their derivative.

The term "Fundamental Principles of the Integral Calculus" appears in Lardner's *An Elementary Treatise on the Differential and Integral Calculus* of 1825, and these include the statement of the evaluation part of the fundamental theorem of calculus. But this statement is one of nine principles that include the fact that the integral is a linear operator as well as many rules for integrating specific functions.

The term "fundamental theorem for integrals" was used to refer to the evaluation part of the fundamental theorem of calculus in Charles de Freycinet's *De L'Analyse Infinitésimal: Étude sur la Métaphysique du haut Calcul* of 1860. de Freycinet (1828–1923) was trained as a mining engineer, was elected to the French senate in 1876, and served four times as prime minister of France. It would be interesting to know if there have been any other heads of state that have written calculus textbooks.

The full modern statement of both parts of the fundamental theorem of calculus with the definite integral defined as a limit in Cauchy's sense, referred to as the "fundamental theorem of integral calculus," can be found in an appendix to an article on trigonometric series published by Paul du Bois-Reymond in 1876. In 1880, he published an extended discussion and proof of this theorem in the widely read journal *Mathematische Annalen*.

The fundamental theorem of integral calculus was popularized in English in the early twentieth century by the publication of Hobson's *The Theory of Functions of a Real Variable and the Theory of Fourier's Series* of 1907. This is a thorough treatment of analysis that was very influential. Hobson gives statements of the fundamental theorem for both the Riemann and Lebesgue integrals. Some evidence that this may be the source of this phrase in English is given by the classic English-language calculus textbook of the first half of the twentieth century, Granville's *Elements of the Differential and Integral Calculus*. Granville does not mention a "fundamental theorem" in his first edition of 1904, but in the second edition of 1911, we do find it. Since Granville defines integration to be the reversal of differentation, his fundamental theorem is that the definite integral is equal to the limit of the approximating summations.

It seems that G. H. Hardy may be responsible for dropping the adjective "integral." In the first edition (1908) of G. H. Hardy's *A Course of Pure Mathematics*, there is no mention of the phrase "fundamental theorem of calculus." It does appear, without the adjective "integral," in the second edition, published in 1914.

Although the term "fundamental theorem of calculus" gained popularity as the twentieth century progressed, it took a while before there was an agreed meaning. Richard Courant's *Differential and Integral Calculus* of 1934 has a section entitled "The Fundamental Theorems of the Differential and Integral Calculus" in which he states Theorems 1.1 and 1.2 as well as several other related results:

- Different indefinite integrals of the same function differ only by an additive constant.
- The integral of a continuous function f is itself a continuous function of the upper limit.
- The difference of two primitives (antiderivatives) of the same function is always a constant.
- Every primitive F of a given function f can be represented in the form $F(x) = c + \int_a^x f(u)\,du$, where c and a are constants.

In the case of Theorem 1.2, Courant's hypothesis is that f is continuous. Theorem 1.1 is stated as being true of any function f with antiderivative F. In fact, this is not quite true. As we shall see, we either need to put some restrictions on f in Theorem 1.1 or abandon the Riemann integral for one that is better-behaved.

Continuity and Differentiability

The fourth big question asks for the relationship between continuity and differentiability. We know that a function that is differentiable at a given value of x must also be continuous at that value, and it is clear that the converse does not hold. The function $f(x) = |x|$ is continuous but not differentiable at $x = 0$. But how nondifferentiable can a continuous function be?

Throughout the first half of the nineteenth century, it was generally believed that a continuous function would be differentiable at most points.[4] Mathematicians recognized that a function might have finitely many values at which it failed to have a derivative. There might even be a sparse infinite set of points at which a continuous function was not differentiable, but the mathematical community was honestly surprised when, in 1875, Gaston Darboux and Paul du-Bois Reymond[5] published examples of continuous functions that are not differentiable at any value.

The question then shifted to what additional assumptions beyond continuity would ensure differentiability. Monotonicity was a natural candidate. Weierstrass constructed a strictly increasing continuous function that is not differentiable at any algebraic number, that is to say, at any number that is the root of a polynomial

---

[4] Although Bernhard Bolzano had shown how to construct a function that is everywhere continuous and nowhere differentiable, his example only existed in a privately circulated manuscript and was not published until 1930.

[5] du Bois-Reymond's example was found by Weierstrass's who had described it in his lectures but never published it.

with rational coefficients. It is not differentiable at $1/2$ or $\sqrt{2}$ or $\sqrt[3]{5} - 2\sqrt[21]{35}$. Weierstrass's function *is* differentiable at π. Can we find a continuous, increasing function that is not differentiable at any value? The surprising answer is *No*. In fact, in a sense that later will be made precise, a continuous, monotonic function is differentiable at "most" values of x. There are very important subtleties lurking behind this fourth question.

Term-by-term Integration

Returning to Fourier series, we saw that the heuristic justification relied on interchanging summation and integration, integrating an infinite series of functions by integrating each summand. This works for finite summations. It is not hard to find infinite series for which term-by-term integration leads to a divergent series or, even worse, a series that converges to the wrong value.

Weierstrass had shown that if the series converges uniformly, then term-by-term integration is valid. The problem with this result is that the most interesting series, especially Fourier series, often do not converge uniformly and yet term-by-term integration is valid. Uniform convergence is sufficient, but it is very far from necessary. As we shall see, finding useful conditions under which term-by-term integration is valid is very difficult so long as we cling to the Riemann integral. As Lebesgue would show in the opening years of the twentieth century, his definition of the integral yields a simple, elegant solution, the Lebesgue dominated convergence theorem.

Exercises

1.1.1. Find the Fourier expansions for $f_1(x) = x$ and $f_2(x) = x^2$ over $[-\pi, \pi]$.

1.1.2. For the functions f_1 and f_2 defined in Exercise 1.1.1, differentiate each summand in the Fourier series for f_2. Do you get the summands in the Fourier series for $2f_1$? Differentiate each summand in the Fourier series for f_1. Do you get the summand in the Fourier series for $f_1'(x)$?

1.1.3. Using the Fourier series expansion for x^2 (Exercise 1.1.1) evaluated at $x = \pi$, show that

$$\sum_{n=1}^{\infty} \frac{1}{n^2} = \frac{\pi^2}{6}.$$

1.1.4. Show that if k is an integer ≥ 1, then

$$\int_{-\pi}^{\pi} \cos(kx)\, dx = \int_{-\pi}^{\pi} \sin(kx)\, dx = 0.$$

Show that if n and k are positive integers, then

$$\int_{-\pi}^{\pi} \sin(kx) \, \cos(nx) \, dx = 0.$$

Show that if n and k are distinct positive integers, then

$$\int_{-\pi}^{\pi} \cos(kx) \, \cos(nx) \, dx = \int_{-\pi}^{\pi} \sin(kx) \, \sin(nx) \, dx = 0.$$

1.1.5. Using the definition of continuity, justify the assertion that the characteristic function of the rationals, Example 1.1, is not continuous at any real number.

1.1.6. Let C be the circumscribed area, I the inscribed area in Newton's illustration, using intervals of length Δx. Newton claims that he has demonstrated that

$$\lim_{\Delta x \to 0} \frac{C}{I} = 1,$$

but what he actually proves is that

$$\lim_{\Delta x \to 0} \left(C - I \right) = 0.$$

Show that since I is monotonically increasing as Δx approaches 0, and C is monotonically decreasing, these two statements are equivalent.

1.1.7. The population of a certain city can be modeled using a population density function, $\rho(x)$, measured in people per square mile, where x is the distance from the center of the city. The density function is valid in all directions for $0 \le x \le 5$ miles. Set up a sum of products that approximates the total population and then convert this sum of products into an integral.

1.1.8. Mass distributed along one side of a balance beam is modeled by a function $m(x)$, where x is the distance from the fulcrum, $0 \le x \le 6$ meters. Set up a sum of products that approximates the total moment resulting from this mass and then convert this sum of products into an integral.

1.1.9. Show that if a function is not bounded on $[a, b]$, then the Riemannn integral on $[a, b]$ cannot exist.

1.1.10. Consider the function $f(x) = 1/\sqrt{|x|}$, $-1 \le x < 0$, $f(0) = 0$. Since this function is not bounded on $[-1, 0]$, the Riemann integral does not exist (see Exercise 1.1.9). Show that, nevertheless, the Cauchy integral of this function over this interval *does* exist.

1.1.11. Explain why it is that if a function is Riemann integrable over $[a, b]$, then it must be Cauchy integrable over that interval.

1.1.12. There are many functions for which there is no simple, closed expression for an antiderivative. The function $\sin(t^2)$ is one such example. Nevertheless, the definite integral of this function can be evaluated to whatever precision is desired, using the definition of the integral as a limit of sums of products. A certain object travels along a straight line with velocity $v(t) = \sin(t^2)$, starting at $x = 3$ at time $t = 0$. Explain how to use the fundamental theorem of calculus (either form) and a definite integral to find the position at time $t = 2$, accurate to six digits.

1.1.13. Work through Poisson's proof of Theorem 1.1 in the specific case $F(x) = \ln(x)$, $f(x) = F'(x) = 1/x$, $a = 1$, $b = 2$. Specifically: What is the value of k? Use the Lagrange remainder theorem to find a bound on $R_j(t)$ that is valid for all n and j. Show that $t^{1+k} \sum_{j=1}^{n} R_j(t)$ approaches 0 as n approaches infinity.

1.1.14. Explain how to use the Lagrange remainder theorem to justify Poisson's assertion that if all derivatives of F exist at every point in $[a, b]$, then

$$\lim_{n \to \infty} t^{1+k} \sum_{j=1}^{n} R_j(t) = 0.$$

1.1.15. Define

$$g(x) = \begin{cases} x, & x \text{ is rational,} \\ 0, & x \text{ is not rational.} \end{cases}$$

For what values of x is g continuous? For what values of x is g differentiable?

1.1.16. Define

$$h(x) = \begin{cases} x^2, & x \text{ is rational,} \\ 0, & x \text{ is not rational.} \end{cases}$$

For what values of x is h continuous? For what values of x is h differentiable?

1.1.17. Prove that if a function is not continuous at $x = a$ then it cannot be differentiable at $x = a$.

1.1.18. Show that

$$\int_0^1 \left(\lim_{n \to \infty} nxe^{-nx^2} \right) dx \neq \lim_{n \to \infty} \int_0^1 \left(nxe^{-nx^2} \right) dx.$$

1.2 Presumptions

In this book, we presume that the reader is familiar with certain notations, definitions, and theorems. The most important of these are summarized here.

Notation

$\{x \in [a, b] \mid f(x) > 0\}$, set notation, to the left of $\mid$ is the description of the general set in which this particular set sits, to the right is the condition or conditions satisfied by elements of this set. Braces are also used to list the elements of the set; thus, $\{1, 2, \ldots, 10\}$ is the set of positive integers from 1 to 10.

$(1/n)_{n=1}^{\infty}$, sequence notation in which the order is important; this sequence could also be written as $(1, 1/2, 1/3, \ldots)$. When it is clear that we are working with a sequence, this may be written without specifying the limits on n: $(a_1, a_2, \ldots) = (a_n)$.

$\mathbb{N}$, the set of positive integers, $\{1, 2, 3, \ldots\}$.

$\mathbb{Q}$, the set of rational numbers.

$\mathbb{R}$, the set of real numbers.

$\mathbb{C}$, the set of complex numbers.

$f_n \to f$, the sequence of functions $(f_n)_{n=1}^{\infty}$ converges (pointwise) to f.

$S \cap T$, the intersection of sets S and T; the set of elements in both S and T.

$S \cup T$, the union of sets S and T; the set of elements in either S or T.

S^C, the complement of S; for the purposes of this book, the complement is always taken in $\mathbb{R}$; S^C is the set of real numbers that are not elements of S.

$S - T$, the set of elements of S that are not in T; $S - T = S \cap T^C$.

$\emptyset$, the empty set; the set that has no elements.

$f(S)$, the image of S; $f(S) = \{f(x) \mid x \in S\}$.

$\lfloor \alpha \rfloor$ denotes the **floor** of α, the greatest integer less than or equal to α; similarly, $\lceil \alpha \rceil$ denotes the **ceiling** of α, the least integer greater than or equal to α.

Definitions

continuity: The function f is continuous at c if for every $\epsilon > 0$ there is a response $\delta > 0$ such that $|x - c| < \delta$ implies that $|f(x) - f(c)| < \epsilon$.

uniform continuity: The function f is uniformly continuous over the set S if for every $\epsilon > 0$ there is a response $\delta > 0$ such that for every $c \in S$, $|x - c| < \delta$ implies that $|f(x) - f(c)| < \epsilon$.

intermediate value property: A function f has the intermediate value property on the interval $[a, b]$ if given any two points $x_1, x_2 \in [a, b]$ and any number N satisfying $f(x_1) < N < f(x_2)$, there is at least one value c between x_1 and x_2 such that $f(c) = N$.

monotonic sequence: a sequence that is either **increasing** (each element is greater than or equal to the previous element) or **decreasing** (each element is less than or equal to the previous element).

monotonic function: either an **increasing function** ($x < y \implies f(x) \leq f(y)$) or a **decreasing function** ($x < y \implies f(x) \geq f(y)$). A function is **piecewise monotonic** on $[a, b]$ if we can partition this interval into finitely many subintervals so that the function is monotonic on each subinterval.

convergence: The sequence $(a_n)_{n=1}^\infty$ converges to A if for every $\epsilon > 0$ there is a response N such that $n \geq N$ implies that $|a_n - A| < \epsilon$. The series $\sum_{k=1}^\infty c_k$ converges to S if the sequence of partial sums, (S_n), $S_n = \sum_{k=1}^n c_k$, converges to S.

pointwise convergence: The sequence of functions (f_n) converges pointwise to F if at each value of x, the sequence $(f_n(x))_{n=1}^\infty$ converges to $F(x)$. Note that the δ response may depend on both ϵ and x.

uniform convergence: The sequence of functions $(f_n)_{n=1}^\infty$ converges uniformly to F over the set S if for every $\epsilon > 0$ there is a response N such that for every $x \in S$ and every $n \geq N$, we have that $\left| f_n(x) - F(x) \right| < \epsilon$. Note that the δ response may depend on ϵ but not on x.

max S: the greatest element in S; **min** S: the least element in S.

least upper bound or **sup** S: the least value that is greater than or equal to every element of S; **greatest lower bound** or **inf** S: the greatest value that is less than or equal to every element of S. We also write

$$\sup_{x \in S} f(x) = \sup\{f(x) \mid x \in S\}, \quad \inf_{x \in S} f(x) = \inf\{f(x) \mid x \in S\}.$$

lim sup ($\overline{\lim}$), **lim inf** ($\underline{\lim}$): For a sequence $(a_n)_{n=1}^\infty$,

$$\overline{\lim_{n \to \infty}} \, a_n = \inf_{n \geq 1} \left(\sup_{k \geq n} a_k \right), \quad \underline{\lim_{n \to \infty}} \, a_n = \sup_{n \geq 1} \left(\inf_{k \geq n} a_k \right).$$

For a function f,

$$\overline{\lim_{x \to c}} \, f(x) = \inf_{\epsilon > 0} \left(\sup \left\{ f(x) \,\middle|\, 0 < |x - c| < \epsilon \right\} \right),$$

$$\underline{\lim_{x \to c}} \, f(x) = \sup_{\epsilon > 0} \left(\inf \left\{ f(x) \,\middle|\, 0 < |x - c| < \epsilon \right\} \right).$$

Cauchy sequence: The sequence (a_n) is Cauchy if for each $\epsilon > 0$ there is a response N such that for every $m, n \geq N$ we have that $|a_m - a_n| < \epsilon$.

nested interval principle: Given any nested sequence of closed intervals in $\mathbb{R}$,

$$[a_1, b_1] \supseteq [a_2, b_2] \supseteq [a_3, b_3] \supseteq \cdots,$$

there is at least one real number contained in all of these intervals,

$$\bigcap_{n=1}^{\infty} [a_n, b_n] \neq \emptyset.$$

vector space: A vector space is a set that is closed under addition, closed under multiplication by scalars from a field such as $\mathbb{R}$, and that satisfies the following conditions where $X, Y, Z, 0$ denote vectors and $a, b, 1$ denote scalars:

1. commutativity: $X + Y = Y + X$,
2. associativity of vectors: $(X + Y) + Z = X + (Y + Z)$,
3. additive identity: $0 + X = X + 0 = X$,
4. additive inverse: $X + (-X) = 0$,
5. associativity of scalars: $a(bX) = (ab)X$,
6. distributivity of scalars: $(a + b)X = aX + bX$,
7. distributivity of vectors: $a(X + Y) = aX + aY$,
8. scalar identity: $1X = X$.

Theorems

The designation ARATRA 3.1 means that this is theorem (or proposition, lemma, or corollary) 3.1 in *A Radical Approach to Real Analysis*.

Theorem 1.3 (DeMorgan's Laws). *Let $\{S_k\}$ be any finite or infinite collection of sets, then*

$$\left(\bigcup_k S_k \right)^C = \bigcap_k S_k^C, \qquad \left(\bigcap_k S_k \right)^C = \bigcup_k S_k^C.$$

Theorem 1.4 (Distributivity). *Let S, T, U be any sets, then*

$$S \cap (T \cup U) = (S \cap T) \cup (S \cap U), \qquad S \cup (T \cap U) = (S \cup T) \cap (S \cup U).$$

Theorem 1.5 (Mean Value Theorem, ARATRA 3.1). *Given a function f that is differentiable at all points strictly between a and x and continuous at all points on* *the closed interval from a to x, there exists a real number c strictly between a and x such that*

$$\frac{f(x) - f(a)}{x - a} = f'(c). \tag{1.9}$$

Theorem 1.6 (Intermediate Value Theorem, ARATRA **3.3).** *If f is continuous on the interval* $[a, b]$*, then f has the intermediate value property on this interval.*

Theorem 1.7 (Darboux's Theorem, ARATRA **3.14).** *If f is differentiable on* $[a, b]$*, then f' has the intermediate value property on* $[a, b]$*.*

Theorem 1.8 (The Cauchy Criterion, ARATRA **4.2).** *A sequence of real numbers converges if and only if it is a Cauchy sequence.*

Theorem 1.9 (Absolute Convergence Theorem, ARATRA **4.4).** *If* $|a_1| + |a_2| + |a_3| + \cdots$ *converges then so does* $a_1 + a_2 + a_3 + \cdots$*.*

Theorem 1.10 (Continuity of Infinite Series, ARATRA **5.6).** *If* $f_1 + f_2 + f_3 + \cdots$ *converges uniformly to F over the interval* (α, β) *and if each of the summands is continuous at every point in* (α, β)*, then the function F is continuous at every point in* (α, β)*.*

Theorem 1.11 (Term-by-term Differentiation, ARATRA **5.7).** *Let* $f_1 + f_2 + f_3 + \cdots$ *be a series of functions that converges at* $x = a$ *and for which the series of deriviatives,* $f_1' + f_2' + f_3' + \cdots$*, converges uniformly over an open interval I that contains a. It follows that*

1. $F = f_1 + f_2 + f_3 + \cdots$ *converges uniformly over the interval I,*
2. *F is differentiable at* $x = a$*, and*
3. *for all* $x \in I$*,* $F'(x) = \sum_{k=1}^{\infty} f_k'(x)$*.*

Theorem 1.12 (Term-by-term Integration, ARATRA **5.8).** *Let* $f_1 + f_2 + f_3 + \cdots$ *be uniformly convergent over the interval* $[a, b]$*, converging to F. If each* f_k *is integrable over* $[a, b]$*, then so is F and*

$$\int_a^b F(x) \, dx = \sum_{k=1}^{\infty} \int_a^b f_k(x) \, dx.$$

Theorem 1.13 (Continuity on $[a, b]$ $\Longrightarrow$ **Uniform Continuity,** ARATRA **6.3).** *If f is continuous over the closed and bounded interval* $[a, b]$*, then it is uniformly continuous over this interval.*

Theorem 1.14 (Continuous $\implies$ Integrable, ARATRA 6.6). *If f is a continuous function on the closed, bounded interval $[a, b]$, then f is integrable over $[a, b]$.*

Theorem 1.15 (Continuity of Integral, ARATRA 6.8). *Let f be a bounded integrable function on $[a, b]$ and define F for x in $[a, b]$ by*

$$F(x) = \int_a^x f(t)\, dt.$$

Then F is continuous at every point between a and b.

Exercises

1.2.1. Give an example of a function and an interval for which the function is continuous but not uniformly continuous on the interval.

1.2.2. Give an example of a sequence that converges but is not monotonic.

1.2.3. Prove or find a counterexample to the statement: Every infinite sequence contains an infinite monotonic subsequence.

1.2.4. Give an example of a sequence of functions and an interval for which the sequence converges pointwise but not uniformly on the interval.

1.2.5. Prove that $\sum_{n=0}^{\infty} 2^{-n} = 2$ by showing how to find a response N for each $\epsilon > 0$.

1.2.6. The lim sup, $\overline{\lim}_{n \to \infty} a_n$, can also be defined as the value A, such that given any $\epsilon > 0$, there is a response N such that $n \geq N$ implies that $a_n < A + \epsilon$, and for every $M \in \mathbb{N}$, there is an $m \geq M$ such that $A - \epsilon < a_m$. Show that this definition is equivalent to the definition

$$A = \inf_{n \geq 1} \left(\sup_{k \geq n} a_k \right).$$

1.2.7. Prove that if $(a_n)_{n=1}^{\infty}$ is bounded, then $\overline{\lim}_{n \to \infty} a_n$ exists.

1.2.8. Prove that if $A = \overline{\lim}_{n \to \infty} a_n$ exists, then we can find a subsequence that converges to A.

1.2.9. Show that the set of all real-valued continuous functions defined on $[0, 1]$ is a vector space.

1.2.10. Use the nested interval principle to prove that every Cauchy sequence converges.

1.2.11. Show that the nested interval principle does not necessarily hold if we replace closed intervals with open intervals.

1.2.12. Justify DeMorgan's laws (Theorem 1.3). Show that

$$x \in \left(\bigcup_k S_k \right)^C \implies x \in \bigcap_k S_k^C \quad \text{and} \quad x \notin \left(\bigcup_k S_k \right)^C \implies x \notin \bigcap_k S_k^C.$$

1.2.13. Justify the distributivity theorem (Theorem 1.4).

1.2.14. Prove that given any two sets F_1 and F_2, if

$$S_1 = F_1 \cap F_2^C \quad \text{and} \quad S_2 = F_1^C \cap F_2,$$

then

$$F_1 = (F_2 \cup S_1) \cap S_2^C.$$

1.2.15. Give an example of a function f and an interval $[a, b]$ such that f is continuous on $[a, b]$, differentiable at all but one point of (a, b), and for which there is no $c \in (a, b)$ for which

$$\frac{f(b) - f(a)}{b - a} = f'(c).$$

1.2.16. Give an example of a function f and an interval $[a, b]$ such that f has the intermediate value property on $[a, b]$ but it is not continuous on this interval.

1.2.17. Use the mean value theorem, Theorem 1.5, to prove the following weaker form of Darboux's theorem: If f' is the derivative of f on an open interval containing c and if $\lim_{x \to c^-} f'(x)$ and $\lim_{x \to c^+} f'(x)$ exist, then these one-sided limits must be equal.

1.2.18. Give an example of a series that converges but does not converge absolutely.

1.2.19. Give an example of a series of continuous functions and an interval such that the series does not converge uniformly over the interval, but it does converge pointwise to a continuous function on this interval.

1.2.20. Give an example of a series of continuous functions and an interval such that the series does not converge to a continuous function on this interval.

1.2.21. Give an example of a series, $f_1 + f_2 + \cdots$, of differentiable functions and an interval such that the series converges uniformly to f over the interval, but

$$f'(x) \neq f_1'(x) + f_2'(x) + \cdots$$

for all x in this interval.

1.2.22. Give an example of a series, $f_1 + f_2 + \cdots$, of integrable functions and an interval $[a, b]$ such that the series does not converge uniformly to f over $[a, b]$ but

$$\int_a^b \left(\sum_{k=1}^{\infty} f_k(x) \right) dx = \sum_{k=1}^{\infty} \int_a^b f_k(x) \, dx.$$

1.2.23. Prove that if f is integrable over $[a, b]$ then there exists $c \in [a, b]$ for which

$$\int_a^c f(x) \, dx = \frac{1}{2} \int_a^b f(x) \, dx.$$

2

The Riemann Integral

Bernard Riemann received his doctorate in 1851, his *Habilitation* in 1854. The habilitation confers recognition of the ability to create a substantial contribution to research beyond the doctoral thesis, and it is a necessary prerequisite for appointment as a professor in a German university. Riemann chose as his habilitation thesis the problem of Fourier series. It was titled *Über die Darstellbarkeit einer Function durch eine trigonometrische Reihe* (On the representability of a function by a trigonometric series), and, strictly speaking, it answered the broader question: When can a function over $(-\pi, \pi)$ be represented as a series of the form $a_0/2 + \sum_{n=1}^{\infty}(a_n \cos(nx) + b_n \sin(nx))$? This is where we find the Riemann integral, introduced in a short section before the main body of the thesis, part of the groundwork that he needed to lay before he could tackle the real problem of representability by a trigonometric series.

Riemann had studied with Dirichlet in Berlin before going to Göttingen to complete his doctorate under the direction of Gauss. In the fall of 1852, Dirichlet visited Göttingen. Shortly afterward, Riemann wrote to his friend Richard Dedekind,

> The other morning Dirichlet stayed with me for about two hours; he gave me the notes necessary for my *Habilitation* so completely that my work has become much easier; otherwise, for some things I would have searched for a long time in the library.[1]

Riemann was almost certainly referring to the extensive introduction to his thesis in which he describes the progress that had been made in understanding Fourier series until that time. But it is also clear that Dirichlet had continued to think about this problem, and he may have had some useful advice.

Riemann's thesis on trigonometric series was not published until 1868, two years after his death at the age of 39. Dedekind was responsible for this publication.

---

[1] Dedekind (1876, p. 578), as quoted in Hochkirchen (2003, p. 261).

Richard Dedekind (1831–1916) and Bernhard Riemann both studied with Gauss at Göttingen and then worked with Dirichlet who succeeded to Gauss's chair. They developed a strong friendship. In 1862, Dedekind took a position at the Brunswick Polytechnikum where he would remain for the rest of his career. Today he is best known for his work in number theory and modern algebra, especially for establishing the theory of the ring of integers of an algebraic number field.

In 1870, three significant papers appeared that built on Riemann's accomplishments: Hermann Hankel's *Untersuchungen über die unendlich oft oscillirendend und unstetigen Funktionen* (Investigations on infinitely often oscillating and discontinuous functions), Eduard Heine's *Über trigonometrische Reihen* (On trigonometric series), and Georg Cantor's *Über einen die trigonometrischen Reihen betreffenden Lehrsatz* (On a theorem concerning trigonometric series). These papers accomplished two important tasks. The first was to clarify the concept of uniform convergence and the related issue of when term-by-term integration is legitimate. The second was to turn the question of integrability of a function to the study of the set of points at which the function is discontinuous, thus opening the way to the development of set theory and a deeper understanding of the structure of the real numbers.

A fourth seminal paper directly inspired by Riemann's thesis was Gaston Darboux's *Mémoire sur les fonctions discontinues* (Memoir on discontinuous functions) of 1875. In 1873, Darboux had published a translation of Riemann's thesis into French. It is clear that he studied it very carefully. His 1875 paper greatly simplified the treatment of the Riemann integral. In discussing the Riemann integral, we shall rely on Darboux's definitions and insights.

Gaston Darboux (1842–1917) studied at the École Normale Supérieur and taught there from 1872 to 1878. He then went to the Sorbonne where, in 1880, he succeeded Michel Chasles as chair of higher geometry. Darboux is best known for his work in differential geometry, but among his many contributions to mathematics, he also edited Fourier's *Collected Works*.

2.1 Existence

Riemann devotes three brief pages to the definition of the definite integral, the definition of an improper integral, and the statement and proof of the necessary and sufficient condition for integrability. He then spends one page describing a function that is discontinuous at every rational number with an even denominator but which is integrable, thus showing that while continuity is a sufficient condition for integrability, it is far from necessary. As Darboux demonstrated, there is a lot to mine from these four pages.

Definition: Integration (Riemann)

A function f is **Riemann integrable** over the interval $[a, b]$ and its integral has the value V if for every error bound $\epsilon > 0$, there is a response $\delta > 0$ such that for any partition $(x_0 = a, x_1, \ldots, x_n = b)$ with subintervals of length less than δ (that is to say, $|x_j - x_{j-1}| < \delta$ for all j) and for any set of tags $x_1^* \in [x_0, x_1]$, $x_2^* \in [x_1, x_2]$, $\ldots, x_n^* \in [x_{n-1}, x_n]$, the corresponding Riemann sum lies within ϵ of the value V:

$$\left| \sum_{j=1}^{n} f(x_j^*)(x_j - x_{j-1}) - V \right| < \epsilon.$$

Given a function f defined on $[a, b]$, we can find a **Riemann sum** approximation to the definite integral $\int_a^b f(x)\,dx$ by choosing a partition of the interval

$$a = x_0 < x_1 < x_2 < \cdots < x_n = b$$

and a set of **tags** $x_1^* \in [x_0, x_1]$, $x_2^* \in [x_1, x_2]$, $\ldots, x_n^* \in [x_{n-1}, x_n]$. The Riemann sum is then given by

$$\sum_{j-1}^{n} f(x_j^*)(x_j - x_{j-1}).$$

Using the Cauchy criterion for convergence, the value V will exist if given any $\epsilon > 0$, there is a response $\delta > 0$ so that any two Riemann sums with intervals of length less than δ will differ by less than ϵ. The value of the integral is denoted by

$$V = \int_a^b f(x)\,dx.$$

The greatest difficulty with this definition is handling the variability in the tags of f since x_j^* can be *any* value in the interval $[x_{j-1}, x_j]$. Darboux saw that the way to do this is to work with the least upper bound[2] (or supremum) and the greatest lower bound (or infimum) of the set $\{f(x) \mid x_{j-1} \le x \le x_j\}$.

Every Riemann sum for this partition lies between the upper and lower Darboux sums (see top of next page). While it may not be possible to find a Riemann sum that actually equals the upper or the lower Darboux sum, we can find Riemann sums for this partition that come arbitrarily close to the Darboux sums.

The function f is Riemann integrable if and only if we can force all Riemann sums to be within ϵ of our specified value $V = \int_a^b f(x)\,dx$ simply by restricting our

---

[2] Actually, Darboux at this time did not make a clear distinction between the supremum and maximum of a set.

Definition: Darboux sums

Given a function f defined on $[a, b]$ and a partition $P = (a = x_0 < x_1 < \cdots < x_n = b)$ of this interval, we define

$$M_j = \sup \{ f(x) \,|\, x_{j-1} \le x \le x_j \} \quad \text{and}$$
$$m_j = \inf \{ f(x) \,|\, x_{j-1} \le x \le x_j \}.$$

The **upper Darboux sum**, is

$$\overline{S}(P; f) = \sum_{j=1}^{n} M_j(x_j - x_{j-1}), \tag{2.1}$$

and the **lower Darboux sum**, is

$$\underline{S}(P; f) = \sum_{j=1}^{n} m_j(x_j - x_{j-1}). \tag{2.2}$$

---

partitions to those with interval length less than an appropriately chosen response δ. This will happen if and only if the upper and lower Darboux sums for these partitions are within ϵ of the specified value V. It follows that f is Riemann integrable if and only if we can make the difference between the upper and lower Darboux sums as small as we wish by controlling the length of the intervals in the partition,

$$x_j - x_{j-1} < \delta \text{ for all } j \implies \sum_{j=1}^{n}(M_j - m_j)(x_j - x_{j-1}) \le \epsilon.$$

In order to guarantee that this sum is less than ϵ, we need some control on the size of $M_j - m_j$, what is called the **oscillation** of the function over the interval $[x_{j-1}, x_j]$. If we can force the oscillation to be as small as we wish by taking sufficiently short intervals, then we have integrability. We choose δ so that $M_j - m_j < \epsilon/(b - a)$. It follows that

$$\sum_{j=1}^{n}(M_j - m_j)(x_j - x_{j-1}) < \frac{\epsilon}{b - a} \sum_{j=1}^{n}(x_j - x_{j-1}) = \frac{\epsilon}{b - a}(b - a) = \epsilon.$$

This implies that every continuous function is integrable (see Exercise 2.1.9).

What about a discontinuous function? If f is discontinuous, then there will be intervals that include the points of discontinuity where the oscillation cannot be made as small as we wish. If our function is integrable and our partition includes intervals where the oscillation is greater than or equal to σ, then the sum of the lengths of these intervals must be less than ϵ/σ. If $\sum^*$ denotes the sum over the intervals on which the oscillation is at least σ, then

$$\epsilon > \sum_{j}^{*}(M_j - m_j)(x_j - x_{j-1}) \geq \sum_{j}^{*}\sigma(x_j - x_{j-1}) \implies \sum_{j}^{*}(x_j - x_{j-1}) < \frac{\epsilon}{\sigma}.$$

If we choose a smaller bound for the difference between the upper and lower Darboux sums, then we get an even smaller bound on the sum of the lengths of the intervals on which the oscillation was at least σ. Since we can force the difference between the upper and lower Darboux sums to be as small as we wish, we can also force the sum of the lengths of the intervals on which the oscillation exceeds σ to be as small as we wish, just by controlling the lengths of the intervals in the partition.

Riemann realized that this also works the other way. If for every $\sigma > 0$, we can force the sum of the lengths of the intervals on which the oscillation exceeds σ to be as small as we wish by restricting the lengths of the intervals in the partition, then we can force the upper and lower Darboux sums to be within any specified ϵ of each other. We define D to be the difference between the least upper bound and the greatest lower bound of $\{f(x) \mid a \leq x \leq b\}$, so that $M_j - m_j \leq D$ for all j. We let $\sigma = \epsilon/2(b - a)$ and choose a limit on the partition intervals so that those on which the oscillation exceeds σ have total length less than $\epsilon/2D$. We split the difference in Darboux sums into $\sum_1$ over those intervals where the oscillation is at least σ and $\sum_2$ over the intervals where the oscillation is strictly less than σ:

$$\sum_{j=1}^{n}(M_j - m_j)(x_j - x_{j-1}) = \sum_{1}(M_j - m_j)(x_j - x_{j-1})$$

$$+ \sum_{2}(M_j - m_j)(x_j - x_{j-1})$$

$$< \sum_{1}D(x_j - x_{j-1}) + \sum_{2}\sigma(x_j - x_{j-1})$$

$$< D\frac{\epsilon}{2D} + \frac{\epsilon}{2(b - a)}(b - a) = \epsilon. \qquad (2.3)$$

We have proven Riemann's criterion for integrability.

Theorem 2.1 (Conditions for Riemann Integrability). *Let f be a bounded function on $[a, b]$. This function is integrable over $[a, b]$ if and only if for any $\sigma > 0$, a bound on the oscillation, and for any $v > 0$, a bound on the sum of the lengths of the intervals where the oscillation exceeds σ, we can find a δ response so that for any partition of $[a, b]$ with subintervals of length less than δ, the subintervals on which the oscillation is at least σ have a combined length that is strictly less than v.*

The Darboux Integrals

In 1881, Vito Volterra showed how to use Darboux sums to create upper and lower integrals that exist for every function. Looking at the upper Darboux sums, we see that as the partition gets finer (including more points), the value of the upper sum gets smaller, decreasing as it approaches the value of the Riemann integral. This suggests taking the greatest lower bound of all the upper Darboux sums. If the Riemann integral exists, it will equal this greatest lower bound. Similarly, if the Riemann integral exists, then it will equal the least upper bound of the lower Darboux sums. Although first described by Volterra, these integrals usually carry Darboux's name because they are defined in terms of his sums.

It is not too hard to see that if f is Riemann integrable, then the upper and lower Darboux integrals must be equal. It will take some work to show the implication in the other direction, that if the upper and lower Darboux integrals are equal, then the function is Riemann integrable. But this work will be worth it, for it produces a very useful test for integrability.

Theorem 2.2 (Darboux Integrability Condition). *Let f be a bounded function on $[a, b]$. This function is Riemann integrable over this interval if and only if the upper and lower Darboux integrals are equal.*

Proof. We take the easy direction first. We leave it as Exercise 2.1.11 to prove that

$$\underline{\int_a^b} f(x)\, dx \le \overline{\int_a^b} f(x)\, dx.$$

It follows that for any partition P, we have

$$\underline{S}(P; f) \le \underline{\int_a^b} f(x)\, dx \le \overline{\int_a^b} f(x)\, dx \le \overline{S}(P; f). \qquad (2.4)$$

Definition: Upper and lower Darboux integrals

Let $\mathcal{P}$ denote the set of all partitions of $[a, b]$. The **upper Darboux integral** of f over $[a, b]$ is defined by

$$\overline{\int_a^b} f(x)\, dx = \inf_{P \in \mathcal{P}} \overline{S}(P; f).$$

Similarly, the **lower Darboux integral** is defined by

$$\underline{\int_a^b} f(x)\, dx = \sup_{P \in \mathcal{P}} \underline{S}(P; f).$$

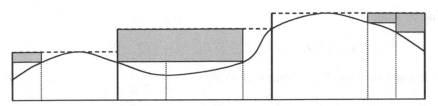

Figure 2.1. Solid vertical bars mark the points of partition P. Dotted vertical bars mark the points of partition P_3. The partition Q consists of all vertical bars, solid or dotted.

If f is Riemann integrable, then we can find a partition for which $\overline{S}(P; f) - \underline{S}(P; f)$ is less than any specified positive value. It follows that the absolute value of the difference between the upper and lower Darboux integrals is also less than any specified positive value, which can only be true if the difference is 0.

In the other direction, if the Darboux integrals are equal, then this common value is our candidate for V, the value of the Riemann integral. Given any $\epsilon > 0$, we can find an upper Darboux sum $\overline{S}(P_1; f)$ and a lower Darboux sum $\underline{S}(P_2; f)$ that are each less than $\epsilon/2$ away from V. If we let P_3 denote the common refinement of P_1 and P_2, $P_3 = P_1 \bigcup P_2$, then

$$\underline{S}(P_2; f) \leq \underline{S}(P_3; f) \leq \overline{S}(P_3; f) \leq \overline{S}(P_1; f),$$

and therefore every Riemann sum for the partition P_3 is within $\epsilon/2$ of V. The same is true for any refinement of P_3. We still need to show that every Riemann sum with sufficiently short intervals differs by at most ϵ from V, even if it shares no points with P_3.

Let D denote the oscillation – the difference between the least upper bound and the greatest lower bound – of f over the entire interval $[a, b]$. Let m denote the number of intervals in P_3. We take any partition P with intervals of length less than $\epsilon/2mD$ (see Figure 2.1). Let $Q = P \bigcup P_3$; Q has at most $m - 1$ more points than P. In Figure 2.1, Q has five more points than P. The difference between the upper Darboux sum for P and the upper Darboux sum for Q is the sum of the areas of the shaded rectangles. There are at most $m - 1$ such rectangles, their lengths are each bounded by the lengths of the intervals in P, which is less than $\epsilon/2mD$, and their heights are bounded by the oscillation of f, which is D. The upper Darboux sums differ by at most

$$\left| \overline{S}(P; f) - \overline{S}(Q; f) \right| < (m - 1) \cdot \frac{\epsilon}{2mD} \cdot D < \frac{\epsilon}{2}.$$

Since Q is a refinement of P_3, we get an upper bound on the upper Darboux sum for P,

$$\overline{S}(P; f) < \overline{S}(Q; f) + \epsilon/2 \leq \overline{S}(P_3; f) + \epsilon/2 \leq V + \epsilon.$$

By a similar argument,

$$\underline{S}(P; f) > \underline{S}(Q; f) - \epsilon/2 \geq \underline{S}(P_3; f) - \epsilon/2 \geq V - \epsilon.$$

Every Riemann sum for P is within ϵ of V.   □

This theorem gives us a simple condition that is equivalent to Riemann integrability. If for each $\epsilon > 0$, we can find just one partition P for which $\overline{S}(P; f) - \underline{S}(P; f) < \epsilon$, then the upper and lower Darboux integrals must be equal. If they are equal, then we can find such a partition for each ϵ.

Corollary 2.3 (One Partition Suffices). *Let f be a bounded function on $[a, b]$. This function is Riemann integrable over this interval if and only if for each $\epsilon > 0$ there is a partition P for which $\overline{S}(P; f) - \underline{S}(P; f) < \epsilon$.*

Improper Integrals

One of the drawbacks of Riemann's definition of the integral is that it only applies to bounded functions on finite intervals, an issue that clearly was of concern to Riemann, for immediately after giving his definition, he explains how to deal with integrals of unbounded functions. Today we refer to these as **improper integrals**.

Strictly speaking, the Riemann integral does not exist in this case. However, there may be a value that can be assigned to such an integral by taking a limit of integrals that are Riemann integrable. For unbounded integrals such as

$$\int_{-1}^{1} \frac{dx}{|x|^{1/2}},$$

we evaluate the integral on intervals for which the function is bounded and then take the limit of these values as the endpoints approach the point at which we have a vertical asymptote:

$$\int_{-1}^{1} \frac{dx}{|x|^{1/2}} = \lim_{\epsilon_1 \to 0^-} \int_{-1}^{\epsilon_1} \frac{dx}{|x|^{1/2}} + \lim_{\epsilon_2 \to 0^+} \int_{\epsilon_2}^{1} \frac{dx}{|x|^{1/2}}$$

$$= \lim_{\epsilon_1 \to 0^-} -2|x|^{1/2}\Big|_{-1}^{\epsilon_1} + \lim_{\epsilon_2 \to 0^+} 2x^{1/2}\Big|_{\epsilon_2}^{1}$$

$$= \lim_{\epsilon_1 \to 0^-} \left(-2|\epsilon_1|^{1/2} + 2\right) + \lim_{\epsilon_2 \to 0^+} \left(2 - \epsilon_2^{1/2}\right)$$

$$= 4.$$

As Riemann went to great pains to point out, the existence of an antiderivative is no guarantee that the improper integral exists. When there is more than one limit,

> **Definition: Improper integral**
>
> An integral is **improper** if either the function that is being integrated or the interval over which the function is integrated is unbounded.

they must be taken independently. For example, the antiderivative of $1/x$ is $\ln |x|$ and $\ln |1| - \ln |-1| = 0 - 0 = 0$, but

$$\int_{-1}^{1} \frac{dx}{x} = \lim_{\epsilon_1 \to 0^-} \int_{-1}^{\epsilon_1} \frac{dx}{x} + \lim_{\epsilon_2 \to 0^+} \int_{\epsilon_2}^{1} \frac{dx}{x}$$

$$= \lim_{\epsilon_1 \to 0^-} \ln |x| \Big|_{-1}^{\epsilon_1} + \lim_{\epsilon_2 \to 0^+} \ln |x| \Big|_{\epsilon_2}^{1}$$

$$= \lim_{\epsilon_1 \to 0^-} \ln |\epsilon_1| - \lim_{\epsilon_2 \to 0^+} \ln |\epsilon_2|.$$

Since neither limit is finite, this function is not integrable over $[-1, 1]$.

Exercises

2.1.1. Explain why if P and Q are partitions of the same interval and Q is a refinement of P, $Q \supseteq P$, and if f is any bounded function on this interval, then

$$\underline{S}(P; f) \leq \underline{S}(Q; f) \leq \overline{S}(Q; f) \leq \overline{S}(P; f).$$

2.1.2. Consider the function f defined by

$$f(x) = \begin{cases} 1, & x = 0, \\ x, & 0 < x < 1, \\ 0, & x = 1. \end{cases}$$

Let P be the partition $(0, 1/4, 1/2, 3/4, 1)$. Find the upper and lower Darboux sums, $\overline{S}(P; f)$ and $\underline{S}(P; f)$.

2.1.3. Using the function f defined in Exercise 2.1.2 and given $\epsilon = 1/2$, find a response δ so that for any partition P into intervals of length less than δ, the difference between $\overline{S}(P; f)$ and $\underline{S}(P; f)$ will be less than $1/2$.

2.1.4. Using the function f defined in Exercise 2.1.2 over the interval $1/2 \leq x \leq 1$, explain why no Riemann sum can equal the upper Darboux sum no matter what partition we choose.

2.1.5. Consider the function

$$g(x) = \sum_{n=1}^{\infty} \frac{1}{2^{2n-1}} \left\lfloor \frac{2^n x + 1}{2} \right\rfloor, \quad 0 \leq x \leq 1,$$

where $\lfloor \alpha \rfloor$ denotes the greatest integer less than or equal to α. Show that this series converges for all $x \in [0, 1]$, that it is monotonically increasing, and that $g(0) = 0$, $g(1) = 1$. Find all points at which g is discontinuous and at these points find the difference between the limit from the left and the limit from the right.

2.1.6. Using the function g defined in Exercise 2.1.5, show that it is Riemann integrable over $[0, 1]$.

2.1.7. Using the function g defined in exercise 2.1.5, find the value of $\int_0^1 g(x)\,dx$. Show the work that leads to your conclusion.

2.1.8. Prove that f is continuous at c if and only if given any $\epsilon > 0$ there is a response δ for which the oscillation of f over $(c - \delta, c + \delta)$ is less than ϵ.

2.1.9. Using the fact that a continuous function on a closed and bounded interval is uniformly continuous on that interval, prove that if f is continuous on $[a, b]$, then f is Riemann integrable over $[a, b]$.

2.1.10. Find the upper and lower Darboux integrals of the characteristic function of the rationals (Example 1.1 on page 3) over the interval $[0, 1]$.

2.1.11. Prove that if f is bounded on $[a, b]$, then

$$\underline{\int_a^b} f(x)\,dx \leq \overline{\int_a^b} f(x)\,dx.$$

2.1.12. Define the function h by

$$h(x) = \begin{cases} x, & x \in [0, 1] \cap \mathbb{Q}, \\ 0, & x \in [0, 1] - \mathbb{Q}. \end{cases}$$

Find the upper and lower Darboux integrals of h over $[0, 1]$.

2.1.13. Define the function k by

$$k(x) = \begin{cases} x, & x \in [-3, 3] \cap \mathbb{Q}, \\ 0, & x \in [-3, 3] - \mathbb{Q}. \end{cases}$$

Find the upper and lower Darboux integrals of k over $[-3, 3]$.

2.1.14. Define the function m by

$$m(x) = \begin{cases} 1, & x = 0, \\ 1/q, & x = p/q \in \mathbb{Q}, \ \gcd(p, q) = 1, \ q \geq 1, \\ 0, & x \notin \mathbb{Q}. \end{cases}$$

Show that m is integrable over $[0, 1]$.

2.1.15. Define the function n by

$$n(x) = \begin{cases} 1, & x = 1/n, n \in \mathbb{N}, \\ 0, & \text{otherwise.} \end{cases}$$

Show that n is integrable over $[0, 1]$ and that $\int_0^1 n(x)\,dx = 0$.

2.1.16. Define the function p by

$$p(x) = \begin{cases} 0, & x = 0, \\ 1/x - \lfloor 1/x \rfloor, & \text{otherwise.} \end{cases}$$

Show that p is integrable over $[0, 1]$.

2.1.17. Find all positive values of α for which the improper integral

$$\int_{-1}^1 \frac{dx}{|x|^\alpha}$$

has a value. Show the work that leads to your conclusion.

2.1.18. Show that for every α, $0 < \alpha < 1$, the improper integral

$$\int_0^1 \left(\left\lfloor \frac{\alpha}{x} \right\rfloor - \alpha \left\lfloor \frac{1}{x} \right\rfloor \right) dx$$

exists and has value $\alpha \ln \alpha$.

2.1.19. Prove that if a function is bounded and Cauchy integrable over $[a, b]$, then it is also Riemann integrable over that interval.

2.2 Nondifferentiable Integrals

What clearly excited Darboux most about Riemann's thesis was his example of a function that has discontinuities at all rational numbers with even denominators and yet is still integrable. Riemann's example appears on the fourth page of his explanation of integration.

Example 2.1. Riemann defined the function

$$((x)) = \begin{cases} x - \lfloor x \rfloor, & \lfloor x \rfloor \le x < \lfloor x \rfloor + 1/2, \\ 0, & x = \lfloor x \rfloor + 1/2, \\ x - \lfloor x \rfloor - 1, & \lfloor x \rfloor + 1/2 < x < \lfloor x \rfloor + 1 \end{cases} \tag{2.5}$$

(see Figure 2.2). He then defined

$$f(x) = \sum_{n=1}^{\infty} \frac{((nx))}{n^2}. \tag{2.6}$$

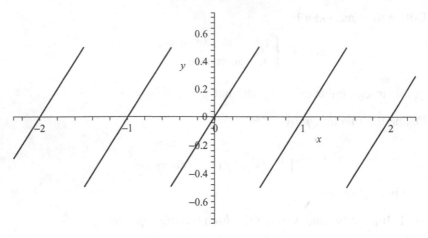

Figure 2.2. Graph of $y = ((x))$.

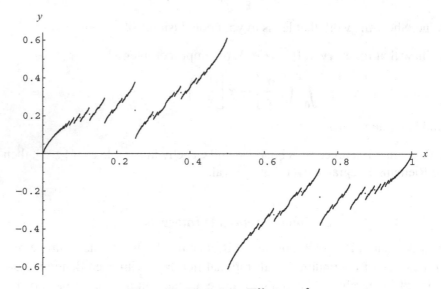

Figure 2.3. Graph of $y = \sum_{n=1}^{\infty} ((nx))/n^2$.

Since $|((nx))| < 1/2$, this series converges for all x. It has a discontinuity whenever nx is half of an odd integer, and that will happen for every x that is a rational number with an even denominator (see Figure 2.3).

Specifically, if $x = a/2b$, where a is odd and a and b are relatively prime, and if n is an odd multiple of b, then

$$\lim_{v \to 0^+} \left(\!\!\left(\frac{na}{2b} + v \right)\!\!\right) - \left(\!\!\left(\frac{na}{2b} \right)\!\!\right) = -1/2 \quad \text{and} \quad \lim_{v \to 0^-} \left(\!\!\left(\frac{na}{2b} + v \right)\!\!\right) - \left(\!\!\left(\frac{na}{2b} \right)\!\!\right) = 1/2.$$

We want to be able to assert that

$$\lim_{v \to 0^+} f\left(\frac{a}{2b} + v\right) - f\left(\frac{a}{2b}\right) = \sum_{\substack{m=1 \\ m \text{ odd}}}^{\infty} \frac{-1/2}{(mb)^2}$$

$$= \frac{-1}{2b^2}\left(1 + \frac{1}{9} + \frac{1}{25} + \cdots\right)$$

$$= \frac{-\pi^2}{16b^2}, \tag{2.7}$$

$$\lim_{v \to 0^-} f\left(\frac{a}{2b} + v\right) - f\left(\frac{a}{2b}\right) = \sum_{\substack{m=1 \\ m \text{ odd}}}^{\infty} \frac{1/2}{(mb)^2}$$

$$= \frac{1}{2b^2}\left(1 + \frac{1}{9} + \frac{1}{25} + \cdots\right)$$

$$= \frac{\pi^2}{16b^2}. \tag{2.8}$$

The first line of these equalities assumes that we can interchange limits, that is

$$\lim_{v \to 0^+} f(x + v) - f(x) = \lim_{v \to 0^+} \left(\sum_{n=1}^{\infty} \frac{(\!(nx + nv)\!) - (\!(nx)\!)}{n^2}\right)$$

$$= \sum_{n=1}^{\infty} \left(\lim_{v \to 0^+} \frac{(\!(nx + nv)\!) - (\!(nx)\!)}{n^2}\right). \tag{2.9}$$

The justification of this interchange rests on the uniform convergence of our series over the set of all x and is left as Exercise 2.2.1.

Our function f has a discontinuity at every rational number with an even denominator, but it is integrable. Given any $\sigma > 0$, there are only finitely many rational numbers between 0 and 1 at which the variation is larger than σ. If the variation is larger than σ at $x = a/2b$, then b must satisfy

$$\frac{\pi^2}{8b^2} > \sigma,$$

which means that b is a positive integer less than $\pi/\sqrt{8\sigma}$.

Given any $\epsilon > 0$, we want to find a bound on the interval length that guarantees that the upper and lower Darboux sums differ by less than ϵ. Choose $\sigma = \epsilon/2$. If there are N rational numbers in $[0, 1]$ with denominators less than $\pi/\sqrt{2\sigma}$, then we choose our interval bound δ so that $N\delta$ is less than $\epsilon/2$.

Darboux's Observation

Darboux observed that if f is integrable over an open interval containing a, if we define a new function F by

$$F(x) = \int_a^x f(t)\, dt,$$

and if $\lim_{x \to a^-} f(x)$ and $\lim_{x \to a^+} f(x)$ exist, then

$$\lim_{h \to 0^-} \frac{F(a + h) - F(a)}{h} = \lim_{x \to a^-} f(x),$$

$$\lim_{h \to 0^+} \frac{F(a + h) - F(a)}{h} = \lim_{x \to a^+} f(x).$$

This follows immediately from the mean value theorem of integral calculus:

$$F(a + h) - F(a) = \int_a^{a+h} f(t)\, dt = h \cdot f(c) \tag{2.10}$$

for some c strictly between a and $a + h$, valid for any $h \neq 0$.

Therefore, if $\lim_{x \to a^-} f(x) \neq \lim_{x \to a^+} f(x)$, then F cannot be differentiable at a. On the other hand, $F(a + h) - F(a)$ can be made arbitrarily small simply by limiting the size of h, and therefore F *is* continuous at every point. The antiderivative of Riemann's function $\sum_{n=1}^{\infty} ((nx))/n^2$ is continuous and not differentiable at rational values with even denominators.

This directly contradicts assertions made by Ampère and by Duhamel that continuity guarantees differentiability, at least at all but a sparse set of values. Our question #4, "What is the relationship between continuity and differentiability?" was now wide open.

Darboux went beyond this to find a continuous function that is not differentiable at *any* value of x.

Example 2.2. Consider

$$g(x) = \sum_{n=1}^{\infty} \frac{\sin((n + 1)! x)}{n!}.$$

This is a uniformly convergent series of continuous functions, and therefore g is continuous for all values of x. The fact that it is not differentiable at any value of x requires a bit more work.[3]

---

[3] Quite a bit more work. Darboux's original justification, published in 1875, had several flaws. He published an addendum in 1879 in which he corrected the justification of his original example and gave a simpler example, $\sum_{n=1}^{\infty} \cos(n!\, x)/n!$ (see Exercises 2.2.4–2.2.9).

That same year of 1875, Paul du Bois-Reymond published Weierstrass's example of an everywhere continuous but nowhere differentiable function.

Example 2.3. Consider the function defined by the uniformly convergent series

$$\sum_{n=0}^{\infty} b^n \cos(a^n \pi x),$$

where $0 < b < 1$ and a is odd integer for which $ab > 1 + 3\pi/2$ (for example, $b = 2/3$, $a = 9$ can be used).[4]

Weierstrass had publicly presented this example to the Berlin Academy in 1872, but it had not appeared in print.

At the same time, Weierstrass produced an example, valid for any bounded, countably infinite set S, of an increasing, continuous function that is not differentiable at any point of S. The set of rational numbers in $[0, 1]$ is an example of a countable set. The set of all algebraic numbers in $[0, 1]$, all roots of polynomials with rational coefficients, is also countable, as we shall see in the next chapter.

Example 2.4. Given our favorite bounded, countably infinite sequence, $(a_1, a_2, a_3, \ldots)$, we define the function

$$h(x) = x + \frac{x}{2} \sin \left(\frac{\ln(x^2)}{2} \right), \quad h(0) = 0.$$

We choose any k strictly between 0 and 1. The Weierstrass function is given by

$$w(x) = \sum_{n=1}^{\infty} k^n h(x - a_n).$$

If $x \neq a_n$, then the derivative of $h(x - a_n)$ with respect to x is

$$\frac{d}{dx} h(x - a_n) = 1 + \frac{1}{2} \sin \left(\frac{1}{2} \ln \left[(x - a_n)^2 \right] \right) + \frac{1}{2} \cos \left(\frac{1}{2} \ln \left[(x - a_n)^2 \right] \right),$$

which always lies strictly between 0 and 2. Since h is an increasing function, so is w. If the set of possible values of x is bounded, so is the set of values of $h(x - a_n)$, and therefore the series that defines w converges uniformly. Since h is a continuous function, so is w.

To show that w is not differentiable at a_N, we rewrite our function as

$$w(x) = k^N h(x - a_N) + \sum_{n \neq N} k^n h(x - a_n).$$

---

[4] For an explanation, see *A Radical Approach to Real Analysis*, 2nd ed., pp. 259–262.

Using the mean value theorem, we see that

$$\left| \frac{\sum_{n \neq N} k^n h(x - a_n) - \sum_{n \neq N} k^n h(a_N - a_n)}{x - a_N} \right| < 2 \sum_{n=1}^{\infty} k^n = \frac{2k}{1 - k}.$$

On the other hand,

$$\frac{h(x - a_N) - h(0)}{x - a_N} = \left(1 + \frac{1}{2} \sin \left(\frac{1}{2} \ln \left[(x - a_N)^2 \right] \right) \right), \qquad (2.11)$$

which oscillates between $1/2$ and $3/2$ as x approaches a_N. For $k < 1/5$, the function w is not differentiable at a_N.

Summary

To summarize the situation with regard to question #4 as it stood in 1875:

- If f is differentiable at a, then it is also continuous at a. Any function that is differentiable at every point in an interval is also continuous over that interval.
- There are functions that are continuous at every point in an interval but differentiable at none of the points in that interval.
- For any countable set of points, we can find an increasing, continuous function that is not differentiable at any point in the set.

What was not known was whether or not it is possible to construct an increasing, continuous function that is not differentiable at any point in the interval. It would take 30 years to find the answer to this question.

With regard to question #3, the fundamental theorem of calculus, we have seen that we can find an integrable function for which

$$\frac{d}{dx} \int_a^x f(t) \, dt$$

does not exist for values of x that are rational numbers with even denominators. Questions that remained open included

- Could $\frac{d}{dx} \int_a^x f(t) \, dt$ exist but not equal $f(x)$?
- Could f be integrable but $\frac{d}{dx} \int_a^x f(t) \, dt$ fail to exist at every point?

Exercises

2.2.1. Prove that $f(x) = \sum_{n=1}^{\infty} ((nx))/n^2$ converges uniformly. Prove that the interchange of limits in Equation (2.9) is allowed.

2.2.2. Use the mean value theorem to prove that if f is continuous on $[a, b]$ and differentiable with a nonnegative derivative at all points of (a, b) except for $c \in (a, b)$ where it is not differentiable, then f is a monotonically increasing function over $[a, b]$.

2.2.3. For the function h defined in Example 2.4, find the supremum and infimum of $\{h'(x) \mid 0 < x < 1\}$. Show that

$$\overline{\lim_{x \to 0}} \, h'(x) \neq \underline{\lim_{x \to 0}} \, h'(x).$$

Exercises 2.2.4–2.2.5 step through Darboux's 1879 proof that

$$\mathcal{F}(x) = \sum_{n=1}^{\infty} \frac{\cos(n! \, x)}{n!}$$

is continuous at all x and not differentiable at any value of x.

2.2.4. Show that $\mathcal{F}$ is a uniformly convergent series of continuous functions and therefore is continuous.

We shall show that

$$\lim_{h \to 0} \frac{\mathcal{F}(x + h) - \mathcal{F}(x)}{h}$$

does not exist at any x, and therefore $\mathcal{F}$ is nowhere differentiable. We fix an $\epsilon > 0$ and, for each $N \in \mathbb{N}$, define $h = \epsilon/N!$. The variable h depends on N.

2.2.5. Show that for every positive integer n,

$$\frac{\cos(n! \, (x + h)) - \cos(n! \, x)}{n! \, h} = - \sin(n! \, x) - \frac{h}{2} \, n! \cos(n! \, c_n) \qquad (2.12)$$

for some c_n between x and $x + h$.

2.2.6. Show that

$$\sum_{n=1}^{N-1} \frac{n!}{N!} \leq \frac{2}{N},$$

and therefore

$$\left| \sum_{n=1}^{N-1} \frac{h}{2} \, n! \cos(n! \, c_n) \right| \leq \frac{\epsilon}{2} \sum_{n=1}^{N-1} \frac{n!}{N!} \leq \frac{\epsilon}{N}. \qquad (2.13)$$

2.2.7. Justify the equality

$$\frac{1}{h} \left(\sum_{n=N+1}^{\infty} \frac{\cos(n! \, (x + h))}{n!} - \sum_{n=N+1}^{\infty} \frac{\cos(n! \, x)}{n!} \right)$$

$$= \sum_{n=N+1}^{\infty} \frac{\cos(n! \, (x + h)) - \cos(n! \, x)}{n! \, h}.$$

Justify each inequality,

$$
\left| \sum_{n=N+1}^{\infty} \frac{\cos(n!\,(x+h)) - \cos(n!\,x)}{n!\,h} \right|
$$

$$
\leq \frac{2}{N!\,h} \left(\frac{1}{N+1} + \frac{1}{(N+1)(N+2)} + \frac{1}{(N+1)(N+2)(N+3)} + \cdots \right)
$$

$$
\leq \frac{4}{\epsilon\,N}. \tag{2.14}
$$

2.2.8. Using Equations (2.12)–(2.14), we see that

$$
\frac{\mathcal{F}(x+h) - \mathcal{F}(x)}{h} = -\left(\sum_{n=1}^{N-1} \sin(n!\,x) \right) + \frac{\cos(N!\,(x+h)) - \cos(N!\,x)}{N!\,h}
$$

$$
+ E(\epsilon, N)
$$

$$
= -\left(\sum_{n=1}^{N-1} \sin(n!\,x) \right) + \frac{\cos(N!\,x + \epsilon)) - \cos(N!\,x)}{\epsilon}
$$

$$
+ E(\epsilon, N), \tag{2.15}
$$

where $\left| E(\epsilon, N) \right| \leq \frac{\epsilon + 4/\epsilon}{N}$. Note that for fixed $\epsilon > 0$, $N = N(h)$ approaches ∞ as h approaches 0. Justify the following statement: If the $\lim_{h \to 0}(\mathcal{F}(x+h) - \mathcal{F}(x))/h$ exists, then

$$
\lim_{N \to \infty} \frac{\cos(N!\,x + \epsilon)) - \cos(N!\,x)}{\epsilon} \tag{2.16}
$$

is independent of ϵ

2.2.9. Show that

$$
\frac{\cos(N!\,x + 2\epsilon) - \cos(N!\,x)}{2\epsilon} - \frac{\cos(N!\,x + \epsilon) - \cos(N!\,x)}{\epsilon}
$$

$$
= \frac{\cos(\epsilon) - 1}{\epsilon}\, \cos(N!\,x + \epsilon). \tag{2.17}
$$

Therefore, the limit in (2.16) is independent of ϵ if and only if

$$
\lim_{N \to \infty} \cos(N!\,x + \epsilon) = 0
$$

regardless of the value of $\epsilon > 0$, and this is not true for any x.

2.3 The Class of 1870

Three important papers appeared in 1870, papers that built on Riemann's work on Fourier series and integration in general. These were written by Eduard Heine, Georg Cantor, and Hermann Hankel. Heinrich Eduard Heine (1821–1881) was a student of Dirichlet in Berlin. In 1848 he took up a position at Halle University where he would remain for the rest of his career. Georg Ferdinand Ludwig Philipp Cantor (1845–1918) was born in St. Petersburg, Russia. The family moved to Germany in 1856. He studied with Kummer and Weierstrass at the University of Berlin, receiving his doctorate in 1867. He joined Heine at Halle in 1869. Cantor's dissertation had been on number theory, but Heine was working on the problem of the uniqueness of the representation of a function as a trigonometric series, and he convinced Cantor to join him in this task.

The traditional approach to Fourier series was to start with a function, calculate its Fourier coefficients,

$$a_k = \frac{1}{\pi} \int_{-\pi}^{\pi} F(x) \cos(kx)\, dx \quad (k \geq 0),$$

$$b_k = \frac{1}{\pi} \int_{-\pi}^{\pi} F(x) \sin(kx)\, dx \quad (k \geq 1),$$

and then study the convergence of the resulting series

$$\frac{a_0}{2} + \sum_{k=1}^{\infty} a_k \cos(kx) + b_k \sin(kx). \tag{2.18}$$

Riemann turned this around by starting with arbitrary trigonometric series in the form of (2.18) and asking what properties such a function must possess. Does a trigonometric series have to be integrable? If it is integrable, then we can calculate its Fourier coefficients. Is this Fourier series always identical to the series with which we started? One outcome of this line of reasoning was the question whether two distinct trigonometric series could converge to the same function. If this were possible, then the difference between these series would be a trigonometric series with some nonzero coefficients that converges to 0. It would have been very surprising if someone had exhibited such a series, but no one could prove that it does not exist.

Uniqueness is easy to prove if the trigonometric series converges uniformly. As Weierstrass had shown in his Berlin lectures of the 1860s, term-by-term integration is valid for uniformly convergent series. Since we begin with a trigonometric series, Fourier's heuristic argument given in Equation (1.4) on page 2 actually proves that the series is unique. The problem is that if a series of continuous functions converges uniformly, then it converges to a continuous function. The most interesting Fourier series of the time converged to discontinuous functions and thus could not be uniformly convergent.

Dirichlet had been able to show that if we start with a continuous function and form the trigonometric series given in Equation (2.18), then that series is uniformly convergent. It follows from his analysis that if we work with a **piecewise**

continuous function, a function that is continuous at all but finitely many points, then its Fourier series is uniformly convergent on any closed interval that does not contain a point of discontinuity. This led Heine to describe a condition that is almost as good as uniform convergence, **uniform convergence in general**. A series with finitely many exceptional points that is uniformly convergent on any closed interval that does not contain one of these points is uniformly convergent in general.

Heine succeeded in proving that if a trigonometric series is uniformly convergent in general and converges to the function identically equal to zero, then all of the coefficients must be zero. That is to say, among the set of trigonometric series that are uniformly convergent in general, no two distinct series converge to the same function. It was Heine who convinced Cantor to take up the question of what happens when the convergence is not uniform in general.

In his 1870 paper, Cantor drew on Riemann's methods to get around the need for uniform convergence on any interval. He proved that if a trigonometric series converges to 0 at all x, then all coefficients of the series must be 0. By 1871, Cantor realized that his proof would work if it is known that the trigonometric series converges to 0 at all but at most finitely many points. He began working on the problem of an infinite number of exceptional points. For what infinite sets S can we conclude that if a trigonometric series converges to 0 at all points not in S, then all of the coefficients must be 0? Cantor was on his way to inventing set theory.

Hankel's Innovations

Early in 1871, Cantor reviewed Hankel's 1870 paper *Untersuchungen über die unendlich oft oszillierenden und unstetigen Funckionen* (Investigations on infinitely often oscillating and discontinuous functions). It spurred his thinking about infinite sets of discontinuities.

Hermann Hankel (1839–1873) took classes with Riemann at Göttingen and Weierstrass in Berlin before earning his doctorate in 1862 at the University of Leipzig where he then taught. His 1870 paper came shortly after his move to Tübingen. In it, he attempted to clarify Riemann's necessary and sufficient conditions for integrability.

We have considered the oscillation of a function over an interval. It is defined as the difference between the least upper bound of the values of the function and the greatest lower bound of those values. Hankel focused this onto a single point. We consider all open intervals that contain that point and look at the oscillation over each of these intervals. As the intervals become smaller, the oscillation can only

---

Definition: Oscillation

Given a function f and an interval I, the **oscillation of f over I** is

$$\omega(f; I) = \sup\{f(x) \mid x \in I\} - \inf\{f(x) \mid x \in I\}.$$

The **oscillation of f at the point c** is

$$\omega(f; c) = \inf_{I \in \mathcal{I}} \omega(f; I),$$

where $\mathcal{I}$ is the set of open intervals containing c. If $\underline{\lim}_{x \to c} f(x) \leq f(c) \leq \overline{\lim}_{x \to c} f(x)$, then this is equivalent to

$$\omega(f; c) = \overline{\lim_{x \to c}} f(x) - \underline{\lim_{x \to c}} f(x).$$

---

decrease. The **oscillation at a point** x is the greatest lower bound over all open intervals that contain x of the oscillation over I.[5]

The following proposition follows immediately from the second definition of oscillation at a point. The equivalence of these definitions is left as Exercise 2.3.14.

Proposition 2.4 (Continuous $\iff \omega = 0$). *The function f is continuous at c if and only if $\omega(f; c) = 0$.*

With this notion, Riemann's criterion for integrability can now be stated in terms of S_σ, the set of points with oscillation at least σ. A function f is integrable over the interval $[a, b]$ if and only if for each $\sigma > 0$, we can put the points of $S_\sigma \bigcap [a, b]$ inside a finite union of intervals, intervals that can be chosen so that the sum of their lengths is less than any predetermined positive amount.

It would take many years before the terminology was fixed, but we see here the beginning of the idea of the **outer content** of a set of points (see definition at top of next page), denoted c_e, e for "exterior."

Any finite set of points has outer content zero. This is because given any $\epsilon > 0$, we can put a small interval around each point so that the sum of the lengths of the intervals is less than ϵ.

If we consider the set $\{1, 1/2, 1/3, 1/4, \ldots\}$, it also has outer content zero. Given any $\epsilon > 0$, we put an interval of length $\epsilon/2$ around 0. That contains all but finitely many points from this set. The remaining points, because they are finite, can be put inside a union of intervals whose lengths add to less than $\epsilon/2$. On the other hand, the set of rational numbers between 0 and 1, $\mathbb{Q} \bigcap [0, 1]$, has outer content 1.

---

[5] Hankel actually defined a different but related concept he called the "jump" of f at c, the largest σ such that inside any interval containing c there is a point x for which $|f(x) - f(c)| > \sigma$.

Definition: Finite cover and outer content

Given a set S, a **finite cover** of S is a finite collection of intervals whose union contains S. The **length** of a cover C, denoted $l(C)$, is the sum of the lengths of the intervals in the cover. The **outer content** of a bounded set S is

$$c_e(S) = \inf_{C \in \mathcal{C}_S} l(C),$$

where $\mathcal{C}_S$ is the set of all covers of S.

Although Hankel did not have the terminology of outer content, he did grasp the idea and turned it into a characterization of when a function is Riemann integrable. Recast into the language of outer content, Hankel's insight is summarized in the following theorem.

Theorem 2.5 (Integrable $\Longleftrightarrow c_e(S_\sigma) = 0$). *Given a bounded function f defined on the interval $[a, b]$, let S_σ be the set of points in $[a, b]$ with oscillation greater than or equal to σ. The function f is Riemann integrable over $[a, b]$ if and only if for every $\sigma > 0$, the outer content of S_σ is zero.*

The first to explicitly use this measure of the size of a set was Otto Stolz (1842–1905) in 1881. The term "content" (*Inhalt*) is due to Cantor in 1884. The distinction between inner and outer content would be made by Guiseppe Peano in 1887 (see Section 5.1). The concept would be popularized in Jordan's *Cours d'analyse* of 1893–1896, because of which it is sometimes referred as Jordan content or Jordan measure when the inner and outer content of a set are the same.

The outer content is the same whether we use open or closed intervals in our finite cover (see Exercise 2.3.15). Because the distinction was not yet recognized as important, mathematicians of this period usually referred simply to intervals without distinguishing whether they were open or closed.

Theorem 2.5 is a profound and very useful result. Hankel did not have the terminology to state this result as we have here, but he did understand it fully. Unfortunately, Hankel used it to reach a faulty conclusion about when a discontinuous function is Riemann integrable. He was led astray because of the paucity of examples of highly discontinuous functions.

Hankel's Types of Discontinuity

It is clear that Hankel was very impressed by Riemann's example of a function (Example 2.1) that is discontinuous at all rational numbers with even denominators, yet *is* integrable. He sought to understand what happens in general. Using a

> **Definition: Dense**
>
> A set S is **dense** in the interval I if *every* open subinterval of I contains at least one point of S.

> **Definition: Totally discontinuous**
>
> A discontinuous function is **totally discontinuous** in an interval if the set of points of continuity is not dense in that interval.

> **Definition: Pointwise discontinuous**
>
> A discontinuous function is **pointwise discontinuous** in an interval if the set of points of continuity is dense in that interval.

technique that he dubbed "condensation of singularities," Hankel showed how to take a function with a singularity at one point, either a discontinuity or an infinite oscillation such as $\sin(1/x)$ near $x = 0$, and use it to construct integrable functions with singularities at every rational number. Cantor would later simplify this method and show how to apply it to any countable set. What is significant for our purposes is that both the set of points at which the function is continuous and the set of points at which the function is discontinuous are **dense**.

The rational numbers are dense in $\mathbb{R}$. The rational numbers with even denominators are also dense. So are the irrational numbers.

Hankel noticed that all of his examples of integrable functions that are discontinuous on a dense set of points have the property that the set of points of continuity is also dense. The examples that we have seen so far of Riemann integrable functions that are discontinuous on a dense set of points include Riemann's function (Example 2.1), the function g in Exercise 2.1.5, and the function m in Exercise 2.1.14. What characterizes all of these examples as well as the others that Hankel found is that the set of points of *continuity* are also dense. This suggested to him that he should separate discontinuous functions into two classes: those for which the points of continuity are not dense and those for which the points of continuity are dense.

Thus, for example, Dirichlet's function (Example 1.1) is totally discontinuous since it is discontinuous at every point.

All of the examples that we have seen so far of Riemann integrable functions that are discontinuous on a dense set of points are pointwise discontinuous. Hankel believed that every pointwise discontinuous function must be Riemann integrable.

Hankel's Error

Hankel's argument for his assertion that every pointwise discontinuous function is Riemann integrable is not unreasonable. We choose an arbitrary $\sigma > 0$. If a function is continuous at one value, then we can find an interval around this value on which the oscillation is less than σ. If the set of points of continuity is dense, then we have succeeded in putting each element of this dense set inside an interval that contains no points of S_σ.

Those points with oscillation larger than σ constitute a very thin set. Between any two points of this set there must be an entire open interval of points not in the set. Hankel believed that such a set must have outer content 0 and therefore must be Riemann integrable. This belief was reinforced by the fact that all of the examples of pointwise discontinuous functions that Hankel knew, examples such as Riemann's function, were integrable.

Hankel's fallacy, and he was not the only prominent mathematician to fall into it, was to assume that such a thin set cannot have positive outer content. Thomas Hawkins has presented evidence that between 1870 and 1875 this was the case for Hankel, for Axel Harnack, and for Paul du Bois-Reymond. But in 1878, when Ulisse Dini published his book on the theory of functions of real variables, *Fondamenti per la teorica delle funczioni di variabili reali*, Dini expressed doubt in the validity of Hankel's claim. As we shall see in Chapter 4, finding the flaw in Hankel's reasoning would greatly advance our understanding of the structure of the real numbers, as it also revealed problems with Riemann's definition of the integral.

Cantor's 1872 Paper

In 1872, Cantor published *Über die Ausdehnung eines Satzes aus der Theorie der trigonometrischen Riehen* (On the extension of a theorem from the theory of trigonometric series) where he proved results on the uniqueness of trigonometric series that converges to zero except possibly at an infinite set of points. The preface to this paper contains Cantor's construction of the irrational numbers, a step he recognized as necessary before he could work with them.

Cantor's discussion of infinite sets to which he could extend his results on uniqueness of the trigonometric series began with the set $\{1, 1/2, 1/3, 1/4, \ldots\}$. This set has the very nice property that if we consider any open interval that contains 0 and remove the points that are in that interval, then we are left with a finite set. The point 0 is called an **accumulation point** of this set, and Cantor designated this set as an infinite set of **type 1**.

Cantor actually defined type 1 sets to be infinite sets for which the derived set is finite, but he was only working with bounded sets. To extend his definition to

---

Definition: Accumulation point

Given a set S, a point x is an **accumulation point** (also known as a **limit point** or **cluster point**) of S if every open interval that contains x also contains infinitely many points of S.

Definition: Derived set, type 1 set

The set of accumulation points of S is called the **derived set** of S, denoted S'. A set is **type 1** if its derived set is nonempty but the derived set of its derived set is empty.

Definition: Type n sets, first and second species

A S set is **type n** if its derived set is type $n-1$ and S itself is not type $n-1$. A set is called **first species** if it is type n for some finite integer $n \geq 0$. A set that is not first species is said to be **second species**.

---

unbounded sets, we count the number of times that we need to take the derived set in order to get to the empty set. A set with no accumulation points is considered to be type 0. The set $\{1, 1/2, 1/3, 1/4, \ldots\}$ is type 1 because its derived set is $\{0\}$ and the derived set of $\{0\}$ is the empty set.

If a derived set is infinite, then we can consider the derived set of its derived set. For example, starting with the set

$$T = \left\{ \frac{1}{m} + \frac{1}{n} \,\middle|\, m, n \in \mathbb{N} \right\},$$

its derived set contains $\{1, 1/2, 1/3, 1/4, \ldots\}$, and with a little work (see Exercise 2.3.11) you can show that the derived set equals $\{1, 1/2, 1/3, 1/4, \ldots\}$. The set T is not type 1, but $T''' = \emptyset$, and we say that T is **type 2**.

A set is of **first species** if we get to the empty set after a finite number of derivations.

The set $\mathbb{Q}$ is not first species. Its derived set is the entire real number line. The derived set of $\mathbb{R}$ is again $\mathbb{R}$. What Cantor was able to prove is that if a trigonometric series converges to 0 at all points except possibly on a set of exceptional points that is first species, then all coefficients of the trigonometric series are zero.

Cantor's identification of sets of first species gave further impetus to the concept of outer content. Any bounded type 1 set, that is to say any set with a finite number of limit points, has outer content zero. We put small intervals around the limit points so that the sum of the lengths of these intervals is less than $\epsilon/2$. As we shall show in the next chapter, any infinite set in a bounded interval has a limit point, so once the limit points have been covered, there can only be finitely many points left.

Any bounded type 2 set has outer content zero. The set of limit points is a bounded type 1 set and so can be covered by intervals of total length less than $\epsilon/2$. The points from the original set that are not covered by these intervals are finite in number.

We now see that, by induction, any bounded first species set has outer content zero. If we have proven that a bounded set of type n must have outer content zero, then so must any bounded set of type $n + 1$ because its derived set has type n. Combining this with Hankel's insights, we see that if S_σ, the set of points at which the oscillation is greater than or equal to σ, is of first species for all $\sigma > 0$, then the function is Riemann integrable. This appears to give even greater credence to Hankel's claim that any pointwise discontinuous function is Riemann integrable. Surely if there is an entire interval of continuity around every point of continuity and the points of continuity are dense, then what is left over must be first species.

It is to Cantor's credit that he realized that what seemed obvious was not so clear. He recognized that before any further progress could be made, he needed to understand the structure of the real number line. Cantor now embarked on a quest that would profitably engage the remainder of his career and mark him as one of the great mathematicians.

Exercises

2.3.1. Find the derived set of each of the following sets. Which of these sets are first species?

1. $\mathbb{Q} \cap [0, 1]$
2. $[0, 1] - \mathbb{Q}$
3. $\left\{ \frac{n}{n+1} \,\middle|\, n \in \mathbb{N} \right\}$
4. $\left\{ \frac{2k-1}{2^n} \,\middle|\, n \in \mathbb{N}, \ 1 \le k \le 2^{n-1} \right\}$
5. $\left\{ \frac{k}{n} \,\middle|\, n \in \mathbb{N}, \ k = 1, 2, \text{ or } 3 \right\}$
6. $(0, 1) \cup (3, 4)$
7. $\bigcup_{n=1}^{\infty} \left(\frac{1}{2n}, \frac{1}{2n-1} \right)$

2.3.2. Find the outer content of each of the sets in Exercise 2.3.1. Justify each answer.

2.3.3. Show that if S has outer content zero and T is any bounded set, then

$$c_e \left(S \cup T \right) = c_e(T).$$

2.3.4. Define the function f by $f(x) = \sin(1/x)$, $x \ne 0$, $f(0) = 0$. What is the oscillation of f at 0? Justify your answer.

2.3.5. Find the oscillation at $x = 1/3$ of the function

$$g(x) = \begin{cases} x, & x \in \mathbb{Q}, \\ -x, & x \notin \mathbb{Q}. \end{cases}$$

Justify your answer.

2.3.6. Which of the following sets are dense in $[0, 1]$? For each set, justify your answer.

1. $\mathbb{Q}$
2. $[0, 1] - \mathbb{Q}$
3. The set of rational numbers with denominators that are a power of 2
4. The set of real numbers that have no 2 in their decimal expansion
5. The set of real numbers that have a 2 somewhere in their decimal expansion
6. The set of rational numbers with denominators less than or equal to 1,000
7. The set of rational numbers with denominators that are prime
8. The set of rational numbers with numerators that are prime

2.3.7. Show that the function m defined in Exercise 2.1.14 is pointwise discontinuous.

2.3.8. Consider the function h defined in Exercise 2.1.12. Is it possible to find an $\epsilon > 0$ so that h is totally discontinuous in the open interval $(-\epsilon, \epsilon)$? Explain why or why not.

2.3.9. Consider the function g in Exercise 2.1.5. Find all values of $x \in [0, 1]$ at which the function is not continuous, and find the oscillation of r at each of these points. Determine whether this function is totally discontinuous or pointwise discontinuous and justify your answer.

2.3.10. Prove that for any set S,

$$S'' \subseteq S',$$

the derived set of the derived set of S is contained in the derived set of S.

2.3.11. Prove that the derived set of $T = \{\frac{1}{m} + \frac{1}{n} | m, n \in \mathbb{N}\}$ is the set $U = \{\frac{1}{n} | n \in \mathbb{N}\} \bigcup \{0\}$. First show that $U \subseteq T'$. To prove that T has no other limit points, show that there are only finitely many points of T in $(1/N, 1/(N-1))$ that are not of the form $1/N + 1/n$.

2.3.12. Prove that

$$T_k = \left\{ \frac{1}{m_1} + \frac{1}{m_2} + \cdots + \frac{1}{m_k} \middle| m_1, m_2, \ldots, m_k \in \mathbb{N} \right\}$$

is type k. It is clear that its derived set includes T_{k-1}. The key is to show that for all k the derived set of T_k does not contain any points other than 0 that are not in T_{k-1}.

2.3.13. Consider the set of zeros of the function f_1 defined by $f_1(x) = \sin(1/x)$. What is the type of this set? What is the type of the set of zeros of $f_2(x) = \sin\left(1/f_1(x)\right)$? What is the type of the set of zeros of $f_3(x) = \sin\left(1/f_2(x)\right)$?

2.3.14. Prove that if $\underline{\lim}_{x \to c} f(x) \le f(c) \le \overline{\lim}_{x \to c} f(x)$, then $\inf_{I \in \mathcal{I}} \omega(f; I) = \overline{\lim}_{x \to c} f(x) - \underline{\lim}_{x \to c} f(x)$, where $\mathcal{I}$ is the set of open intervals containing c.

2.3.15. Given a set S, let $\mathcal{O}$ be the set of finite covers of S by open sets and let $\mathcal{C}$ be the set of finite covers of S by closed sets. Show that if $O \in \mathcal{O}$ is any finite cover of S by open intervals, then adding the endpoints to each of these open intervals gives an element of $\mathcal{C}$, a finite cover by closed sets. Show that if $C \in \mathcal{C}$ is an finite cover of S by closed sets, then removing the endpoints of each of these closed intervals covers all of S except possibly for a finite set of points. Use these observations to prove that it does not matter whether we use open or closed (or even half-open) intervals in defining the outer content of a set.

2.3.16. Given a function f defined over some interval, we use S_σ to denote the set of points at which the oscillation of f is least σ. Give an example of a function for which $c_e(S_\sigma) = 0$ for all $\sigma > 0$ but for which

$$c_e\left(\bigcup_{n=1}^{\infty} S_{1/n}\right) = 1.$$

2.3.17. Prove that if the function f is Riemann integrable on $[0, 1]$, then it is either continuous or pointwise discontinuous on $[0, 1]$.

3

Explorations of $\mathbb{R}$

Like most students entering college, mathematicians of the midnineteenth century thought they understood real numbers. In fact, the real number line turned out to be much subtler and more complicated than they imagined. As Weierstrass, Dedekind, Cantor, Peano, Jordan, and many others would show, the real numbers contain many surprises and can be quite unruly. Until they were fully understood, it would be impossible to come to a solid understanding of integration.

The complexity of real numbers illustrates a recurrent theme of mathematics. The real number line is a human construct, created by extrapolation from the world we experience, employing a process of mental experiments in which choices must be made. In one sense, the choices that have been made in formulating the properties of the real number line are arbitrary, but they have been guided by expectations built from reality.

The complexity of the real numbers arises from the superposition of two sets of patterns: the geometry of lines and distances on the one hand, and the experience of discrete numbers – integers, rationals, algebraic numbers – on the other. This is a template for much of mathematical creation. Patterns that arise in one context are recognized as sharing attributes with those of a very different genesis. As these patterns are overlaid and the points of agreement are matched, a larger picture begins to emerge. The miracle of mathematics lies in the fact that this artificial creation does not appear to be arbitrary. Repeatedly throughout the history of mathematics, this superposition of patterns has led to insights that are useful. In Wigner's phrase, we are privileged to witness the "unreasonable effectiveness of mathematics." We have tapped into something that does appear to have a reality beyond the constructions with which we began. That it entails unexpected subtleties should come as no surprise.

3.1 Geometry of $\mathbb{R}$

The real number line is, above all else, a line. While true lines may not exist in the world of our senses, we do see them at the intersection of flat or apparently flat surfaces such as the line of the horizon when looking across a sea or prairie. To imagine the line as infinite is easy, for that is simply imagining the absence of an end.

For the line as a geometric construct, the natural operation is demarcation of distance. There are two critical properties of distances that become central to the nature of real numbers:

1. However small a distance we measure, it is always possible to imagine a smaller distance.
2. Any two distances are commensurate. However small one distance might be and large the other, one can always use the smaller to mark out the larger.

The second property is known as the **Archimedean principle**,[1] that given any two distances, one can always find a finite multiple of the smaller that exceeds the larger.

Let us now take our line and mark a point on it, the **origin**. We conduct a mental experiment. We stretch the line, doubling distances from the origin. What does the line now look like? It cannot have gotten any thinner. It did not have any width to begin with. A point that was a certain distance from the origin is now twice as far, but the first property tells us that the line itself should look the same. No gaps or previously unseen structures are going to appear as we stretch it. No matter how many times we double the length, what we see does not change.

An Infinite Extension

We now kick our mental experiment up a level and imagine stretching the line by an *infinite* amount. What happens to our line? There are two reasonable answers, and a choice must be made. One reasonable answer is that it still looks like a line. This answer builds on the human expectation that whatever has never changed, never will change. Every time we have doubled the length, the line has remained unchanged. Is stretching by an infinite factor so different?

It *is* different. Go back to our original line and identify one of the points other than the origin. What happens to that point as we magnify by an infinite factor? No matter how large our field of vision, that point has moved outside of it. All points other than the origin have moved outside the field of vision. All that is left is the point at the origin. Infinite magnification has turned our line into a single point.

I said that we have a choice. We could hold onto our instinctive answer that the infinitely magnified line is still a line. That choice contradicts the Archimedean principle, and so it is not the generally accepted route to construction of the real numbers, but I do want to go down that road a little way.

---

[1] Also known as the Archimedean axiom or the continuity axiom. It predates Archimedes, appearing as definition 4 of book 5 of Euclid's *Elements*.

Consider one of the points other than the origin on the infinitely magnified line. Where was it before the magnification? It certainly was not any measurable distance away from the origin, otherwise it would have sailed off to infinity when we magnified the line. It was not on top of the origin because then it would have stayed put. It must have been off the origin, but less than any measurable distance away from the origin. Its distance from the origin must have been so small that it is incommensurate with any measurable distance. No matter how many of these tiny distances we take, we cannot fill any measurable distance, no matter how small. The tiny distances are known as **infinitesimals**. Leibniz used them to explain his development of the calculus. It *is* possible to develop analysis using the real number line with infinitesimals, though the full complexity of infinitesimals is much greater than what is suggested by this simple thought experiment. This approach to calculus through infinitesimals is called **nonstandard analysis** and was developed in the early 1960s, beginning with the work of Abraham Robinson at Princeton. Initially, the logical underpinnings required to work with infinitesimals were daunting. Since then, these foundations have been greatly simplified, and nonstandard analysis has vocal proponents.

If the real number line includes infinitesimals, then every point on the real line must be surrounded by a cloud of points that are an infinitesimal distance away. The infinitely magnified line, if it really is a replica of the original line, must also contain infinitesimals, and they must be stretched from points that were infinitesimally small with respect to the original infinitesimals. Thus every infinitesimal is surrounded by a cloud of points whose distance is infinitesimally infinitesimal, and these by third-order infinitesimals, and so on.

"And so on" is a wonderful human phrase. We actually can imagine this unimaginable construction, or at least imagine enough that we are prepared to accept it and work with it. There is a real choice to be made. That choice was made in the early nineteenth century. Looking back, the decision to hold to the Archimedean principle appears inevitable. As I said earlier, the real number line is, above all else, a line. We are working with distances, and distances in our everyday experience are commensurate. Those seeking foundations for analysis preferred to stay to the simpler, surer, and more intuitive ground of commensurate distances.

Topology of ℝ

The only geometry on a line is the measurement of distance. There is an entire branch of mathematics that is built on the concept of distance and its generalizations: topology. Topology gets much more interesting in higher dimensions, but there is a lot going on in just one dimension.

We begin with some basic definitions. The basic building block of topology is the ϵ-**neighborhood** of a point a. Objects such as neighborhoods are best visualized

---

Definition: ϵ-Neighborhood

Given $\epsilon > 0$, the ϵ-**neighborhood** of a point a, $N_\epsilon(a)$, is the set of all points whose distance from a is strictly less than ϵ:

$$N_\epsilon(a) = \{x \mid |x - a| < \epsilon\}. \tag{3.1}$$

Definition: Open set

We say that a set S is **open** if for each $x \in S$, there is an ϵ-neighborhood of x that is contained entirely inside S: $x \in S$ implies that there is an $\epsilon > 0$ such that $N_\epsilon(x) \subseteq S$.

---

in two- or three-dimensional space, but you also need to visualize them as they live on the real number line.

On the real number line, this is the open interval $(a - \epsilon, a + \epsilon)$. In the plane $\mathbb{R}^2$, it is a disc centered at a (without the bounding edge), and in $\mathbb{R}^3$ it is the solid ball centered at a (without the bounding surface). It is also often useful to work with a **deleted** or **punctured neighborhood** of a point a, consisting of a neighborhood with the point a removed, $\{x \mid 0 < |x - a| < \epsilon\}$.

Any open interval is open. Any union of open intervals is open. The empty set is open. (Since there is no $x \in S$, this statement is true about every x that is in S). The entire real line $\mathbb{R}$ is open. In two dimensions, the inside of any polygon, without the boundary, is open.

Theorem 3.1 (Equivalent Definition of Continuity). *A function is continuous over its domain if and only if the inverse image of every open set in the range is an open set in the domain.*

Consider the function defined by $f(x) = x^2$. The inverse image of $(1, 4)$ consists of all points of $\mathbb{R}$ whose square lies strictly between 1 and 4. This is $(-2, -1) \bigcup (1, 2)$. The inverse image of $(-1, 4)$ is $(-2, 2)$. Think about why.

Proof. We begin with the ϵ-δ definition of continuity. A function f is continuous at a if and only if given any $\epsilon > 0$, there exists a response δ such that $|x - a| < \delta$ implies that $|f(x) - f(a)| < \epsilon$.

We first translate this statement into the language of ϵ-neighborhoods. A function f is continuous at a if and only if given any ϵ-neighborhood of $f(a)$, there is a δ-neighborhood of a such that

$$f\left(N_\delta(a)\right) \subseteq N_\epsilon\left(f(a)\right).$$

This containment is equivalent to $N_\delta(a) \subseteq f^{-1}(N_\epsilon(f(a)))$. Thus the ϵ-δ definition of continuity can be restated as: Given a point a in the domain of f and any $\epsilon > 0$, there exists a response δ such that

$$N_\delta(a) \subseteq f^{-1}\left(N_\epsilon\left(f(a)\right)\right).$$

We now translate this into the language of open sets. Start by assuming that f is continuous and let T be an open set in the range. If $x \in f^{-1}(T)$, then $f(x) \in T$, so we can find an ϵ-neighborhood $N_\epsilon(f(x)) \subseteq T$. We have seen that continuity implies that there is a $\delta > 0$ for which $N_\delta(x) \subseteq f^{-1}(N_\epsilon(f(x))) \subseteq f^{-1}(T)$, so $f^{-1}(T)$ is open.

If we assume that $f^{-1}(T)$ is open for every open set T in the range, then pick any $f(x) \in T$ and any ϵ for which $N_\epsilon(f(x)) \subseteq T$. We know that $N_\epsilon(f(x))$ is open, and therefore by our assumption so is $f^{-1}(N_\epsilon(f(x)))$. Since this set is open, we can find a $\delta > 0$ for which $N_\delta(x) \subseteq f^{-1}(N_\epsilon(f(x)))$. $\square$

Be careful. If f is continuous, it does not necessarily follow that if S is open, so is $f(S)$ (consider any nonmonotonic continuous function, see Exercise 3.1.2).

More Definitions

A closed interval is closed. Any finite set of points is closed. Any set of points that forms a convergent sequence, taken together with its limit, is a closed set. The empty set and the entire real line $\mathbb{R}$ are closed. These are the only two sets that are both open and closed (see Exercise 3.1.8). Many sets such as $(0, 1]$ and $\{1, 1/2, 1/3, 1/4, \ldots\}$ (without the limiting value 0) are neither open nor closed.

If every neighborhood of x contains a point of S other than x itself, then every neighborhood of x contains infinitely many points of S (see Exercise 3.1.7), and therefore x is an accumulation point of S (recall the definition on page 46). We can characterize closed sets in terms of accumulation points.

Proposition 3.2 (Characterization of Closed Sets). *A set S is closed if and only if it contains all of its accumulation points.*

Proof. If S is closed, then its complement is open. No accumulation point of S could be in the complement (see Exercise 3.1.11), so S contains all of its accumulation

---

Definition: Closed set

A set S is **closed** if and only if its complement (S^C, the set of points that are not in S) is open.

Definition: Interior, closure, and boundary

The **interior** of S is the union of all open sets contained in S. The **closure** of S is the intersection of all closed sets that contain S. The **boundary** of S, ∂S, consists of all points in the closure that are not in the interior.

---

points. If S contains all of its accumulation points, then any point in S^C sits inside some ϵ-neighborhood entirely inside S^C (see Exercise 3.1.12). Therefore, S^C is open and so S is closed. $\square$

It follows that any closed set contains its derived set. If S is closed, then its elements might or might not be accumulation points. Thus, the set

$$S = \{0, 1/2, 1/3, 1/4, \ldots, 1/n, \ldots\}$$

is closed, but 0 is the only accumulation point of S. On the other hand, if $T = [0, 1]$, then T is closed and every point in T is an accumulation point of T. This closed set is equal to its derived set. Any derived set is closed (see Exercise 3.1.21).

It took a long time for the mathematical community to recognize the importance of open and closed sets. Closed sets, which contain their own derived set, are the older notion and can be traced back to Cantor in 1884. Mathematicians working in analysis in the late nineteenth and even early twentieth centuries would fail to make it clear whether the intervals that they described were to include endpoints or not. Often it made no difference. Sometimes it was critically important. One of the most infamous examples was Borel's 1905 *Leçons sur les fonctions de variable réelles* (Lectures on real variable functions).[2] As late as the early twentieth century, some mathematicians including the Youngs used the term "open set" to mean a set that is not closed. The current meaning can be traced to Baire in 1899. It was Lebesgue who popularized the current definition.

The interior of an open set is itself. The interior of the closed interval $[a, b]$ is the open interval (a, b). The interior of a finite set of points is the empty set. The closure of a closed set is itself, while the closure of the open interval (a, b) is $[a, b]$. The half-open, half-closed interval $[-1, 1)$ has closure $[-1, 1]$, interior $(-1, 1)$, and a boundary that consists of two points, $\{-1, 1\}$. The closure of $\{1/n \mid n \in \mathbb{N}\}$ is $\{1/n \mid n \in \mathbb{N}\} \bigcup \{0\}$. Its interior is the empty set. The boundary is $\{1/n \mid n \in \mathbb{N}\} \bigcup \{0\}$.

The interior, closure, and boundary can also be described in terms of ϵ-neighborhoods.

---

[2] Renfro speculates that this might have been the fault of Maurice Fréchet, then a young graduate student, who transcribed these lectures.

Proposition 3.3 (Characterizations of Interior, Closure, and Boundary). *A point x is in the interior of S if and only if there is some ϵ-neighborhood of x that is completely contained in S. A point x is in the closure of S if and only if every ϵ-neighborhood of x contains a point that is in S. A point x is in ∂S if and only if every ϵ-neighborhood of x contains a point that is in S and a point that is not in S.*

Proof. The characterization of the interior of S is left for Exercise 3.1.22.

From DeMorgan's laws (Theorem 1.3, the complement of the intersection of the closed sets that contain S is the union of the open sets that are contained in S^C. Therefore, x is in the closure of S if and only if it is not in the interior of S^C, which holds if and only if there is an ϵ-neighborhood of x that is not completely contained in S^C.

The characterization of points in the boundary follows from the definition of the boundary as the set of points that are in the closure and not in the interior.   □

Note that a set S is dense (recall the definition on p. 45) in T if, for all $x \in T$, every ϵ-neighborhood of x contains infinitely many points of S. The set of rational numbers is dense in $\mathbb{R}$. We can take a much smaller set and still be dense in $\mathbb{R}$. The set of rational numbers whose denominators are even (recall Riemann's function, Example 2.1) is dense. Even if we restrict ourselves to the set of rational numbers whose denominators are powers of 2, this also is a dense subset of $\mathbb{R}$. If a set S is dense in $\mathbb{R}$, then every real number is an accumulation point of S. In this case, the closure of S is the entire real number line.

Exercises

3.1.1. Prove that every constant function is continuous.

3.1.2. Give an example of a continuous function f and an open set S in the domain of f such that $f(S)$, the image of S, is not open.

3.1.3. Define the function f over the domain $[-1/\pi, 1/\pi]$ by

$$f(x) = \sin(1/x), \ x \neq 0, \quad f(0) = 0.$$

Describe $f^{-1}(N_{1/2}(0))$, the inverse image of $N_{1/2}(0)$. Is this inverse image open, closed, or neither?

3.1.4. Define the function g over the domain $[-1/\pi, 1/\pi]$ by

$$g(x) = x \sin(1/x), \ x \neq 0, \quad g(0) = 0.$$

Describe $g^{-1}(N_{1/2}(0))$, the inverse image of $N_{1/2}(0)$. Is this inverse image open, closed, or neither?

3.1.5. For each of the following sets in $\mathbb{R}^2$, state whether the set is open, closed, or neither. Then find the closure, the interior, and the boundary of the set.

1. $\{(x, y) \mid x > 0\}$
2. $\{(x, y) \mid x + y > 1 \text{ and } y \leq 1\}$
3. $\{(x, y) \mid x + y \geq 1, \ x > 0, \text{ and } y > 1\}$
4. $\{(x, y) \mid y = x^2\}$
5. $\{(x, y) \mid x = 1, y < 2\}$
6. $\{(x, y) \mid 1 \leq x \leq 2, y \in \mathbb{N}\}$
7. $\{(x, y) \mid 1 \leq x \leq 2, y \in \mathbb{Q}\}$

3.1.6. For each of the following sets in $\mathbb{R}$, state whether the set is open, closed, or neither. Then find the closure, the interior, and the boundary of the set.

1. $\mathbb{Q}$
2. $\mathbb{R} - \mathbb{Q}$
3. $\displaystyle\bigcup_{n=1}^{\infty} (1/2n, 1/2n - 1)$
4. $\displaystyle\bigcup_{n=1}^{\infty} [1/2n, 1/2n - 1]$
5. the set of all rational numbers with denominators that are less than 1,000
6. the set of all rational numbers with denominators that are powers of 2
7. the set of all rational numbers with numerators that are powers of 2

3.1.7. Prove that every neighborhood of x contains at least one point of S other than x itself if and only if every neighborhood of x contains infinitely many points of S.

3.1.8. Prove that if a set is both open and closed, then it is either $\mathbb{R}$ or the empty set.

3.1.9. Give an example of a set with an accumulation point that is not a boundary point.

3.1.10. Give an example of a set with a boundary point that is not an accumulation point.

3.1.11. Prove that if S^C, the complement of S, is open, then S^C cannot contain an accumulation point of S.

3.1.12. Prove that if $x \in S^C$ is not an accumulation point of S, then there is an $\epsilon > 0$ for which $N_\epsilon(x) \subseteq S^C$.

3.1.13. Prove that any union of open sets is open.

3.1.14. Prove that any finite intersection of open sets is open.

3.1.15. Give an example of an infinite intersection of open sets that is not open.

3.1.16. Give an example of an infinite intersection of open sets that *is* open.

3.1.17. Prove that any intersection of closed sets is closed.

3.1.18. Prove that any finite union of closed sets is closed.

3.1.19. Give an example of an infinite union of closed sets that is not closed.

3.1.20. Give an example of an infinite union of closed sets that *is* closed.

3.1.21. Let S be a derived set (the set of accumulation points of some set T). Show that S^C is open, and thus S is closed.

3.1.22. Prove that a point x is in the interior of S if and only if there is some ϵ-neighborhood of x that is completely contained in S.

3.2 Accommodating Algebra

The real number line entails more than distance. We want to assign values to its points. We choose a point that will represent the origin, 0, and then pick a second point, label it "1," and use the distance between the origin and 1 as our basic linear unit. We can locate points whose distances from the origin correspond to integers and rational numbers. We can even locate points that correspond to irrational lengths such as $\sqrt{2}$, the length of the diagonal of a unit square. Taking mirror images across the origin, we locate the negatives of these numbers. We have imposed a system of a discrete set of objects – integers, rational numbers, algebraic numbers – onto our continuum of distances. But this system does not account for all points in $\mathbb{R}$. In 1844, Joseph Liouville proved that there are points on the real line that are not algebraic.

It is a very small and self-evident step to believe that every point on the continuum of the real line corresponds to a number, but this step carries enormous repercussions, for from now on we will be using the geometric notion of distance to inform our concept of number that, until now, had been restricted to quantities arising from algebraic constructions.

Implications of the Bolzano–Weierstrass Theorem

Although some mathematicians resisted expanding the notion of number beyond algebraic numbers, most recognized the need to do so. Foremost among them were Richard Dedekind and Karl Weierstrass.

Beginning in the academic year 1857–1858 and then every 2 years until the 1880s, Weierstrass lectured on analysis at the University of Berlin. In these lectures, he developed and expounded many of the basic principles of analysis. His students would work through these ideas, refine them, and eventually publish them. It was in these lectures that Weierstrass first explained what today we call the Bolzano–Weierstrass theorem. His proof rested on the nested interval principle. The shared attribution with Bernhard Bolzano arises from Weierstrass's acknowledgment of his indebtedness to Bolzano's 1817 proof that the convergence of every Cauchy sequence implies that every bounded, increasing sequence has a limit.

Theorem 3.4 (Bolzano–Weierstrass Theorem). *Any bounded infinite set S has an accumulation point.*

Proof. Weierstrass's proof proceeds as follows. If we have an infinite set of points contained in the bounded interval $[a, b]$, we consider the two half-intervals, $[a, (a + b)/2]$ and $[(a + b)/2, b]$, and choose one that has infinitely many points from our set. We then divide this interval in half and choose an interval of length $(b - a)/4$ that contains infinitely many points from our set. We continue in this way, generating an infinite sequence of nested intervals of arbitrarily small length, each of which contains infinitely many points from our set. The nested interval principle guarantees a point α that lies in all of these intervals. Every neighborhood of α contains one of these nested intervals, and therefore every neighborhood of α contains infinitely many points from our set. The point α is an accumulation point of S. $\qquad\qquad\square$

The nested interval principle implies what we can call the Bolzano–Weierstrass principle, that every bounded infinite set has a limit point, but the Bolzano–Weierstrass principle also implies the nested interval principle. Given a nested sequence of intervals, we create an infinite set S by choosing one point from each interval so that no point duplicates any of the previously chosen points. By Bolzano–Weierstrass, the set S has a limit point, and since every neighborhood of this limit point contains infinitely many elements of S, this limit point must be inside every interval.

How do we define and assign values to those points on the real number line that are predicted by Bolzano–Weierstrass but are neither algebraic nor roots of common functions? Is there a way we can start with arithmetic and build to all of the values represented by the real number line? Several notable mathematicians wrestled with this question. Beginning in the late 1850s, Richard Dedekind, Karl Weierstrass, Georg Cantor, and H. Charles Mèray each found his own solution. Dedekind and Cantor published their solutions in 1872. The details of these solutions are less important than what they have in common, all drawing on a fundamental observation about the structure of $\mathbb{R}$: The set of rationals is dense on the real number line.

Every open interval, no matter how far we may have zoomed in, contains at least one and therefore infinitely many rational numbers (see Exercise 3.2.1). Any point on the real number line is uniquely determined by reference to these rational numbers. To Dedekind, each irrational number is described by considering all rationals less than the point in question and all rationals that are greater. Given two such sets with the property that every element of the first is strictly less than every element of the second and their union consists of all rational numbers, such a "Dedekind cut" defines a unique point in $\mathbb{R}$. Weierstrass used series. Cantor and Mèray used sequences of rational values that could be forced as close as desired to the point in question by taking sufficiently many terms or going out sufficiently far in the sequence. Cantor identified each point in $\mathbb{R}$ with the collection of rational Cauchy sequences that converge to this point. Each point on the real number line is identified by an appropriate collection of Cauchy sequences.

We are using a modified form of Cantor's definition when we identify the elements of $\mathbb{R}$ with all possible decimals to infinitely many places. Such a decimal expansion represents a choice of a particular Cauchy sequence. For example, we identify π with a Cauchy sequence that begins $(3, 3.1, 3.14, 3.141, 3.1415, 3.14159, \ldots)$. When we write "$\pi = 3.14159\ldots$," there are two observations that we need to make. The first is that, actually, we have not specified the location of π. The information supplied by $3.14159\ldots$ tells us nothing about the digit that follows 9. There are an infinitude of points on the real number line that begin with this particular decimal expansion. Giving a thousand or a million or a billion digits gets us no further in the sense that there are still infinitely many different points whose expansions start with those digits.

The second observation is that the statement "$\pi = 3.14159\ldots$" nevertheless does tell us something very important. It implies that there *is* a sequence of rational approximations to π that begins $(3, 3.1, 3.14, 3.141, 3.1415, \ldots)$ and that eventually will enter and stay within any open interval containing π, no matter how small that interval might be. We may not know how to find an arbitrary term of this sequence, but we are asserting its existence. Such a statement tells us that π is a point on the real number line and that it is located within the interval $[3.14159, 3.14160]$.

There are, of course, many explicit Cauchy sequences that represent π. One of them is given by

$$\left(4, 4 - \frac{4}{3}, 4 - \frac{4}{3} + \frac{4}{5}, 4 - \frac{4}{3} + \frac{4}{5} - \frac{4}{7}, 4 - \frac{4}{3} + \frac{4}{5} - \frac{4}{7} + \frac{4}{9}, \ldots\right).$$

The point of Cantor's construction is not that we can find a Cauchy sequence for each real number, but that it exists.

Completeness

Dedekind based his construction on the assumption that every nonempty bounded set should have a least upper bound. Cantor based his construction on the

---

Definition: Completeness

A set of numbers is called **complete** if it has any of the four equivalent properties:

- Every sequence of closed, nested intervals has a nonempty intersection that belongs to the set.
- Every bounded subset has a least upper bound in the set.
- Every Cauchy sequence converges to a point in the set.
- Every infinite bounded subset has a limit point in the set.

---

assumption that every Cauchy sequence should converge. The nested interval principle is another equivalent assumption. These are different but equivalent ways of making precise what we mean when we describe the real number line as a continuum. This property of $\mathbb{R}$ would eventually come to be known as **completeness**.

In particular, the set of all real numbers is complete. The set of all rational numbers is not complete. Today, rather than attempting to define the set of real numbers so as to justify their completeness, it is common to simply assert as axiomatic that $\mathbb{R}$ contains all rational numbers and is complete.

I have already explained Weierstrass's proof that the nested interval principle and the Bolzano–Weierstrass principle are equivalent. The equivalence of the remaining statements is left for you in Exercises 3.2.8–3.2.10.

The nineteenth century witnessed an increasing sense of paradox from the interplay of algebra and geometry on the real number line. It reached a peak in the 1880s. One of the curious phenomena that was discovered and debated in that decade began with the observation that the rational numbers are denumerable or **countable**.

Informally, a set is countable if it is possible to list its elements in order: first, second, third, ... The rational numbers in [0, 1] can be listed by ordering them by the size of the denominator (when reduced) and among those of equal denominator by the numerator:

$$\frac{0}{1}, \frac{1}{1}, \frac{1}{2}, \frac{1}{3}, \frac{2}{3}, \frac{1}{4}, \frac{3}{4}, \frac{1}{5}, \frac{2}{5}, \frac{3}{5}, \frac{4}{5}, \frac{1}{6}, \frac{5}{6}, \frac{1}{7}, \dots \tag{3.2}$$

The fact that the set of rational numbers is countable leads to an important characterization of open sets in $\mathbb{R}$.

Theorem 3.5 (Characterization of Open Sets). *Every open set in* $\mathbb{R}$ *is a countable union of disjoint open intervals.*

Proof. Let U be an open set, and choose any $t \in U$. Let $a = \inf\{x \mid (x, t] \subseteq U\}$. The point a cannot be in U, otherwise there would be a neighborhood of a contained in U

> **Definition: Countable**
>
> A set is **countable** or **denumerable** if it is finite or if it is in one-to-one correspondence with $\mathbb{N}$, the set of positive integers. A set that is countable and not finite is called **countably infinite**.

and we could find an $x < a$ for which $(x, t] \subseteq U$. Similarly, let $b = \sup\{x \mid [t, x) \subseteq U\}$. The point b cannot be in U, but $(a, b) \subseteq U$. (Note that we might have $a = -\infty$ and/or $b = \infty$.) We call this interval $I(t) = (a, b)$.

If s and t are two points in U, then either $I(s) = I(t)$ or $I(s) \cap I(t) = \emptyset$, so U is a union of disjoint open intervals. To see that there are at most countably many open intervals, we observe that we can find a distinct rational number inside each interval. $\qquad\square$

Harnack's Mistake

We take the ordering of the rational points in $[0, 1]$ given in (3.2) and call them ($a_1 = 0, a_2 = 1, a_3 = 1/2, a_4 = 1/3, \ldots$). We choose any positive ϵ and let I_k be the open interval of length $\epsilon/2^k$ that is centered at a_k: $I_k = (a_k - \epsilon/2^{k+1}, a_k + \epsilon/2^{k+1})$. Does the union of these intervals contain all points in $[0, 1]$? In other words, can we put the closed interval $[0, 1]$ inside a countable union of open intervals whose lengths add up to ϵ? This was a problem first posed by Axel Harnack in 1885. He convinced himself that the answer is "yes."

Axel Harnack (1851–1888) was the younger twin brother of the German theologian Adolf von Harnack. Axel earned his doctorate at Erlangen-Nürnberg University in 1875, working under the direction of Felix Klein. He is best known for his work in harmonic analysis and the theory of algebraic curves.

In essence, what Harnack did was to ask himself, "What is the complement of a countable union of intervals?" He believed that it must also be a countable union of intervals. Think about this. The intervals might be open or closed or half-open/half-closed, and a closed interval might be a single point. It is certainly true that the complement of any finite union of intervals is a finite union of intervals. It is not obvious that the same would not be true for countable unions. But if Harnack was right, then the complement of $\bigcup I_k$ is a countable union of intervals. The intervals in the complement must be single points, otherwise they would contain rational numbers between 0 and 1. We now put each of these countably many points inside intervals whose lengths add up to ϵ, and we now have all of $[0, 1]$ contained within a union of countably many intervals whose lengths add up to 2ϵ.

In fact, Harnack's basic premise, that the complement of a countable union of intervals is a countable union of intervals, is wrong. The complement can be an uncountable union. Georg Cantor and others understood this. The flaw in Harnack's reasoning underscores some of the complexity of the real number line as a set of numbers, and we shall treat it in full detail in the next section. But that does not prove that his answer was wrong. This was accomplished by Émile Borel in 1895.

Borel's Series

A year after earning his doctorate, Borel published *Sur quelques points de la théorie des fonctions* (On some points in the theory of functions), a paper that dealt with a question from complex analysis. Specifically, he studied analytic continuation across a boundary on which we have a countable dense set of poles (points for which the function is unbounded in every neighborhood). We shall focus on a very special case of the type of function he studied,[3] the series

$$\sum_{n=1}^{\infty} \frac{A_n}{|x - a_n|}, \qquad \text{where } A_n > 0, \quad \sum_{n=1}^{\infty} A_n^{1/2} = A < \infty,$$

and the points $\{a_n\}$ are dense in $[0, 1]$. For example, we could take $\{a_n\}$ to be the set of rational numbers in $[0, 1]$.

Does this series converge at any points in $[0, 1]$? At first glance, it may appear that the answer is obviously "no." After all, this function is unbounded in *every* interval. But take a closer look. Choose any constant $c > 0$. If we can find an $x \in [0, 1]$ so that

$$|x - a_n| \geq c A_n^{1/2}, \quad \text{for all } n \geq 1,$$

then the series converges,

$$\sum_{n=1}^{\infty} \frac{A_n}{|x - a_n|} \leq \sum_{n=1}^{\infty} \frac{A_n^{1/2}}{c} = \frac{A}{c}.$$

For any x in

$$E_c = \left\{ x \in [0, 1] \,\big|\, |x - a_n| \geq c A_n^{1/2} \text{ for all } n \geq 1 \right\},$$

our series converges. But are there any points in E_c?

Borel considered the complement of E_c. It is a union of open intervals, $|x - a_n| < c A_n^{1/2}$, the nth of which has length $2c A_n^{1/2}$. The sum of the lengths of these intervals is $2cA$. If we choose $c < 1/2A$, then these intervals have total combined length less than 1. To prove his result about analytic continuation, Borel needed to know that there is at least one point in E_c.

Borel proved that if the sum of the lengths of the open intervals is strictly less than 1, then there must be points – in fact, uncountably many points – that are

---

[3] The interested reader can find a fuller description in Hawkins's *Lebesgue's Theory of Integration*, pp. 97–106.

not in any of these intervals. For $c < 1/2A$, the set E_c contains uncountably many points in $[0, 1]$. In a note appended to this paper, he remarked that his proof actually demonstrated a stronger statement, a theorem that today we call the Heine–Borel theorem. In stating and proving this theorem, Borel was interested only in the case of a countable collection of open intervals. We give it in its most general form.

Theorem 3.6 (Heine–Borel Theorem). *If $\{U_k\}$ is any countable or uncountable collection of open sets whose union contains $[a, b]$, then there is a finite subcollection, $\{U_{k_1}, U_{k_2}, \ldots, U_{k_n}\}$, whose union also contains $[a, b]$.*

This theorem will become one of our most useful tools for proving results about measure theory and Lebesgue integration. It implies that if we have a collection of open intervals for which the sum of the lengths is strictly less than 1, then the union of those intervals cannot contain $[0, 1]$. If it did, then by Heine–Borel there would be a finite subcollection that also contained $[0, 1]$, and this is impossible. In Exercise 3.2.3, you are asked to prove that a finite union of intervals of total length less than 1 cannot cover $[0, 1]$.

Henri Lebesgue in his 1904 book *Leçons sur l'Intégration et la Recherche des Fonctions Primitives* (Lectures on integration and the search for antiderivatives) gave the following proof, which is valid for any collection of open intervals, including an uncountable collection.

Proof. Consider the set S of points $x \in [a, b]$ for which $[a, x]$ is contained in a finite union of intervals from $\{U_k\}$. Since $a \in S$, we know that this set is not empty. Since the set has an upper bound, it must have a least upper bound, call it $\beta = \sup S$. The point β is contained in one of these open sets, say $U_j = (\alpha_j, \beta_j)$, $\alpha_j < \beta < \beta_j$. If we take the finite union of intervals that contain $[a, \beta)$ and add to it the open set U_j, we have a finite union of open sets that contains $[a, \beta_j)$. This contradicts the assumption that $\beta = \sup S$ unless $\beta = b$. □

Compactness

This property of closed, bounded intervals is so important that it has a name, **compactness**. The Heine–Borel theorem tells us that any closed, bounded interval, $[a, b]$, is compact. We are interested in all sets that share this property.

Corollary 3.7 (Compact $\Longleftrightarrow$ Closed and Bounded). *A set is compact if and only if it is closed and bounded.*

Proof. We leave it as Exercises 3.2.4 and 3.2.5 to show that if a set is not closed or not bounded, then it is not compact.

> **Definition: Cover and compactness**
>
> A collection of open sets, $\mathcal{C}$, whose union contains the set S is called an **open cover** of S. Given a cover $\mathcal{C}$ of S, a **subcover** is any subcollection of sets from $\mathcal{C}$ whose union also contains S. A set S is said to be **compact** if every open cover of S contains a finite subcover.

We assume that S is closed and bounded. Since S is bounded, we can find a closed interval $[a, b]$ that contains S. Since S is closed, its complement S^C is open. Let $\mathcal{C}$ be an open cover of S. If we add the set S^C to this collection, we get a collection of open sets, call it $\mathcal{C}'$ whose union is all of $\mathbb{R}$. It follows that $\mathcal{C}'$ is an open cover of $[a, b]$ and so, by the Heine–Borel theorem, it has a finite subcover of $[a, b] \supseteq S$. If S^C is in this finite subcover, we remove it to get a finite subcollection of $\mathcal{C}$ that will still cover S. $\qquad\square$

Proposition 3.8 (Continuous Image of Compact Is Compact). *If S is compact and f is continuous on S, then $f(S)$ is compact.*

Proof. We first prove that $f(S)$ is bounded. Let (I_n) be a sequence of intervals of length 1 for which $f(S) \subseteq \bigcup I_n$. Since f is continuous, $f^{-1}(I_n)$ is open and $\left(f^{-1}(I_n)\right)$ is an open cover of S. Since S is compact, we can find a finite subcollection $\left(f^{-1}(I_{n_k})\right)$ that contains all of S. It follows that $\bigcup I_{n_k}$ contains $f(S)$, and this is a finite union of intervals of length 1.

We now prove that $f(S)$ is closed. Let y be any accumulation point of $f(S)$ and choose a sequence $(y_j)_{j=1}^{\infty} \subseteq f(S)$ that converges to y. Since $\left(f^{-1}(y_j)\right)_{j=1}^{\infty}$ is contained in the bounded set S, the Bolzano–Weierstrass theorem promises us an accumulation point, x_0, of this sequence. It follows that there is a subsequence that converges to this accumulation point, $f^{-1}(y_{j_k}) \to x_0$. Since S is closed, x_0 must be in S. By continuity, y_{j_k} converges to $f(x_0)$, and therefore $y = f(x_0) \in f(S)$. $\qquad\square$

Two Corollaries

There are two immediate corollaries of the Heine–Borel theorem that are historically intertwined. They predate Borel's theorem of 1895. The first corollary is the Bolzano–Weierstrass theorem, Theorem 3.4. Since S is bounded, it is contained in a closed interval $[a, b]$. If there are no limit points, then for each point of $[a, b]$ we can find a neighborhood that contains only finitely many points of S. The collection of these neighborhoods is an open cover of $[a, b]$, and so there is a finite subcover of $[a, b] \supseteq S$. But this finite subcover contains only finitely many points of S.

Since the Heine–Borel theorem follows from the assumption that every bounded set has a least upper bound, and it in turn implies the Bolzano–Weierstrass theorem, we see that the Heine–Borel theorem is yet another equivalent statement of what it means to say that $\mathbb{R}$ is complete.

The second corollary is Theorem 1.13: A continuous function on a closed and bounded interval is uniformly continuous on that interval. In 1904, Lebesgue observed that Heine–Borel implies "a pretty demonstration of the uniformity of continuity." In his words,

> Let f be a continuous function at all points of $[a, b]$. Each point of $[a, b]$ is, by definition, inside an interval Δ on which the oscillation of f is less than ϵ. One can cover $[a, b]$ with a finite number of them. Let l be the length of the shortest interval that is used. In each interval of length l, the oscillation of f is at most 2ϵ since such an interval overlaps at most two of the intervals Δ. The continuity is uniform.[4]

How Heine's Name Got Attached to This Theorem

Theorem 3.6 was discovered by Émile Borel in 1895. Eduard Heine died in 1881. Why is this result called the Heine–Borel theorem? Pierre Dugac has told the story[5] behind this theorem, which he, tongue-in-cheek, refers to as the "Dirichlet–Heine–Weierstrass–Borel–Schoenflies–Lebesgue theorem." I shall summarize the high points. The essence of the answer is that Heine became known as the first person to prove Theorem 1.13, and his proof looks suspiciously like the proof of Theorem 3.6.

When Cauchy, in 1823, first proved that every continuous function is integrable, he needed more than ordinary continuity. He needed uniform continuity.[6] There is some debate about whether Cauchy meant continuity or uniform continuity when he talked about continuous functions. He never made a clear distinction. Dirichlet was the first to recognize this distinction and to realize that over a closed and bounded interval it did not matter because any continuous function would also be uniformly continuous.

In 1852, Dirichlet gave a course on integration at the University of Berlin, in which he gave a proof of Theorem 1.13 that I shall loosely paraphrase. Given a function f, continuous on $[a, b]$, and given any $\epsilon > 0$, we need to find a response δ so that for any pair of values $x_1, x_2 \in [a, b]$, $|x_1 - x_2| < \delta$ implies that $|f(x_1) - f(x_2)| < \epsilon$. Consider the set

$$S_a = \left\{ t \in [a, b] \,\middle|\, a \leq x \leq t \Rightarrow |f(x) - f(a)| < \epsilon \right\}.$$

---

[4] Lebesgue (1904, p. 113, fn). The notation for closed intervals was not standard at that time. Instead of using $[a, b]$, Lebesgue refers to it as "(a, b) including a and b."

[5] See Dugac (1989, pp. 91f).

[6] See *A Radical Approach to Real Analysis*, 2nd ed., pp. 241–243.

Clearly $a \in S_a$, so the the set is not empty. Since it is bounded, it has a least upper bound. Set $c_1 = \sup S_a$. Dirichlet observes that if $c_1 \neq b$, then $|f(a) - f(c_1)| = \epsilon$ (see Exercise 3.2.6). We now define

$$c_2 = \sup\{t \in [c_1, b] \mid c_1 \leq x \leq t \Rightarrow |f(x) - f(c_1)| < \epsilon\}.$$

We continue in this way, obtaining a sequence of values $a < c_1 < c_2 < c_3 < \cdots \leq b$ with the property that $|f(c_{k+1}) - f(c_k)| = \epsilon$ and if $c_k < x < c_{k+1}$, then $|f(x) - f(c_k)| < \epsilon$. If there are only a finite number of values of c_k before we get to b, then we have uniform continuity (see Exercise 3.2.7). We only need to rule out the possibility that $(a < c_1 < c_2 < \cdots)$ is an infinite sequence, converging to some $c < b$.

If there is such a limit c, we know that f is continuous at c, and therefore we can find a response δ so that $c - \delta < x < c + \delta$ implies that $|f(x) - f(c)| < \epsilon/2$. Since the sequence (c_k) converges to c, we can find two consecutive elements of the sequence that lie inside this δ-neighborhood, say c_j and c_{j+1}. But then

$$\epsilon = |f(c_j) - f(c_{j+1})| \leq |f(c_j) - f(c)| + |f(c) - f(c_{j+1})| < \epsilon.$$

This concludes Dirichlet's proof.

Heine had been one of Dirichlet's students in Berlin. In 1872 he published an important paper that solidified many of the key concepts of analysis, *Die Elemente der Functionenlehre* (Elements of function theory). Among these was uniform continuity. Heine stated Theorem 1.13 and proved it using precisely Dirichlet's argument. Heine did not credit Dirichlet. The notes from Dirichlet's 1852 lectures would be published in 1904, but Heine's paper marked the first time that this argument appeared in print.

In succeeding years, specific cases of the Heine–Borel theorem reappeared in various contexts. In 1880, Weierstrass proved that if for each $x \in [a, b]$ a series converges uniformly in some neighborhood of x, then the series converges uniformly over $[a, b]$. In 1882, Salvatore Pincherle proved that if for each $x \in [a, b]$ there is a neighborhood in which f stays bounded, then f must be bounded over $[a, b]$.

Arthur Schönflies, in 1900, was the first to point out that the Heine–Borel theorem applied equally well to covers consisting of an uncountable number of open sets. He was also the first to connect Heine's name to this result, describing it as an extension by Borel of a theorem of Heine and designating it as the Heine–Borel theorem. This name gained popularity when it was picked up by William Henry Young in his 1902 paper, *Overlapping Intervals*.

Lebesgue was particularly incensed that Heine's name had been attached to this theorem, and he campaigned for the designation Borel–Schönflies. Both Paul

Montel and Giuseppe Vitali questioned whether Schönflies had any special claim to this theorem. They referred to it as the Borel–Lebesgue theorem, acknowledging Lebesgue's priority in proving the general case. Borel himself would come to call it the "first fundamental theorem of measure theory," a name with a great deal of merit that, unfortunately, has not stuck.

In 1904, Oswald Veblen pointed out that Theorem 3.4, the Bolzano–Weierstrass theorem, follows from the Heine–Borel theorem and, in fact, is equivalent to it. The key to proving Heine–Borel is that every bounded set must have a least upper bound, but this is precisely the property of completeness and thus equivalent to the Bolzano–Weierstrass theorem.

Exercises

3.2.1. Show that if every open interval contains at least one point of S, then every open interval contains infinitely many points of S. Show that no matter how many points of S we have found, if it is a finite number, then we can always find one more.

3.2.2. Every point in $\mathbb{R}$ can be represented with an infinite decimal expansion. Thus, 3 can be written as 3.000000. .... Explain why 3 can also be written as 2.9999999. .... Which points of $\mathbb{R}$ have more than one infinite decimal expansion? Are there any points of $\mathbb{R}$ that have more than two infinite decimal expansions?

3.2.3. Prove that the interval $[0, 1]$ cannot be contained in a finite union of open intervals whose lengths sum to less than 1.

3.2.4. Show that if a set S is not bounded, then we can find an infinite collection of open sets whose union contains S and such that no finite subcollection will contain S.

3.2.5. Show that if a set S is not closed, then we can find an infinite collection of open sets whose union contains S and such that no finite subcollection will contain S. Use the fact that S is not closed if and only if there is a limit point of S that is not in S.

3.2.6. Prove that if f is continuous on $[a, b]$ and $s = \sup\{t \leq b \mid a \leq x \leq t \Rightarrow |f(x) - f(a)| < \epsilon\}$ and $s \neq b$, then $|f(a) - f(s)| = \epsilon$.

3.2.7. Clean up Dirichlet's proof of Theorem 1.13. Assume that for every $\epsilon > 0$, there is a finite sequence of values, say $a = c_0 < c_1 < \cdots < c_{n-1} < c_n = b$, such that if $c_k < x < c_{k+1}$ then $|f(x) - f(c_k)| < \epsilon$. Explain how to use this to find a δ response to any challenge of an $\epsilon > 0$.

3.2.8. Prove that the Bolzano–Weierstrass principle, that every infinite bounded set has a limit point, implies that every Cauchy sequence converges.

3.2.9. Prove that if every Cauchy sequence converges, then every bounded set has a least upper bound.

3.2.10. Prove that if every bounded set has least upper bound, then every sequence of closed, nested intervals has a nonempty intersection.

3.2.11. In 1880, Weierstrass published a proof that if a series $\sum f_n$ converges uniformly in some neighborhood of each x in the closed and bounded interval $[a, b]$, then the series converges uniformly over the entire interval. An outline of the proof – translated into modern terminology – is given below. Justify each of these statements. Identify where and how Weierstrass used the completeness of $\mathbb{R}$.

1. For each $x \in [a, b]$, define

$$R(x) = \sup \left\{ \epsilon > 0 \,\middle|\, \text{convergence is uniform over } N_\epsilon(x) \right\}.$$

   If $x, y \in [a, b]$ and $D = |x - y|$, then $R(x) \le D + R(y)$ and $R(y) \le D + R(x)$; therefore,

$$R(x) - D \le R(y) \le R(x) + D.$$

2. For $x \in [a, b]$, R is a continuous function.
3. The minimum value of R over $[a, b]$ is strictly positive.
4. Choose an integer n so that $(b - a)/n$ is strictly less than the minimum value of R. Convergence is uniform over each of the intervals

$$\left[a + (j - 1)\frac{b - a}{n}, a + j\frac{b - a}{n} \right], \quad 1 \le j \le n.$$

5. Convergence is uniform over $[a, b]$.

3.2.12. Let $(A_n)_{n=1}^\infty$ be a sequence of positive numbers. Show that there exists a positive sequence $(b_n)_{n=1}^\infty$ such that

$$\sum_{n=1}^\infty \frac{A_n}{b_n} \quad \text{and} \quad \sum_{n=1}^\infty b_n$$

both converge if and only if $\sum_{n=1}^\infty A_n^{1/2}$ converges.

3.2.13. Consider the series $\sum_{n=1}^\infty A_n/|x - a_n|$, where the a_n are the rational numbers in $(0, 1)$ with denominators that are powers of 2:

$$a_1 = \frac{1}{2}, \; a_2 = \frac{1}{4}, \; a_3 = \frac{3}{4}, \; a_4 = \frac{1}{8}, \; a_5 = \frac{3}{8}, \; a_6 = \frac{5}{8}, \; a_7 = \frac{7}{8}, \; a_8 = \frac{1}{16}, \ldots.$$

For each $a_n = b/2^m$, b odd, define A_n to be 2^{-4m}. Show that

$$\sum_{n=1}^\infty A_n^{1/2} = \sum_{m=1}^\infty \frac{1}{2^{m+1}} = \frac{1}{2}.$$

Find a value strictly between 0 and 1 that is not in

$$\bigcup_{n=1}^{\infty} (a_n - A_n^{1/2}, a_n + A_n^{1/2}) = \bigcup_{m=1}^{\infty} \bigcup_{k=1}^{2^{m-1}} \left(\frac{2k-1}{2^m} - \frac{1}{4^m}, \frac{2k-1}{2^m} + \frac{1}{4^m} \right).$$

3.2.14. Consider the bounded infinite set of numbers of the form

$$\sum_{m=1}^{M} \frac{\pm 1}{3^m}, \quad M \in \mathbb{N}.$$

Show that 1/3 is not an accumulation point. Find three accumulation points of this set.

3.3 Set Theory

The first mathematician to really exploit the strangeness of the algebraic overlay of the real number line was Dirichlet when he exhibited the nowhere continuous function of Example 1.1,

$$f(x) = \begin{cases} 1, & x \in \mathbb{Q}, \\ 0, & x \notin \mathbb{Q}. \end{cases}$$

What is significant about this function is that it treats each point of the real continuum as a discrete object that can be examined, tested, and categorized. Although earlier mathematicians had defined functions in terms of their action at each real value, no one before Dirichlet had pushed that notion of function to its logical conclusion. Once this door was opened, truly strange functions began to emerge.

The key to the construction of pathological functions was an understanding of the variety of possible subsets of $\mathbb{R}$. Following Cantor's 1872 paper on trigonometric series, many mathematicians assumed that any infinite subset must be either dense in some interval or one of Cantor's sets of first species. If true, this would have implied a simplicity in the structure of $\mathbb{R}$ that would have made the development of analysis much easier. Cantor was less certain that this was the whole story.

Over the period 1872–1874, Cantor devoted his attention to the structure of subsets of $\mathbb{R}$. The first paper that emerged from this study marked the birth of set theory. It made precise Cantor's realization that the algebraic numbers, those arising from the algebraic overlay, constitute not just a minority but a negligible wisp among the points of the real continuum. Georg Cantor had begun the study of transfinite cardinals.

Cardinality

To lay the foundations for an explanation of Cantor's work, it will be helpful to borrow concepts and language that would not come into being until much later, specifically the concept of **cardinality**. We shall not attempt a formal definition of cardinality. That would take us too far afield. Informally, it is a property that every set possesses and that, in some sense, measures the size of the set. What is important for our purposes is that two sets have the same cardinality if there is a one-to-one and onto correspondence between them.

Thus, the sets $\{1, 2, 3, 4, 5\}$, $\{a, b, c, d, e\}$, and $\{2, 7, 10, 15, 38\}$ share the same cardinality, which is denoted by 5. Cardinality is most interesting when applied to infinite sets. Finite sets are not good examples from which to build an intuitive understanding of cardinality. The "smallest" infinite set is $\mathbb{N}$, the positive integers, $\{1, 2, 3, \ldots\}$. Proper containment need not change cardinality. The sets $\{2, 3, 4, \ldots\}$ and $\{2, 4, 6, 8, \ldots\}$ have the same cardinality as $\mathbb{N}$. In the first case, the one-to-one and onto mapping is $n \to n + 1$; in the second it is $n \to 2n$. Even if we expand to $\mathbb{Z}$, the set of all integers, we still have the same cardinality because it is possible to list the integers so that we establish a correspondence with $\mathbb{N}$: $\{0, 1, -1, 2, -2, 3, -3, \ldots\}$. The cardinality of these sets is designated as $\aleph_0$, read **aleph null**. The countable sets are precisely those with cardinality that is finite or $\aleph_0$.

A word of warning about establishing that a set S is countable: The key is that we must be able to list the elements of S so that if you give me any positive integer, I can find the unique element of S to which it corresponds, and if I give you any element of S, you can find the unique positive integer to which it corresponds. Thus, listing $\mathbb{Z}$ as $(\ldots, -3, -2, -1, 0, 1, 2, 3, \ldots)$ or as $(0, 1, 2, 3, \ldots, -1, -2, -3 \ldots)$ does *not* establish this one-to-one and onto relationship. In both there is no well-defined positive integer to which -1 corresponds.

If we consider $\mathbb{Q}$, the set of all rational numbers, we still have not changed the cardinality. To keep life a littler simpler, we first list just the positive rational numbers. We cannot list all the positive integers, then all the halves, then the thirds, and so on. There must be a *finite* integer that corresponds to 3/2. We consider rational numbers only in reduced form (we write integers with denominator 1), and we order them by the sum of the numerator and denominator. For those with the same sum of numerator and denominator, we order them by the numerator:

$$\frac{1}{1}, \frac{1}{2}, \frac{2}{1}, \frac{1}{3}, \frac{3}{1}, \frac{1}{4}, \frac{2}{3}, \frac{3}{2}, \frac{4}{1}, \frac{1}{5}, \ldots$$

A similar trick can be used to show that any countable union of countable sets is again countable. For each $m \in \mathbb{N}$, let $(a_{m,1}, a_{m,2}, a_{m,3}, \ldots)$ be a countable sequence of values. We can order the union of all of these sequences, ordering by the sum of the subscripts and by the first subscript when the sums are equal (see Figure 3.1).

Once we have ordered the positive rational numbers,

$$a_1 = 1, a_2 = 1/2, a_3 = 2, a_4 = 1/3, \ldots,$$

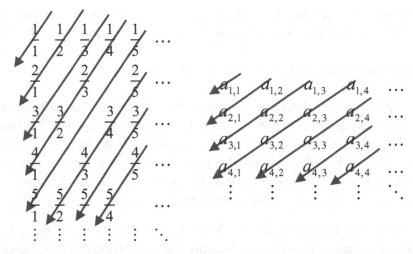

Figure 3.1. Ordering the positive rational numbers (rational numbers not in reduced form have been removed). Ordering any countable collection of countable sets.

we can order all rationals by listing them as

$$0, a_1, -a_1, a_2, -a_2, \ldots.$$

Cantor's 1874 paper contained two important insights:

- The set of all algebraic numbers is countable.
- The set of all real numbers is not countable.

The implication is that the size of the set of real numbers is a bigger infinity than the size of the set of algebraic numbers. Liouville had shown that there are points on the real number line that are not algebraic. Cantor's insights imply that virtually all points on the real number line are not algebraic. We shall talk through the first insight and then describe Cantor's original proof for the second.

Every algebraic number is the root of a polynomial with integer coefficients. To avoid duplication, we can divide by the coefficient of the highest power of x and deal with monic[7] polynomials with rational coefficients. Polynomials of degree n have at most n distinct roots. Each polynomial of degree n is determined by the choice of n rational numbers. Since the set of rational numbers is countable, so is the number of polynomials of degree n. Since each polynomial has at most n roots, for each n the number of algebraic numbers that are roots of polynomials of degree n is again countable. We have a countable number of possible degrees, and for each degree a countable number of roots, so the total number of algebraic numbers is countable. There is, of course, much duplication in this counting, but we know that

---

[7] The coefficient of the highest power is 1.

rational numbers are algebraic, and so the cardinality of the algebraic numbers can be neither more nor less than $\aleph_0$. The set of algebraic numbers has cardinality $\aleph_0$.

Theorem 3.9 ($|\mathbb{R}| \neq \aleph_0$). *The cardinality of* $\mathbb{R}$ *is not* $\aleph_0$.

The proof presented here is a recasting into modern language of Cantor's original 1874 proof, interesting because of its use of the nested interval principle. The more commonly known proof is based on an argument made by Cantor in 1891. You will see it later in this section. The 1874 proof is important not just because it was first. It also shows that if a set satisfies the nested interval principle and between any two distinct elements there always lies a third, then the set cannot be countable.

Proof. We need to prove that we cannot order the real numbers as $(r_1, r_2, r_3, \ldots)$, so we assume that we can and look for an absurdity that arises from this asssumption.

We pick any closed interval on the real line, say $[0, 1]$, and find the first two real numbers in our list that are inside this interval. Since all real numbers are in our list, there are at least two inside $[0, 1]$. Call them $a_0 < b_0$. All of the numbers in the open interval (a_0, b_0) are also inside $[0, 1]$, so we have not yet encountered any of them. We continue down the list until we find the first two real numbers in our list that are inside (a_0, b_0), call them $a_1 < b_1$. We have not yet encountered any numbers from the open interval (a_1, b_1), so we continue until we find the first two in this interval.

We are generating a nested sequence of closed intervals,

$$[0, 1] \supseteq [a_0, b_0] \supseteq [a_1, b_1] \supseteq [a_2, b_2] \supseteq \cdots,$$

with $0 < a_0 < a_1 < a_2 < \cdots$ and $1 > b_0 > b_1 > b_2 > \cdots$. By the nested interval principle, there is at least one real number contained in all of these intervals, call it r_m. By the strict inclusion of these intervals, r_m is not equal to any of the endpoints.

Now we have a problem because a_n and b_n are preceded by at least $2n$ elements from our sequence $(r_1, r_2, r_3, \ldots)$. This means that we can find an n for which a_n and b_n come after r_m in our sequence. This contradicts the fact that $a_n < b_n$ are the first two real numbers in the list that lie within the open interval (a_{n-1}, b_{n-1}). $\quad\square$

Theorem 3.9 says that the continuum, the set of points in $\mathbb{R}$, has a cardinality strictly larger[8] than $\aleph_0$, a cardinality that is denoted by the letter $\mathfrak{c}$. Are there any subsets of $[0, 1]$ whose cardinality is larger than $\aleph_0$ and less than $\mathfrak{c}$? Cantor believed that it was not possible, a belief known as the continuum hypothesis.

---

[8] It is not clear that cardinalities can always be ordered. A discussion of what it means for one cardinal number to be larger than another can be found in Section 5.4.

The Continuum Hypothesis

The cardinality of $\mathbb{R}$ is denoted by c (for the continuum). One might well ask, why not $\aleph_1$ as the next cardinality after $\aleph_0$? The problem is that we do not know that c is the next cardinality after $\aleph_0$. Cantor believed that it is, but he could not prove it. The statement that $c = \aleph_1$ is known as the **continuum hypothesis**. For much of the twentieth century, proving the continuum hypothesis remained one of the great unsolved challenges of mathematics.

The solution, achieved through work of Kurt Gödel and Paul Cohen, gives us a surprising insight into the real numbers: We get to choose whether or not the continuum hypothesis is true for the real number line. In other words, the assumptions that have been made about the structure of the real number line are consistent with having a subset whose cardinality lies between $\aleph_0$ and c, and they are also consistent with not having such subsets. The real number line is not as definitively determined as we might have thought.

Kurt Gödel (1906–1978) is perhaps best known for his incompleteness theorem, that any system of mathematical axioms that is sufficiently complex to include arithmetic will have propositions that can be neither proven nor disproven within that system. Gödel taught at the University of Vienna. After the outbreak of the Second World War, fearing that he might be conscripted into the German army, Gödel left for Princeton where he eventually became a member of the Institute for Advanced Study. Einstein was one of his closest friends.

Paul Cohen (1934–2007) studied at Brooklyn College and the University of Chicago, writing his doctoral dissertation on *Topics in the Theory of Uniqueness of Trigonometric Series*. For most of his career, he taught at Stanford University. In 1966, he received the Fields Medal,[9] for his work on the foundations of set theory.

The continuum hypothesis is not the only property of the real number line over which we have a choice. In the early twentieth century, mathematicians recognized the importance of a certain subset S of $[0, 1]$, with the property that every real number is a rational distance away from exactly one element in S. Initially, it was assumed that such a set must exist. We can construct it first by picking one rational number. We then take the irrational numbers and divide them into classes. Two irrationals are in the same class if they are a rational distance apart. We select one irrational from each class. Surely, this gives us our set. A few mathematicians

---

[9] Until 2002, the Fields Medal was the highest honor any mathematician could win. Awarded only every four years but to up to four people, it is restricted to mathematicians under the age of 40. In 2002, the Abel Prize was created by the Norwegian government to honor Niels Henrik Abel. Similar to the Nobel Prize, it is awarded each year.

quibbled that this means making an uncountable number of selections, and it is not clear how that can be done.

The ability to choose one element from each equivalence class and so define this subset is a consequence of what came to be known as the **axiom of choice**. It appears in many proofs, but it also has surprising and disturbing consequences such as the Banach–Tarski paradox. As it turns out, the truth of the existence of this set is also a matter of choice, a fact also proven by Gödel and Cohen. This axiom will come to play an important role in Section 5.4, where we shall explore it in greater detail and say something about the Banach–Tarski paradox.

The problem arises from thinking of $\mathbb{R}$ as a set of values that in some sense are equivalent to algebraic values. The status of the continuum hypothesis shows just how strange it really is to impose the concept of sets of numbers onto the points of a continuous line. This does not mean that we should not think of $\mathbb{R}$ as a set of numbers. That is an extremely useful construction that will lie at the heart of our eventual solution of all of the problems regarding integration and the fundamental theorem of calculus. But it is a reminder that we must tread very carefully. We are now in a realm where intuition can no longer be trusted.

Power Sets

If A and B are sets, we use A^B to denote the set of all mappings from B to A. The reason for this notation is best explained through an example. Let

$$A = \{a, b, c\}, \qquad B = \{1, 2, 3, 4, 5\}.$$

A mapping from B to A assigns one of the letters from A to each of the numbers in B. There are three possible images of 1 (we can have $1 \to a$ or $1 \to b$ or $1 \to c$), three possible images of 2 (no reason that we cannot use the same image more than once), and so on for a total of 3^5 possible mappings. We see that for finite sets, the cardinality of A^B is the cardinality of A raised to the cardinality of B.

We now extend this idea to infinite cardinalities. In general $\mathcal{A}^{\mathcal{B}}$ means the cardinality of the set of mappings from a set with cardinality $\mathcal{B}$ to a set with cardinality $\mathcal{A}$.

For example $\aleph_0^2$ denotes the cardinality of the set of mappings from $\{1, 2\}$ to $\mathbb{N}$. Each mapping is uniquely determined by a pair of positive integers, $\{(i, j) \mid i, j \in \mathbb{N}\}$. The first coordinate is the image of 1; the second coordinate is the image of 2. We have seen (see Figure 3.1) that the cardinality of such pairs is again $\aleph_0$, and therefore

$$\aleph_0^2 = \aleph_0.$$

It is easy to extend this to any finite positive integer n: $\aleph_0^n = \aleph_0$.

What about $2^{\aleph_0}$? This is the cardinality of the set of mappings from $\mathbb{N}$ to $\{0, 1\}$. We are looking at an infinite sequences of 0s and 1s, such as $1010010001000010\ldots$ There is a natural correspondence between such sequences and the set of real numbers between 0 and 1. We just put a decimal point in front of the sequence and read this sequence in base 2:

$$0.1010010001\ldots_2$$
$$= \frac{1}{2} + \frac{0}{2^2} + \frac{1}{2^3} + \frac{0}{2^4} + \frac{0}{2^5} + \frac{1}{2^6} + \frac{0}{2^7} + \frac{0}{2^8} + \frac{0}{2^9} + \frac{1}{2^{10}} + \cdots.$$

We do have times when two different sequences represent the same real number, for example,

$$0.00100\overline{111} = 0.00101\overline{000},$$

but there are a countable number of these duplications. Exercises 3.3.5–3.3.7 establish that $2^{\aleph_0}$ is the cardinality of $\mathbb{R}$,

$$2^{\aleph_0} = \mathfrak{c}. \tag{3.3}$$

The **power set** of S is the collection of all subsets of S. It is easy to establish a correspondence between the power set and the set of all mappings from S to $\{0, 1\}$. For each mapping, $a \to 1$ if and only if a is in the subset. The mapping from $\mathbb{N}$ to $\{0, 1\}$ given by $1010010001000010\ldots$ corresponds to the subset that contains $\{1, 3, 6, 10, \ldots\}$. For this reason, the power set of any set S is usually denoted by 2^S.

It was in 1891, in an address to the first congress of the German Mathematical Association, that Cantor stated and proved what is now known as Cantor's theorem, that the cardinality of S can never equal the cardinality of 2^S, its power set.

Theorem 3.10 (Cantor's Theorem). *For any set S, the cardinality of S is not the same as the cardinality of the power set of S.*

Proof. We assume that S and 2^S have the same cardinality and look for a contradiction. If the cardinalities were the same, then we would have a one-to-one and onto mapping from S to the collection of subsets of S, $\psi : S \to 2^S$. We construct a subset $T \subseteq S$ according to the following rule: $a \in T$ if and only if $a \notin \psi(a)$. Since T is a subset of S, it is an element of 2^S. Since ψ is a one-to-one and onto mapping, we can find an element $b \in S$ for which $\psi(b) = T$. Is b in T? If b is in $T = \psi(b)$, then $b \in \psi(b)$, so b is not an element of T. That is a contradiction, so b cannot be in T. But if b is not in $T = \psi(b)$, then $b \notin \psi(b)$, so b is in T. Having assumed a one-to-one and onto mapping, we are led to a contradiction whether or not b is in T. The mapping cannot exist. $\qquad\qquad\qquad\qquad\qquad\qquad\qquad\qquad\qquad\qquad\square$

Note that, since $\mathbb{R}$ has the cardinality of the power set of $\mathbb{N}$, this theorem implies that $\mathbb{R}$ is not countable. In fact, a variation on this argument has become the common proof that $\mathbb{R}$ is not countable. Assume that the set of real numbers in $[0, 1]$ is countable, and write down their decimal expansions in order

$$0.a_1a_2a_3a_4a_5\ldots$$
$$0.b_1b_2b_3b_4b_5\ldots$$
$$0.c_1c_2c_3c_4c_5\ldots$$
$$0.d_1d_2d_3d_4d_5\ldots$$
$$\vdots$$

Now choose any decimal whose digits do not include 0 or 9 (exercise 3.3.9 asks you to explain why we avoid those digits) and whose first digit is not a_1, whose second digit is not b_2, whose third digit is not c_3, whose fourth digit is not d_4, and so on. This number does not appear in the list, and so we were wrong when we claimed that we could list them in order. We see now that $\mathfrak{c}$ is not the largest possible cardinality; $2^{\mathfrak{c}}$ is bigger; $2^{(2^{\mathfrak{c}})}$ is even bigger than that.

We can go even further, taking the union of countably many sets of which the first set has cardinality $\mathfrak{c}$ and the nth set is the power set of the $n-1$st set: $|S_1| = \mathfrak{c}$, $S_n = 2^{S_{n-1}}$, $T_1 = \bigcup_{n=1}^{\infty} S_n$. This union has cardinality strictly larger than any of the cardinalities in the sequence. We can then take the power set of this union, $T_2 = 2^{T_1}$. We can now restart with this power set, continue through countably many power sets, and take the union of these sets: $U_1 = \bigcup_{n=1}^{\infty} T_n$. This is only the second iteration of this process. We can do it countably many times: $V_1 = \bigcup_{n=1}^{\infty} U_n$, $W_1 = \bigcup_{n=1}^{\infty} V_n, \ldots$. We have still only described countably many cardinalities. There are sets of cardinalities that are themselves uncountable, even sets of cardinalities that have cardinality $2^{\mathfrak{c}}$. In fact, for every cardinality $\mathfrak{b}$, there is a set of cardinalities that itself has cardinality $\mathfrak{b}$.

Exercises

3.3.1. Show that if $(b_1, b_2, \ldots)$ is any sequence, if $x < b_n$ for every n, and if $y \geq b_n$ for some n, then $x < y$.

3.3.2. Explain how to establish a one-to-one correspondence between $\mathbb{N}$ and the set of all rational numbers.

3.3.3. Explain how to establish a one-to-one correspondence between $\mathbb{R}$ and the set of all irrational numbers.

3.3.4. Describe a one-to-one and onto mapping between each of following pairs of sets:

1. $\mathbb{N}$ and the set of integer multiples of 3
2. $\mathbb{R}$ and set of real numbers in [0, 1]
3. $\mathbb{Q}$ and the set of rational numbers in [0, 1]
4. $\mathbb{N}$ and the set of all pairs of rational numbers in $\mathbb{R}^2$
5. $\mathbb{R}$ and the set all pairs of numbers $(a, b) \in \mathbb{R}^2$

3.3.5. Prove that the set of real numbers in [0, 1] that have more than one representation in base 2 is a countable set.

3.3.6. Prove that $c + \aleph_0 = c$. That is to say, find a one-to-one mapping from $\mathbb{R}$ to $\mathbb{R}$ for which the image omits countably many elements of $\mathbb{R}$. Hint: If $x = b/2^n$, $n \geq 0$, b odd, then $x \to x/2$. Otherwise, $x \to x$.

3.3.7. Prove that $\aleph_0 \cdot c = c$. That is to say, find a one-to-one mapping from $\{(a, b) \mid a \in \mathbb{N}, b \in \mathbb{R}\}$ onto $\mathbb{R}$.

3.3.8. Explain the connection between the proof of Cantor's theorem, Theorem 3.10, and the proof that the set of real numbers in [0, 1] is not countable.

3.3.9. In the common proof that $\mathbb{R}$ is not countable, why do we avoid the digits 0 and 9?

3.3.10. Prove that if $A \subseteq B$ and there is a mapping from A onto B, then A and B have the same cardinality.

3.3.11. Describe the set $A^{\mathbb{N}}$, where $A = \{1, 2, 3\}$. Find the cardinality of this set and justify your answer.

3.3.12. Describe the set $\mathbb{N}^A$, where $A = \{1, 2, 3\}$. Find the cardinality of this set and justify your answer.

3.3.13. Describe the set $\mathbb{N}^{\mathbb{N}}$. Find the cardinality of this set and justify your answer.

3.3.14. Describe the sets $\mathbb{N}^{\{1\}}$ and $\{1\}^{\mathbb{N}}$. Find the cardinality of these sets and justify your answers.

3.3.15. What is the meaning of $S^{\emptyset}$ and $\emptyset^S$ when S is a finite nonempty set? Does the rule $\left| A^B \right| = |A|^{|B|}$ still hold?

3.3.16. Consider the set of pairs (x, y) that are roots of polynomials in x and y with rational coefficients; for example, $x^2 + xy + y^2$. What is the cardinality of the set of all such pairs? Justify your answer.

3.3.17. What is the cardinality of the set of all rational polynomials in π (all expressions of the form $a_n \pi^n + a_{n-1} \pi^{n-1} + \cdots + a_1 \pi + a_0$, where $a_0, a_1, \ldots, a_{n-1}$, $a_n \in \mathbb{Q}$)?

3.3.18. Find a nested sequence of intervals with rational endpoints, $[a_1, b_1] \supseteq [a_2, b_2] \supseteq \cdots$, so that no rational number is contained in

$$\bigcap_{n=1}^{\infty} [a_n, b_n].$$

4

Nowhere Dense Sets and the Problem with the Fundamental Theorem of Calculus

This chapter will focus on the types of sets that confused Hankel and Harnack and many other mathematicians of the late nineteenth century. Consider again what is left over when we order the rational numbers between 0 and 1, say $a_1 = 0, a_2 = 1, a_3 = 1/2, \ldots$, and remove all numbers within $1/8$ of a_1, within $1/16$ of $a_2, \ldots$, within $1/2^{n+2}$ of a_n. As Borel showed, there is something left over. In fact, if we let S denote the set of points that are not eliminated, then $c_e(S) \geq 1/2$ (see Exercise 4.1.1).

This set also gives a counter-example to Hankel's contention that every pointwise discontinuous function is Riemann integrable. If we define $\chi_S(x) = 1$ if $x \in S, = 0$ if $x \notin S$, then χ_S is continuous at every rational number and so is pointwise discontinuous. But no matter how we partition the interval $[0, 1]$, the subintervals that contain points of S – and thus have oscillation 1 – must have total length at least $1/2$. This function is not Riemann integrable over $[0, 1]$.

Although S seems like a very sparse set, it does not fall into the category of any of our characterizations of sparse sets. As we shall see in this chapter, it is not countable. It does not have outer content 0. Since the outer content is not zero, it cannot be first species. We need a new term to describe the way in which this set is sparse.

A set T is dense in (a, b) if every open interval in (a, b) contains at least one point of T. The set S is nowhere dense in (a, b) if and only if every open interval in (a, b) contains an open subinterval with no points of S (see Exercise 4.1.3). Finite sets are nowhere dense and so are **discrete sets**, sets such as $\mathbb{N}$ for which each element is contained in a neighborhood that has no other elements of that set. But there are also nowhere dense set that are not discrete.

The confusion exhibited by Hankel and many of his contemporaries arises from the attempt to connect the intuitive idea of a "sparse" set with any of the precise definitions that were beginning to emerge. In some sense, to be countable is to be sparse. Such a set, though infinite, is of a smaller order of infinity than the

---

Definition: Nowhere dense

A set is **nowhere dense** if there are no open intervals in which S is dense. In other words, S is nowhere dense if its derived set does not contain any open intervals.

---

continuum, $\mathbb{R}$. But as we know with the rationals, a countable set can still be dense. Mathematicians of the 1870s and into the 1880s would conflate the concepts of discrete, nowhere dense, first species, and outer content zero, and often throw in the assumption that such a set must be countable. It would take a while to straighten these out and clarify how they are related.

The problem arises from the use of three incomparable ways of measuring the size of a set: cardinality, density, and measure (which, for the moment, means outer content). We shall straighten these out in Section 4.1. In Section 4.2 we shall explain the disturbing implications for the fundamental theorem of calculus that arise from the existence of nowhere dense sets with positive outer content. Volterra's example of a bounded, pointwise discontinuous function that is not Riemann integrable is built using such a set. In Section 4.3, we shall rely on our improved understanding of nowhere dense sets to explore Osgood's justification of term-by-term integration for any bounded series of continuous functions that converge to a continuous function. Nowhere dense sets lie at the heart of these first three sections, but as Osgood's proof makes very clear, they are an inconvenient tool with which to explore analysis. I have included Osgood's proof for a specific pedagogical purpose. In struggling with it, the reader is prepared to appreciate – as did analysts of the early twentieth century – the incredible simplicity and clarity of Lebesgue's approach. In Section 4.4 we shall explore Baire's insights into the gulf that separates nowhere dense sets from intervals, insights that will restrict the possibilities for the set of discontinuities of a function. His work marks the culmination of our understanding of nowhere dense sets, preparing the ground for and inspiring Lebesgue's development of measure theory.

4.1 The Smith–Volterra–Cantor Sets

We have seen that any first species set will have outer content zero. What about sets that are not first species? Cantor would prove in 1883 that given any set S, its derived set is a union of a countable set and a perfect set, a set that is its own derived set.

The only perfect sets we have seen so far are the empty set, all of $\mathbb{R}$, closed and bounded proper intervals (not single points), and finite unions of closed, bounded, and proper intervals. Since it contains its accumulation points, any perfect set is closed. Not all closed sets are perfect.

Definition: Perfect

A set is **perfect** if it is equal to its derived set. In other words, S is perfect if and only if every point of S is an accumulation point of S, and all accumulation points of S are in S.

There are two big questions that we shall answer in this section:

1. Can a nonempty perfect set be nowhere dense? If so, then we would have a set that is second species and nowhere dense.
2. Can a nowhere dense set have positive outer content? If so, then Hankel's proof that pointwise discontinuous functions are Riemann integrable collapses. It should be possible to find a function for which S_σ has positive outer content even though it is nowhere dense.

As we shall see, bounded, perfect, nowhere dense sets can be constructed with any outer content we wish, provided only that the outer content is strictly less than the length of interval that contains our set.

The first construction of a perfect, nowhere dense set was by the British mathematician Henry J. S. Smith (1826–1883) in 1875. Smith, who taught at Balliol College in Oxford and was appointed Savilian professor of geometry in 1860, is known primarily for his work in number theory. Not many mathematicians were aware of Smith's construction, a fate that was shared by some of his other groundbreaking work. Most of the exciting mathematics was happening in Germany and France, and that is where attention was focused. In 1881, Vito Volterra showed how to construct such a set, but Volterra was still a graduate student, and he published in an Italian journal that was not widely read. Again, little notice was paid. Finally, in 1883, Cantor rediscovered this construction for himself, and suddenly everyone knew about it. Cantor's example is known as the Cantor ternary set. We shall use this term to refer to Cantor's specific example, but the family of examples of perfect, nowhere dense sets exemplified by the work of Smith, Volterra, and Cantor will be referred to as the Smith–Volterra–Cantor sets, or SVC sets.

The Cantor Ternary Set

We shall build a perfect, nowhere dense set with outer content zero that is contained in [0, 1]. We begin with this interval and remove the middle third, leaving us with $[0, 1/3] \cup [2/3, 1]$. We now remove the middle third from each of these intervals, leaving us with $[0, 1/9] \cup [2/9, 1/3] \cup [2/3, 7/9] \cup [8/9, 1]$. We remove the middle third from each of these, leaving us with eight intervals, each of length $1/27$. We remove the middle third from each of these and continue this process

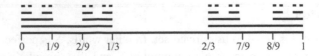

Figure 4.1. Construction of the Cantor ternary set by removal of middle thirds.

indefinitely (see Figure 4.1). We shall call the set of values that remain the **Cantor ternary set**. What is left?

We clearly still have many of the rational numbers between 0 and 1 whose denominators are powers of 3, the endpoints of the intervals we kept. It may seem that that is all that we have, but there is more.

The easiest way to see what is left is to consider the base 3 expansion of the real numbers between 0 and 1. This uses the digits 0, 1, and 2, for example,

$$0.210201_3 = \frac{2}{3} + \frac{1}{3^2} + \frac{0}{3^3} + \frac{2}{3^4} + \frac{0}{3^5} + \frac{1}{3^6} = \frac{586}{729}.$$

When we eliminate all values between $1/3$ and $2/3$, we are eliminating those numbers between 0.1_3 and 0.2_3. In other words, we take out all values with a 1 in the third's place and a nonzero digit after that. When we remove the intervals $(1/9, 2/9)$ and $(7/9, 8/9)$, we are removing the values between 0.01_3 and 0.02_3 and the values between 0.21_3 and 0.22_3. In other words, we remove those values with a 1 in the ninth's place and a nonzero digit somewhere after that. As we continue our removal, what we are eliminating are the numbers with a 1 anywhere in the base 3 expansion, provided the 1 is eventually followed by a nonzero digit.

We can simplify the description of the Cantor ternary set. Base 3 representations that terminate can also be written with repeating 2s. Thus, we have

$$0.1_3 = 0.022222\ldots_3,$$
$$0.021_3 = 0.02022222\ldots_3,$$
$$0.0220201_3 = 0.022020022222\ldots_3.$$

We can define the elements of the Cantor ternary set as those numbers that can be written in base 3 without using the digit 1.

It is now easy to find elements of the Cantor ternary set that are not rational numbers with denominators that are powers of 3. For example,

$$0.0202\overline{02}_3 = \frac{2}{3^2} + \frac{2}{3^4} + \frac{2}{3^6} + \cdots = \frac{2}{9} \cdot \frac{1}{1 - 1/9} = \frac{2}{9} \cdot \frac{9}{8} = \frac{1}{4}.$$

Proposition 4.1 (Properties of the Cantor Ternary Set). *The Cantor ternary set, C, is perfect, nowhere dense, and has outer content zero.*

Proof. The complement of C is a union of open intervals, and therefore C is closed. By Proposition 3.2, C contains all of its accumulation points. To show that every point in C is an accumulation point, we begin with any $a \in C$ and any $\epsilon > 0$. We can find another element of C that is in the ϵ-neighborhood of a by finding a k for which $3^{-k} < \epsilon/2$ and then switching the kth digit of a. If there is a 0 in the 3^{-k} place of a, we change it to a 2. If there is a 2 in that place, we change it to a 0. The number with the switched digit is in C and is equal to $a \pm 2 \cdot 3^{-k}$, and so it differs from a by less than ϵ. We thus see that every element of C is an accumulation point of C, and therefore C is a perfect set.

Since this set is perfect, it can be dense in some open interval only if it already contains that open interval. But between any two numbers, we can find a number that requires the digit 1 somewhere in its base 3 representation, so C cannot contain any open intervals. It is nowhere dense.

Taking open intervals that contain and are just little larger than $[0, 1/3]$ and $[2/3, 1]$, we can get the Cantor ternary set inside a finite union of open intervals whose total length is less than $2/3 + \epsilon$ for any positive ϵ. Using the four intervals of length $1/9$ that contain C, we can get this set inside a finite union of open intervals of total length less than $4/9 + \epsilon$. Using the eight intervals of length $1/27$, we see that the outer content is less than $8/27 + \epsilon$. In general, for any $n \in \mathbb{N}$, we can get C inside 2^n open intervals whose total length is less than $(2/3)^n + \epsilon$. But this bound can be brought as close to 0 as we wish, so the outer content of C is zero. $\qquad\square$

The SVC sets consist of those sets that are constructed by starting with a closed interval and removing an open subinterval. One then removes an open subinterval from each of the remaining subintervals and continues through an infinite sequence of such removals, choosing the subintervals that are removed so that every open subinterval of the original set overlaps with at least one of the subintervals that are removed. The SVC set is the intersection of this countably infinite collection of the sets that remain after each iteration. Every SVC set is closed and nowhere dense.

A particular family of SVC sets consists of those formed by, at the kth iteration, removing an open interval of length $1/n^k$ from the center of each of the remaining closed intervals. We shall denote the resulting set SVC(n), $n \geq 3$. The Cantor ternary set will, from now on, be referred to as SVC(3).

The Devil's Staircase

The SVC sets lead to many strange constructions. One of the strangest is a function often referred to as the **Lebesgue singular function**, but we shall use a more descriptive label, **the devil's staircase**, DS(x).

Example 4.1. The function DS is a mapping from $[0, 1]$ to $[0, 1]$. If x is in SVC(3), then we take the base 3 expansion of x without 1s, replace each digit 2 in this expansion by the digit 1, and read the resulting number as a base 2 expansion. For example,

$$\frac{2}{3} = 0.2_3 \longrightarrow 0.1_2 = \frac{1}{2},$$

$$\frac{20}{81} = 0.0202_3 \longrightarrow 0.0101_2 = \frac{1}{4} + \frac{1}{16} = \frac{5}{16},$$

$$\frac{1}{3} = 0.02222\ldots_3 \longrightarrow 0.01111\ldots_2 = \frac{1}{4} + \frac{1}{8} + \frac{1}{16} + \cdots = \frac{1}{2},$$

$$\frac{1}{4} = 0.02\overline{02}_3 \longrightarrow 0.01\overline{01}_2 = \frac{1}{4} + \frac{1}{16} + \frac{1}{64} + \cdots = \frac{1}{3}.$$

This function maps SVC(3) onto all of $[0, 1]$. Since SVC(3) is also a subset of $[0, 1]$, this onto mapping implies that SVC(3) must have the same cardinality as $[0, 1]$, the cardinality c.

Note that if a and b are elements of SVC(3), $a < b$, then $DS(a) \leq DS(b)$. Equality occurs if and only if a and b are the endpoints of one of the open intervals that was removed to create SVC(3). Assume that a and b are the endpoints of one of these intervals, and assume that they agree in the n digits, d_1 through d_n. We can represent these values by

$$a = 0.d_1 d_2 \ldots d_n 1_3 = 0.d_1 d_2 \ldots d_n 0222 \ldots_3, \qquad b = 0.d_1 d_2 \ldots d_n 2_3.$$

Their images under the function DS are the same:

$$DS(b) = 0.e_1 e_2 \ldots e_n 1_2, \qquad (e_j = d_j/2),$$
$$DS(a) = 0.e_1 e_2 \ldots e_n 0111 \ldots_2 = 0.e_1 e_2 \ldots e_n 1_2.$$

We now extend this function to all of $[0, 1]$ by mapping the points not in SVC(3) to the same value as the image of the endpoints of the removed open interval in which the point lies. In other words, every point in $[1/3, 2/3]$ is mapped to $1/2$. Every point in $[1/9, 2/9]$ is mapped to $1/4$; every point in $[7/9, 8/9]$ is mapped to $3/4$ (see Figure 4.2).

We thus get a continuous, increasing function with a graph that connects $(0, 0)$ and $(1, 1)$ but with horizontal steps of lengths that add up to 1. Between any two steps, no matter how close they may be, there are infinitely many other steps. This function has the curious property that it is a nonconstant function with a derivative that exists and is zero at every point in $[0, 1]$ except at values in SVC(3), a set with outer content zero. Even though the derivative of DS does not exist at all values in $[0, 1]$, every interval in $[0, 1]$ does contain a point at which the derivative exists. It is possible to define the Riemann integral of the derivative if we simply restrict the

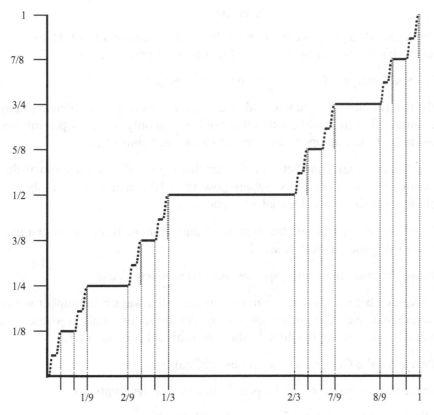

Figure 4.2. The devil's staircase.

x_i^* at which we evaluate DS′ to those points where it exists. But then the integral of the derivative of DS is a constant function, not DS.

This suggests that for the evaluation part of the fundamental theorem of calculus,

$$F'(x) = f(x) \implies \int_a^b f(x)\,dx = F(b) - F(a),$$

we want to insist that the derivative of F must exist at all points in $[a, b]$. But this is not the end of our troubles. As we shall see in the next section, we can use the SVC sets to create examples of functions that *are* differentiable at every point in $[a, b]$, $F'(x) = f(x)$, and yet

$$\int_a^b f(x)\,dx \neq F(b) - F(a).$$

Exercises

4.1.1. Prove that if $(a_n)_{n=1}^{\infty}$ is any order of the rational numbers in $[0, 1]$, then the set of points that are not within $(^1/_2)^{n+2}$ of a_n has outer content at least $1/2$.

4.1.2. Give an example of a closed set that is not perfect.

4.1.3. By definition, if S is nowhere dense in (a, b), then it is not dense in any interval contained in (a, b). Show that this holds if and only if every open interval contained in (a, b) has an open subinterval that has no points of S.

4.1.4. Explain the connection between the fact that the set of rational numbers that can be written with denominators that are powers of 10 is dense in $\mathbb{R}$ and the fact that every real number has a decimal expansion.

4.1.5. Prove that every real number between 0 and 1 can be represented in a base 3 expansion using the digits 0, 1, and 2.

4.1.6. Prove that a set that is first species must be nowhere dense.

4.1.7. Prove that between any two numbers, there will always be a number with a 1 somewhere in its base 3 representation. Consider the set of numbers whose base 3 expansion requires using the digit 1. Show that this set is dense in $\mathbb{R}$.

4.1.8. Prove that the Cantor ternary set has cardinality $\mathfrak{c}$.

4.1.9. Prove that no finite set can be perfect unless it is the empty set.

4.1.10. Prove that no countably infinite set can be perfect.

4.1.11. Prove that a set is nowhere dense if and only if its derived set is nowhere dense.

4.1.12. Prove that the devil's staircase is continuous by explaining how to find a response $\delta > 0$ to the challenge $\epsilon > 0$ so that if $|x - y| < \delta$, then $|DS(x) - DS(y)| < \epsilon$.

4.1.13. Using the definition of the derivative, justify the assertion that the derivative of the devil's staircase, DS, is not defined at any point of the Cantor ternary set, SVC(3).

4.1.14. Let F denote the set of values in $[0, 1]$ that can be written in base 5 without the use of the digits 1 or 3. Thus, $1/5 = 0.1_5 = 0.0444\ldots_5$ is in F but $7/25 = 0.12_5$ is not. Describe the open intervals that are removed from $[0, 1]$ to create F. Find the outer content of F.

4.1.15. For the set F defined in Exercise 4.1.14, define a function, DSF, that takes each point in F, $x = 0.d_1d_2\ldots_5$, to $y = 0.e_1e_2\ldots_3$, where $e_i = d_i/2$. Thus, $1/5 = 0.0444\ldots_5$ is mapped to $0.0222\ldots_3 = 0.1_3 = 1/3$. Show that if (a, b) is one of the open intervals that is removed to form F, then $DSF(a) = DSF(b)$. We can extend DSF to all of [0, 1] by defining it to equal $DSF(a) = DSF(b)$ on each of removed intervals (a, b). Sketch the graph of DSF.

4.2 Volterra's Function

The Cantor ternary set, SVC(3), is an example of a perfect, nowhere dense set with outer content zero. Vito Volterra was interested in a perfect, nowhere dense set with positive outer content because it would enable him to construct a function with a bounded derivative that exists everywhere, but this derivative would not be Riemann integrable over any closed, bounded interval.

Volterra was born in 1860 in Ancona and earned his doctorate in physics at the University of Pisa in 1882, studying under the direction of Enrico Betti. He taught at Pisa and then Torino before being appointed to the chair of mathematical physics at the University of Rome in 1900. In 1931, all university faculty in Italy were required to take an oath of allegiance to the Fascist government. Volterra was one of eleven in all of Italy who refused.

Since the Riemann integral is, strictly speaking, defined only for bounded functions, it is fairly easy to find examples of functions whose derivative cannot be integrated.

Example 4.2. We consider the function

$$f(x) = x^2 \sin(x^{-2}), \quad x \neq 0, \quad f(0) = 0.$$

If $x \neq 0$, then the derivative of f is given by (see Figure 4.3)

$$f'(x) = 2x \sin(x^{-2}) - 2x^{-1} \cos(x^{-2}).$$

To find the derivative of f at $x = 0$, we need to rely on the definition

$$f'(0) = \lim_{x \to 0} \frac{f(x) - f(0)}{x - 0} = \lim_{x \to 0} x \sin(x^{-2}) = 0.$$

So f is a function that is differentiable over the interval $[-1, 1]$, but f' is not a bounded function on this interval. The Riemann integral exists only for bounded functions. Notice that if we treat $\int_{-1}^{1} f'(x) \, dx$ as an improper integral, then it does satisfy the fundamental theorem of calculus

$$\lim_{\epsilon \to 0^+} \int_{\epsilon}^{1} 2x \sin(x^{-2}) - 2x^{-1} \cos(x^{-2}) \, dx = \lim_{\epsilon \to 0^+} x^2 \sin(x^{-2}) \Big|_{\epsilon}^{1} = \sin(1). \quad (4.1)$$

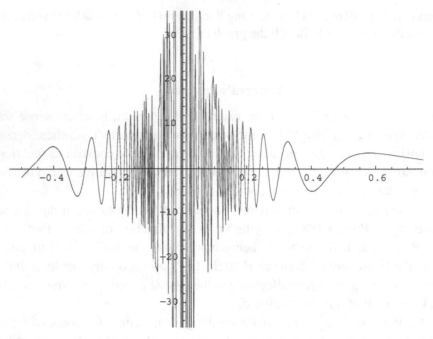

Figure 4.3. Graph of the derivative defined by $f'(x) = 2x \sin(x^{-2}) - 2x^{-1} \cos(x^{-2})$.

Similarly,

$$\lim_{\epsilon \to 0^-} \int_{-1}^{\epsilon} 2x \sin(x^{-2}) - 2x^{-1} \cos(x^{-2}) \, dx = -\sin(1). \qquad (4.2)$$

The improper integral of f' from -1 to 1 exists and is equal to 0.

But is it possible to find a counterexample to the evaluation part of the funda-mental theorem of calculus that does not rely on unbounded functions? Can we find a function f for which f' exists and is bounded over an interval, say $[0, 1]$, but f' is not Riemann integrable over $[0, 1]$?

In 1878, Dini observed that if f is a nonconstant function that has a bounded derivative over $[a, b]$, and if f' is zero on a dense subset of $[a, b]$, then f' cannot be integrable on $[a, b]$ (see Exercise 4.2.3), but he could not produce an example of such a function. This is precisely the type of function that Volterra constructed in 1881.

SVC(4)

The set SVC(3) was created by removing an interval of length $1/3$, then two intervals of length $1/3^2$, then four of length $1/3^3$, and so on. To create the set SVC(4), we remove an open interval of length $1/4$ from the middle of $[0, 1]$,

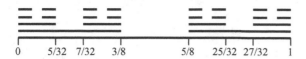

Figure 4.4. Construction of the set SVC(4).

then an open interval of length $1/4^2$ from each of the two remaining pieces, then intervals of length $1/4^3$ from each of the four intervals that remain, and so on.

We do not have a nice characterization of SVC(4) comparable to the base 3 description of SVC(3), but we still wind up with a perfect, nowhere dense set (see Exercises 4.2.4 and 4.2.5). Any finite collection of open sets that covers SVC(4) must have lengths that add up to at least

$$1 - \frac{1}{4} - \frac{2}{4^2} - \frac{2^2}{4^3} - \frac{2^3}{4^4} - \cdots = 1 - \frac{1}{4}\left(1 + \frac{1}{2} + \frac{1}{2^2} + \cdots\right) = \frac{1}{2}.$$

We can find finite open covers of SVC(4) for which the sum of the lengths of the intervals comes as close as we wish to $1/2$, and therefore the outer content of this set is $1/2$ (see Figure 4.4).

Our next function has a derivative that exists and is bounded but is not continuous at $x = 0$.

Example 4.3. Consider the function

$$g(x) = x^2 \sin(x^{-1}), \quad x \neq 0, \quad g(0) = 0.$$

This is very much like our previous function, but the derivative is now bounded,

$$g'(x) = 2x \sin(x^{-1}) - \cos(x^{-1}), \quad x \neq 0, \quad g'(0) = 0.$$

Recall that a function is Riemann integrable if and only if for every $\sigma > 0$, the set of points for which the oscillations exceeds σ has outer content zero. For the function g', the oscillation at $x = 0$ is 2. We have only a single point at which the oscillation is positive, so this function is Riemann integrable. But what if we could construct a function for which the oscillation at every point of SVC(4) is 2? That would imply that the set of points at which the oscillation is greater than 1 does not have zero content, and so the function cannot be Riemann integrable. Our basic idea is to take copies of g and paste them into each of the intervals that have been removed. The behavior of our new function at each of the endpoints of the removed intervals will look just like the behavior of g at $x = 0$.

Example 4.4. (Volterra's Function). We craft our function with some care. To find the piece of the function that will go into the interval of length $1/4$, we start

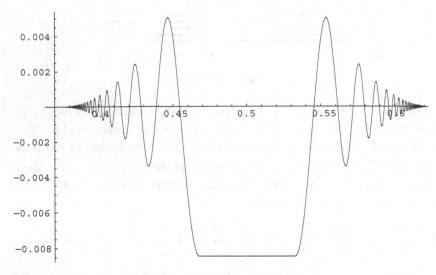

Figure 4.5. Graph of $h_1(x)$, $0.35 \leq x \leq 0.7$.

with our function g and find the largest value less than $1/8$ at which g' is zero, call it a_1. We now define the function $h_1(x)$ by (see Figure 4.5)

$$h_1(x) = \begin{cases} 0, & x < 3/8, \\ g(x - 3/8), & 3/8 \leq x \leq 3/8 + a_1, \\ g(a_1), & 3/8 + a_1 < x < 5/8 - a_1, \\ g(5/8 - x), & 5/8 - a_1 \leq x \leq 5/8, \\ 0, & x > 5/8. \end{cases} \qquad (4.3)$$

We have constructed this function so that it is differentiable at every point in $[0, 1]$, and the oscillation of h_1' is 2 at both $3/8$ and $5/8$.

We now define h_2, which will be nonzero in the two intervals of length $1/16$. We first find a_2, the largest value less than $1/32$ at which g' is zero. We then have (see Figure 4.6)

$$h_2(x) = \begin{cases} 0, & x < 5/32, \\ g(x - 5/32), & 5/32 \leq x \leq 5/32 + a_2, \\ g(a_2), & 5/32 + a_2 < x < 7/32 - a_2, \\ g(7/32 - x), & 7/32 - a_2 \leq x \leq 7/32, \\ 0, & 7/32 < x < 25/32, \\ g(x - 25/32), & 25/32 \leq x \leq 25/32 + a_2, \\ g(a_2), & 25/32 + a_2 < x < 27/32 - a_2, \\ g(27/32 - x), & 27/32 - a_2 \leq x \leq 27/32, \\ 0, & x > 27/32. \end{cases} \qquad (4.4)$$

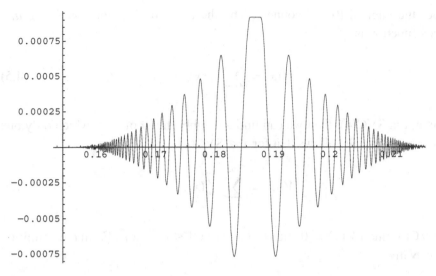

Figure 4.6. The graph of $h_2(x)$, $0.15 \leq x \leq 0.22$.

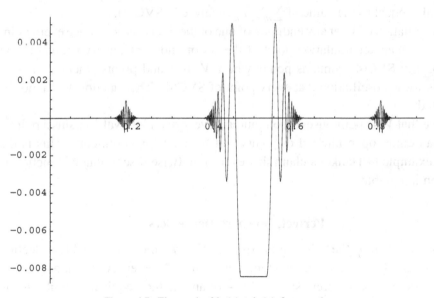

Figure 4.7. The graph of $h_1(x) + h_2(x)$, $0 \leq x \leq 1$.

The derivative of $h_1 + h_2$ has oscillation 2 at 5/32, 7/32, 3/8, 5/8, 25/32, and 27/32 (see Figure 4.7).

We continue in this way. For each n we find a_n, the largest value less than $1/2^{2n+1}$ at which g' is zero. We construct a function h_n that is nonzero only inside the intervals of length 4^{-n}, and in each of those intervals it is two mirrored copies of g over the interval $[0, a_n]$ connected by the constant function equal to $g(a_n)$. Volterra's function is

$$V(x) = \sum_{n=1}^{\infty} h_n(x). \tag{4.5}$$

If x is not in SVC(4), then we can find a neighborhood of x on which only one of the h_n is nonzero. It follows that for $x \notin$ SVC(4),

$$V'(x) = \sum_{n=1}^{\infty} h'_n(x). \tag{4.6}$$

If $x \in$ SVC(4), then $V(x) = 0$ and $|V(x) - V(y)| \leq (y - x)^2$. From the definition of the derivative,

$$\lim_{y \to x} \frac{V(x) - V(y)}{x - y} = 0. \tag{4.7}$$

This is also equal to the value of $\sum_{n=1}^{\infty} h'_n$ at any x in SVC(4).

The oscillation of V' at any endpoint of one of the intervals is 2. Since every point of SVC(4) is an accumulation point of the set of endpoints, every neighborhood of a point in SVC(4) contains points where V' is 1 and points where V' is -1, and thus we get oscillation 2 at every point of SVC(4). The function V' cannot be integrated.

Notice that V' *is* pointwise discontinuous. Every open interval contains a point – in fact an entire open interval of points – at which V' is continuous. This is the counterexample to Hankel's claim that every pointwise discontinuous function is Riemann integrable.

Perfect, Nowhere Dense Sets

The basic idea behind the SVC sets is to progressively remove a countable collection of open intervals so that from each remaining interval we remove yet another open interval. As the next theorem shows, this – or any method equivalent to it – is the only way to obtain a perfect, nowhere dense set.

Theorem 4.2 (Characterization of Perfect, Nowhere Dense Sets). *If S is a bounded, perfect, nowhere dense, nonempty set, $a = \inf S$, $b = \sup S$, then there is a countably infinite collection of open intervals, each contained in $[a, b]$ such that S is the derived set of the set of endpoints of these intervals. Furthermore, the cardinality of S is $\mathfrak{c}$.*

Proof. Since S contains its accumulation points, it is closed and points a and b are in S. By Theorem 3.5, the complement of S in $[a, b]$, $S^C \cap [a, b]$, consists of a countable union of disjoint open intervals. Since every point of S is an accumulation point of S, a cannot be a left endpoint of one of these open intervals, b cannot be a right endpoint of one of these open intervals, and no right endpoint of any of these intervals is a left endpoint of another interval. If $S^C \cap [a, b]$ consisted of only finitely many intervals, then S would contain a closed interval of positive length. Since S is nowhere dense, the number of intervals in $S^C \cap [a, b]$ must be countably infinite.

We now show that S is the derived set of the set of endpoints of the disjoint open intervals whose union is $S^C \cap [a, b]$. Let $(I_1, I_2, I_3, \ldots)$ be an ordering of these disjoint open intervals. Let $L(I_j)$ be the left endpoint of interval I_j and $R(I_j)$ the right endpoint. The set of endpoints is contained in S and, since S is perfect, its derived set is also in S. Since S is nowhere dense, if s is any element of S, every neighborhood of s has a nonempty intersection with at least one of the intervals I_j, and therefore an endpoint of this interval is in this neighborhood of s. It follows that s is an accumulation point for the set of endpoints, and, therefore, S is the derived set of the set of endpoints.

To prove that S is not countable, it is enough to find a one-to-one mapping from $2^{\mathbb{N}}$ into S. Given an infinite sequence of 0s and 1s $(m_1, m_2, m_3, \ldots)$ where $m_j = 0$ or 1, we create a sequence of nested intervals starting with $[a_0, b_0] = [a, b]$. Given $[a_{j-1}, b_{j-1}]$, we find the first open interval in our ordered sequence, I_{n_j}, contained in $[a_{j-1}, b_{j-1}]$. We define

$$[a_j, b_j] = \begin{cases} [a_{j-1}, L(I_{n_j})], & \text{if } m_j = 0, \\ [R(I_{n_j}), b_{j-1}], & \text{if } m_j = 1. \end{cases}$$

We see that a_j is always a or a right endpoint and b_j is always b or a left endpoint. It follows that $a_j \neq b_j$ and I_{n_j} is strictly contained within (a_j, b_j). The right-hand endpoints of our nested intervals, $b_1 > b_2 > \cdots$, form a bounded, decreasing sequence that converges to some $\beta \in S$. This β is the image to which our sequence is mapped,

$$(m_1, m_2, m_3, \ldots) \to \beta.$$

To prove that two distinct sequences map to distinct elements of S, we let k be the first position at which the sequences differ. If $m_k = 0$, then the image of this sequence is strictly less than $L(I_{n_k})$. If $m_k = 1$, then the image of this sequence is strictly greater than $R(I_{n_k})$. The images are distinct. $\square$

Theorem 4.2 has implications for the continuum hypothesis. If a perfect set is dense in an interval (a, b), then it must contain every point in $[a, b]$. Therefore, every nonempty perfect set has cardinality c. When this is combined with Cantor's result (see p. 82) that every derived set is the union of a countable set and a perfect set, we see that every derived set, and therefore every closed set, has cardinality that is finite, equal to $\aleph_0$, or equal to $\mathfrak{c}$. Thus, if there is a subset of $\mathbb{R}$ with cardinality strictly between $\aleph_0$ and $\mathfrak{c}$, then it is not closed.

In 1903, Young proved that a subset of $\mathbb{R}$ with cardinality strictly between $\aleph_0$ and $\mathfrak{c}$ cannot be the intersection of a countable collection of open sets. In 1914, Hausdorff extended this to exclude sets that are the union of a countable collection of sets that are the intersection of a countable collection of open sets. Finally, in 1916, Hausdorff and Alexandrov showed that a set with such an intermediate cardinality cannot be a Borel set (see definition on p. 127).

SVC(n)

While there are many ways of constructing the countable disjoint open intervals that constitute the complement of our perfect, nowhere dense set, the method used for SVC(3) and SVC(4) will work for finding perfect, nowhere dense sets in [0, 1] whose outer content comes as close to 1 as we wish. We define SVC(n) as the set that remains after removing an interval of length $1/n$ centered at $1/2$, then an interval of length $1/n^2$ from the center of each of the two remaining intervals, then intervals of length $1/n^3$ from the centers of each of the remaining four intervals, and so on, leaving a set with outer content

$$1 - \frac{1}{n} - \frac{2}{n^2} - \frac{2^2}{n^3} - \cdots = 1 - \frac{1}{n} \cdot \frac{1}{1 - 2/n} = \frac{n-3}{n-2}.$$

Exercises

4.2.1. For each of the following combinations, either give an example of a bounded, nonempty set with these properties or explain why such a set cannot exist.

1. nowhere dense, first species, and outer content 0
2. nowhere dense, first species, and positive outer content
3. nowhere dense, second species, and outer content 0
4. nowhere dense, second species, and positive outer content
5. dense in some interval, first species, and outer content 0
6. dense in some interval, first species, and positive outer content
7. dense in some interval, second species, and outer content 0
8. dense in some interval, second species, and positive outer content

4.2.2. Find a perfect, nowhere dense subset of [0, 1] with outer content 9/10.

4.2.3. Show that if f is a nonconstant function that has a bounded derivative over $[a, b]$, and if f' is zero on a dense subset of $[a, b]$, then f' cannot be integrable on $[a, b]$.

4.2.4. Show that SVC(4) is closed and that every point is an accumulation point of this set.

4.2.5. Prove that every open interval contained in [0, 1] contains a subinterval with no points of SVC(4), and therefore SVC(4) is nowhere dense.

4.2.6. For the function g in Example 4.3, prove that the oscillation of g' at $x = 0$ is 2.

4.2.7. Find the values of a_1 and a_2 to 10-digit accuracy, where a_1 is the largest number less than $1/8$ for which g' is zero and a_2 is the largest number less than $1/32$ for which g' is zero.

4.2.8. Show that even though $V'(x) = \sum_{n=1}^{\infty} h'_n(x)$, this series does not converge uniformly.

4.2.9. Show that if $S = $ SVC(3) and the intervals are ordered so that longer intervals precede shorter, then the mapping described in the proof of Theorem 4.2 takes $(0, 1, 0, 1, 0, 1, \ldots)$ to

$$\frac{2}{9} + \frac{2}{81} + \frac{2}{729} + \cdots = \frac{2}{9}\frac{1}{(1 - 1/9)} = \frac{1}{4}.$$

Explain why this mapping is independent of how we choose to order intervals of the same length.

4.2.10. Show that the derived set of the set described in Exercise 3.2.14 (p. 71) is perfect and nowhere dense. Describe the countable union of open sets for which this derived set is the complement in $[-1/2, 1/2]$.

4.2.11. Of the types of sets listed in Exercise 4.2.1 that do exist, which can be countable?

4.2.12. Give an example of a bounded, countable, nowhere dense set that has positive outer content.

4.2.13. Let S be a bounded set with exactly one accumulation point, a. Define S_n to be the set of points in S that are at least $1/n$ away from a. Use the fact that $S - \{a\} = \bigcup_{n=1}^{\infty} S_n$ to prove that S is countable.

4.2.14. Using induction on the type, prove that any first species set is countable. It is possible to mimic the proof from Exercise 4.2.13 and define S_n to be the set of points in S that are at least $1/n$ from any of the accumulation points of S, but the statement

$$S - S' = \bigcup_{n=1}^{\infty} S_n$$

now requires proof.

4.2.15. One assumption that is sufficient to make the evaluation part of the fundamental theorem of calculus correct for Riemann integrals is to assume that F' is continuous: If $F' = f$ where f is continuous, then $\int_a^b f(t)\,dt = F(b) - F(a)$. Explain why this assumption eliminates Volterra's counter-example.

4.2.16. Show that while the assumption $F' = f$ is continuous may be sufficient to imply the evaluation part of the fundamental theorem of calculus (see Exercise 4.2.15), it is not a necessary condition.

4.2.17. Following Weierstrass, we can modify the definition of the Riemann integral by taking our Riemann sums, $\sum_{j=1}^n f(x_j^*)(x_j - x_{j-1})$, with x_j^* restricted to be a point of continuity of f in the interval $[x_{j-1}, x_j]$. Show that even with this modified definition, the derivative of Volterra's function is not Riemann integrable.

4.3 Term-by-Term Integration

Despite Volterra's function that revealed a disturbing exception to the fundamental theorem of calculus, no one fully realized that the Riemann integral was inadequate until Lebesgue described his own integral and demonstrated how many of the difficulties that had been associated with the Riemann integral now evaporated. Nowhere was this more evident than in the question of term-by-term integration. A question that is extremely difficult to answer in the context of the Riemann integral – what are the conditions that allow us to integrate an infinite series by integrating each summand? – would suddenly have a simple and direct answer. This more than anything else convinced mathematicians that the Lebesgue integral was the correct approach to integration.

To fully appreciate the simplicity that Lebesgue made possible, it is necessary to spend some time wrestling with term-by-term integration in the setting of the Riemann integral.

In the 1860s, Weierstrass proved that if a series of integrable functions converges uniformly over the interval $[a, b]$, then we can integrate the series by summing the integrals,

$$\int_a^b \left(\sum_{k=1}^\infty f_k(x) \right) dx = \sum_{k=1}^\infty \left(\int_a^b f_k(x)\,dx \right).$$

Heine popularized this result and clarified the distinction between pointwise and uniform convergence in *Die Elemente der Functionenlehre* (The Elements of Function Theory) published in 1872. Uniform convergence is sufficient for term-by-term integration, but it was clear that it was not necessary. Many series that do not converge uniformly still allow for term-by-term integration. As we saw in Section 2.3, Heine tried to work around the nonuniform convergence by isolating a small set that was problematic and focusing on its complement where convergence would be uniform.

As would be realized eventually, this problematic set has to be closed and nowhere dense. In the early 1870s, it was hopefully believed that this meant it was a small set. But as the work of Smith, Volterra, and Cantor showed, a closed, nowhere dense set can be very large and can in fact have outer content as close as desired to the length of the entire interval in which it lies.

In the 1880s, Paul du Bois-Reymond tackled the problem of term-by-term integration, proving in 1883 that any Fourier series of an integrable function can be integrated term by term. In 1886, he published results on the general problem and focused attention on those values of x for which convergence is uniform inside some neighborhood of x. His approach was picked up in 1896 by William F. Osgood whose work we will study in detail.

Paul du Bois-Reymond (1831–1889) was German. His father had moved to Germany from Neuchâtel in francophone Switzerland. He received his doctorate under the direction of Ernst Kummer at the University of Berlin in 1853 and held positions at a succession of universities including Heidelberg, Freiburg, Tübingen, and finally at the Technische Hochschule Charlottenberg (Charlottenberg Institute of Technology) in Berlin. Otto Hölder whom we will meet later was one of his doctoral students in Tübingen.

William F. Osgood (1864–1943) is one of the few Americans to feature in this story. He went to Germany for his graduate work, studying with Max Noether at Erlangen and earning his doctorate in 1890. He returned to spend his career teaching at Harvard.

To clarify the problems with which du Bois-Reymond and Osgood had to deal, it is useful to consider some examples. Rather than working with series, it is simpler if we work with sequences of integrable functions, S_n, and ask whether

$$\int_a^b \left(\lim_{n \to \infty} S_n(x) \right) dx = \lim_{n \to \infty} \left(\int_a^b S_n(x)\, dx \right).$$

This is equivalent to working with series for we can always define

$$S_n(x) = \sum_{k=1}^n f_k(x), \quad \text{or} \quad f_k(x) = S_k(x) - S_{k-1}(x).$$

We also assume that this sequence of functions converges pointwise to 0,

$$\lim_{n \to \infty} S_n(x) = 0, \text{ for all } x.$$

If $S_n(x) \to f(x)$ and if f is integrable, then we can replace our sequence $S_n \to f$ by $(S_n - f) \to 0$. The limit of integrable functions needs not be integrable, a fact that was illustrated in an example by René Baire in 1898.

Example 4.5. (Baire's Sequence). We define

$$f_n(x) = \begin{cases} 1, & \text{if } x = p/q, \ q \le n, \\ 0, & \text{otherwise.} \end{cases} \qquad (4.8)$$

Each f_n is discontinuous on a finite set of points, so each f_n is integrable. The function these approach is Dirichlet's function (Example 1.1), which is not Riemann integrable.

For our purposes, we shall assume that the functions in the sequence as well as the limiting function are all continuous. In this case, we lose no generality if we restrict ourselves to sequences that converge to 0. Finally, we assume that the interval over which we integrate is [0, 1].

What Can Happen

Example 4.6. We begin with the example (see Figure 4.8)

$$A_n(x) = nxe^{-nx^2}.$$

Since $\lim_{n \to \infty} A_n(x) = 0$ at every x, we see that

$$\int_0^1 \left(\lim_{n \to \infty} A_n(x) \right) dx = 0.$$

However, we also have that

$$\int_0^1 nxe^{-nx^2} \, dx = \frac{1 - e^{-n}}{2},$$

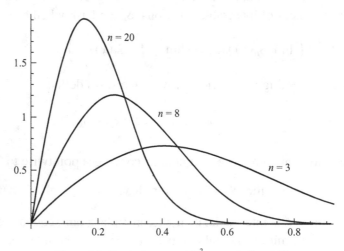

Figure 4.8. Graph of $y = nxe^{-nx^2}$ at $n = 3$, 8, and 20.

and therefore

$$\lim_{n \to \infty} \left(\int_0^1 A_n(x)\, dx \right) = \lim_{n \to \infty} \frac{1 - e^{-n}}{2} = \frac{1}{2}.$$

The integral of the limit is not equal to the limit of the integral. If we convert this into a series,

$$A_n(x) = \sum_{k=1}^{n} \left(kxe^{-kx^2} - (k-1)xe^{-(k-1)x^2} \right),$$

we have an example where term-by-term integration yields the wrong value.

The sequence (A_n) does not converge uniformly, so the ability to interchange limit and integral is not guaranteed, but there are examples of sequences for which the convergence is not uniform, and yet the integral of the limit is equal to the limit of the integral.

Example 4.7. We next consider (see Figure 4.9)

$$B_n(x) = \frac{n^2 x}{1 + n^3 x^2}.$$

Again we have $\lim_{n \to \infty} B_n(x) = 0$ at every x, and therefore,

$$\int_0^1 \left(\lim_{n \to \infty} B_n(x) \right) dx = 0.$$

The convergence of (B_n) is not uniform. In fact, the maximum value of B_n in $[0, 1]$ is $n^{1/2}/2$, which occurs at $x = n^{-3/2}$. We cannot force all values of

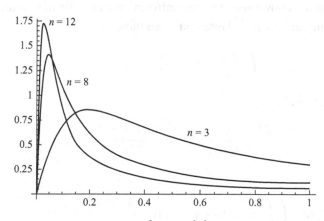

Figure 4.9. Graph of $y = n^2 x / (1 + n^3 x^2)$ at $n = 3, 8,$ and 12.

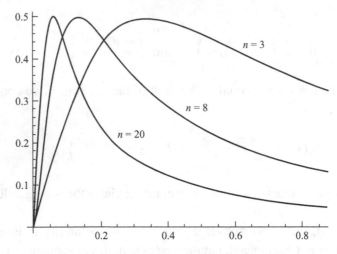

Figure 4.10. Graph of $y = nx/(1 + n^2x^2)$ at $n = 3$, 8, and 20.

B_n within ϵ of 0 by taking n sufficiently large. Nevertheless, in this case

$$\int_0^1 \frac{n^2x}{1 + n^3x^2} \, dx = \frac{\ln(1 + n^3)}{2n},$$

and therefore

$$\lim_{n\to\infty} \left(\int_0^1 B_n(x) \, dx \right) = \lim_{n\to\infty} \frac{\ln(1 + n^3)}{2n} = 0. \tag{4.9}$$

Example 4.8. Our last example is (see Figure 4.10)

$$C_n(x) = \frac{nx}{1 + n^2x^2}.$$

Again, we see that the convergence is not uniform, because the maximum value of C_n is $1/2$, occurring at $x = n^{-1}$. Here again we have that

$$\int_0^1 \left(\lim_{n\to\infty} C_n(x) \right) dx = 0.$$

In this case, the integral is

$$\int_0^1 \frac{nx}{1 + n^2x^2} \, dx = \frac{\ln(1 + n^2)}{2n},$$

and, therefore,

$$\lim_{n\to\infty} \left(\int_0^1 C_n(x) \, dx \right) = \lim_{n\to\infty} \frac{\ln(1 + n^2)}{2n} = 0.$$

Interchanging limits and integrals works in these last two cases despite the fact that convergence is not uniform. What is different about them?

We shall not be able to explain why the interchange works for B_n until we have the tools of Lebesgue integration in hand, but we can tackle C_n now. What is most noticeable about the sequence (C_n) is that these functions stay bounded. In the case of the A_n, we were able to get convergence to 0 and still have the area under the graph of $y = A_n(x)$ increase toward $1/2$ because the maximal values of the functions were increasing. This did not interfere with the convergence to 0, because the location of that maximum kept moving left, approaching $x = 0$. Whatever positive value of x we might choose, eventually the maximum will occur to the left of it, and from then on the sequence approaches 0. But if our functions are all bounded, then we cannot use that trick.

Preserving Some Uniformity

We now return to Osgood who, following du Bois-Reymond, separated those points that lie within neighborhoods within which we have uniform convergence from those points that do not. In the three examples, 4.6–4.8, if we take any $x > 0$, then the convergence *is* uniform in some neighborhood of x. The problem point is $x = 0$. No matter what neighborhood of 0 we choose, these sequences do not have uniform convergence in that neighborhood. Osgood called these Γ-**points** (read "gamma points").

Note that convergence is not uniform in any neighborhood of a point in Γ_α. In all three of our examples, $x = 0$ is the only Γ-point. In the first two examples, $\Gamma_\alpha = \{0\}$ for any $\alpha > 0$, no matter how large. In the third example, 0 is an element of Γ_α, provided $0 < \alpha < 1/2$.

Proposition 4.3 (Characterization of Γ_α). *For any sequence of continuous functions, $(f_1, f_2, f_3, \ldots)$, that converges pointwise to 0 and for any $\alpha > 0$, the set Γ_α is closed and nowhere dense.*

Proof. By Proposition 3.2, Γ_α is closed if it contains its accumulation points. Let x_0 be an accumulation point of Γ_α, and let $N_\delta(x_0)$ be an arbitrary neighborhood

---

Definition: Γ-Points

Given a sequence of functions, $(f_1, f_2, f_3, \ldots)$, that converges pointwise to 0, and any $\alpha > 0$, we define Γ_α to be the set of x such that given any integer m and any neighborhood of x, $N_\delta(x)$, there is an integer $n \geq m$ and a point $y \in N_\delta(x)$ for which $|f_n(y)| > \alpha$. We call x a Γ-**point** if it is an element of Γ_α for some $\alpha > 0$.

of x_0. Since there is an element $x \in \Gamma_\alpha$ that lies in the open set $N_\delta(x_0)$, we may choose a neighborhood of x, $N_{\delta'}(x)$, which is entirely contained within $N_\delta(x_0)$. For any integer m, there is an $n \geq m$ and a point $y \in N_{\delta'}(x) \subseteq N_\delta(x_0)$ for which $|f_n(y)| > \alpha$. Therefore x_0 is also in Γ_α.

To prove that Γ_α is nowhere dense, we assume that it is dense in some open interval (a_1, b_1) and look for a contradiction. We can find an integer n_1 and a point $y_1 \in (a_1, b_1)$ for which $|f_{n_1}(y_1)| > \alpha$. Since f_{n_1} is continuous, there must be an open interval (a_2, b_2) containing y_1 and contained in (a_1, b_1) over which the absolute value is larger than $\alpha/2$. Since Γ_α is dense in (a_2, b_2), there is an integer $n_3 > n_2$ and a point $y_2 \in (a_2, b_2)$ for which $|f_{n_2}(y_2)| > \alpha$. We can find a neighborhood of y_2, (a_3, b_3), $a_2 < a_3 < b_3 < b_2$, over which $|f_{n_3}(y)| > \alpha/2$.

Continuing in this way, we generate an increasing sequence of integers, $(n_1 < n_2 < n_3 < \cdots)$, and a sequence of nested intervals, $[a_1, b_1] \supseteq [a_2, b_2] \supseteq [a_3, b_3] \supseteq \cdots$, for which $y \in [a_k, b_k]$ implies that $|f_{n_k}(y)| \geq \alpha/2$. Since we have strict containment, $a_{k-1} < a_k < b_k < b_{k-1}$, we can consider the closed intervals,

$$(a_1, b_1) \supseteq [a_2, b_2] \supseteq [a_3, b_3] \supseteq \cdots,$$

where $y \in [a_{k+1}, b_{k+1}]$ implies that $|f_{n_k}(y)| \geq \alpha/2$.

By the nested interval principle, there is a point c contained in all of these intervals. Since $f_n(c)$ converges to 0, there is an N such that $n \geq N$ implies that $|f_n(c)| < \alpha/2$. Choose any $n_k \geq N$. This gives our contradiction because c is in $[a_{k+1}, b_{k+1}]$ and, therefore, $|f_{n_k}(c)| \geq \alpha/2$. $\square$

Is Boundedness Sufficient?

As du Bois-Reymond knew by the time he was working on the problem of term-by-term integration, closed nowhere dense sets can be quite large. As we saw in the last section, there are closed and nowhere dense subsets of $[0, 1]$ with outer content as close to 1 as we wish. Because of the difficulty of working with such sets, du Bois-Reymond was never certain whether or not uniform boundedness, combined with continuity, would be enough to allow term-by-term integration. He died before Osgood answered this question. Unbeknownst to either du Bois-Reymond or Osgood, a mathematician at the University of Bolgna, Cesare Arzelà (1847–1912), had proven in 1885 that, for any convergent sequence of integrable and uniformly bounded functions, the limit of the integrals is equal to the integral of the limit. Arzelà, who had studied with Ulisse Dini in Pisa, anticipated many of the results in analysis that others would discover, but his work was not widely known. We shall follow Osgood's proof that relies on continuity because it is simpler and ties directly to our study of perfect, nowhere dense sets. In fact, the only place that Osgood used continuity was in the proof of Proposition 4.3.

Osgood's proof is more difficult to read than it needs to be because he did not have access to the Heine–Borel theorem. Borel had published that result a year earlier, but it would be another five years before Heine–Borel would be recognized as the powerful tool that it is. Because of this, Osgood relied on the nested interval principle that, as we have seen, is equivalent to the Heine–Borel theorem, but is less well suited for Osgood's needs. To simplify matters, we shall use Heine–Borel at the critical points in this proof.

The first time we use Heine–Borel is to prove Osgood's lemma. In the statement of this lemma he assumed that the set G is closed, bounded, and nowhere dense. In fact, he did not need the assumption that G is nowhere dense, so we shall prove a more general form of his lemma. Recall that c_e is the outer content (p. 44).

Lemma 4.4 (Osgood's Lemma). *Let G be a closed, bounded set and let $G_1, G_2, \ldots$ be subsets of G such that*

$$G_1 \subseteq G_2 \subseteq G_3 \subseteq \cdots \qquad \text{and} \qquad \bigcup_{k=1}^{\infty} G_k = G.$$

It follows that

$$\lim_{k \to \infty} c_e(G_k) = c_e(G).$$

As Osgood points out, we really need G to be closed and bounded. For example, if $G = \mathbb{Q} \cap [0, 1]$, we can let G_k be the set of rational numbers between 0 and 1 with denominators less than or equal to k. In this case, $c_e(G_k) = 0$ for all k, but $c_e(G) = 1$.

Proof. Given an arbitrary $\delta > 0$, we must show that there is response N so that $n \geq N$ implies that $c_e(G_n) \leq c_e(G) < c_e(G_n) + \delta$. The first inequality follows from the fact that $G_n \subseteq G$. Since $c_e(G_n) \geq c_e(G_N)$, the second inequality will follow if we can show that $c_e(G) < c_e(G_N) + \delta$.

Let U_k be a finite union of disjoint, open intervals that contains G_k and such that the sum of the lengths of the intervals is strictly less than $c_e(G_k) + \delta/2^k$. The collection $\{U_k\}_{k=1}^{\infty}$ is an open cover of G. By the Heine–Borel theorem, it has a finite subcover. Let N be the largest subscript in this finite subcover.

If U_k is in the finite subcover of G, $k < N$, then we divide it into two disjoint sets: $U_k^{(1)} = U_k \cap U_N$ and $U_k^{(2)} = U_k - U_k^{(1)}$, the part of U_k in U_N and the part not contained in U_N. Since both U_k and U_N are finite unions of open intervals, $U_k^{(2)}$ is a finite union of intervals. (They might be open, closed, or half open.) Since both U_N and U_k contain G_k, the sum of the lengths of the intervals in $U_k^{(1)}$ is at least $c_e(G_k)$, and therefore the sum of the lengths of the intervals in $U_k^{(2)}$ is strictly less than $\delta/2^k$. Because $U_k^{(2)}$ is a finite union of intervals, if any of the intervals are not open, we can replace them by slightly larger open intervals and still keep the sum of the lengths strictly less than $\delta/2^k$. We denote this finite union of open intervals containing $U_k^{(2)}$ by V_k.

We let U denote the union of U_N with all of the V_k, $k < N$, for which U_k is in the finite subcover of G. The set U is still a finite union of open intervals that contains G, and we have the desired bounds,

$$c_e(G) \le c_e(U) < c_e(G_N) + \delta/2^N + \sum_{k=1}^{N-1} \delta/2^k < c_e(G_N) + \delta. \qquad \Box$$

The Arzelà–Osgood Theorem

Theorem 4.5 (Arzelà–Osgood Theorem). *Let* $(f_1, f_2, f_3, \ldots)$ *be a sequence of continuous, uniformly bounded functions on* $[0, 1]$ *that converges pointwise to* 0. *It follows that*

$$\lim_{n\to\infty} \left(\int_0^1 f_n(x)\,dx \right) = \int_0^1 \left(\lim_{n\to\infty} f_n(x) \right) dx = 0.$$

As a consequence of this theorem, given any series of continuous functions that converges pointwise to a continuous function, if the partial sums stay within a uniformly bounded distance of the value of the series, then we may integrate the series by integrating each summand. In the proof that follows we shall use Darboux sums. Osgood did not use them in his proof, but their use clarifies Osgood's argument.

Proof. We need to show that for any $\alpha > 0$, we can find an N such that $n \ge N$ implies that

$$\left| \int_0^1 f_n(x)\,dx \right| < \alpha.$$

To do this, we shall separate the points in $\Gamma_{\alpha/2}$ from those that are not. The proof presented here is a modified version of Osgood's proof, recast so as to take maximum advantage of the Heine–Borel theorem.

From the definition of outer content, we can find a finite union of open intervals that contains $\Gamma_{\alpha/2}$ and for which the sum of the lengths of the intervals is as close as I wish to the outer content of $\Gamma_{\alpha/2}$. We shall call the union of these open intervals U. The complement of U is a finite union of closed intervals, some of which might be single points. We shall use C to denote the intersection of this complement with $[0, 1]$, still a finite union of closed intervals. The theorem now breaks into two parts, limiting the size of the integral over U and limiting the size of the integral over C. First, we need to specify our choice of U. For each $g \in \Gamma_{\alpha/2}$, we know that $\lim_{n\to\infty} f_n(g) = 0$, so we can find an i_g so that $n \geq i_g$ implies that $|f_n(g)| < \alpha/2$. Define G_i to be the set of $g \in \Gamma_{\alpha/2}$ for which $n \geq i$ implies that $|f_n(g)| < \alpha/2$. We see that

$$G_1 \subseteq G_2 \subseteq G_3 \subseteq \cdots \quad \text{and} \quad \bigcup_{i=1}^{\infty} G_i = \Gamma_{\alpha/2}.$$

Let B be a uniform bound on the f_n, $|f_n(x)| \leq B$ for all $n \geq 1$ and all $x \in [0, 1]$. By Lemma 4.4, we can find an integer K and a finite union of open intervals, U, so that U contains $\Gamma_{\alpha/2}$ and the sum of the lengths of the intervals in U is less than $c_e(G_K) + \alpha/(2B)$. We define C to be the complement of U in $[0, 1]$. The set C is a finite union of disjoint closed intervals, and therefore it also is closed and bounded.

We first bound the integral over C. For each x in C, x is not in $\Gamma_{\alpha/2}$, so there is a neighborhood, say $N_{\delta_x}(x)$, and an integer, say A_x, so that if $y \in N_{\delta_x}(x)$ and $n \geq A_x$, then $|f_n(y)| < \alpha/2$. The set of intervals $N_{\delta_x}(x)$ taken over all $x \in C$ is an open cover of C. By the Heine–Borel theorem, we can find a finite number of these neighborhoods whose union contains C,

$$N_{\delta_{x_1}}(x_1) \cup N_{\delta_{x_2}}(x_2) \cup \cdots \cup N_{\delta_{x_n}}(x_n) \supseteq C.$$

Let $\mathcal{A} = \max_{1 \leq i \leq n} A_{x_i}$. For all x in C, if $n \geq \mathcal{A}$, then $|f_n(x)| < \alpha/2$, and therefore for any $n \geq \mathcal{A}$, the integral over C of f_n is bounded by

$$\left| \int_C f_n(x)\,dx \right| \leq \int_C |f_n(x)|\,dx < \frac{\alpha}{2} c_e(C). \tag{4.10}$$

We now consider the integral over U of any f_n with $n \geq K$. We know that the sum of the lengths of the intervals in U is less than $c_e(G_K) + \alpha/2B$. If we take any partition P of U, the sum of the lengths of the intervals that contain points in G_K must be at least $c_e(G_K)$. The infimum of the values of $|f_n|$ on these intervals is strictly less than $\alpha/2$. On all other intervals in this partition, the value of $|f_n|$ is still bounded by B. This implies that we have the following bound on the lower Darboux sum for this partition:

$$\underline{S}(P; f_n) \leq \frac{\alpha}{2} c_e(G_K) + B(\alpha/2B).$$

Since this inequality holds for all lower Darboux sums, and we know that the function is Riemann integrable, this also provides an upper limit for the Riemann integral of $|f_n|$ over U:

$$\left| \int_U f_n(x)\,dx \right| \leq \int_U |f_n(x)|\,dx \leq \frac{\alpha}{2} c_e(G_K) + B\frac{\alpha}{2B}. \tag{4.11}$$

Combining equations (4.10) and (4.11) and using the fact that $c_e(C) + c_e(G_K) = c_e([0, 1]) = 1$, we see that if $n \geq \max\{\mathcal{A}, K\}$, then

$$\left| \int_0^1 f_n(x)\,dx \right| \leq \int_0^1 |f_n(x)|\,dx$$

$$< \frac{\alpha}{2} c_e(C) + \frac{\alpha}{2} c_e(G_K) + B\,\frac{\alpha}{2B}$$

$$\leq \frac{\alpha}{2} + \frac{\alpha}{2} = \alpha. \tag{4.12}$$

$\square$

Exercises

4.3.1. Prove that for any $k \geq 1$,

$$\lim_{n \to \infty} \frac{\ln(1 + n^k)}{n} = 0.$$

4.3.2. Evaluate $\int_0^1 \left(kxe^{-kx^2} - (k-1)xe^{-(k-1)x^2} \right) dx$, $k \geq 1$, and then show that

$$\sum_{k=1}^{\infty} \left(\int_0^1 \left(kxe^{-kx^2} - (k-1)xe^{-(k-1)x^2} \right) dx \right) = \frac{1}{2}.$$

4.3.3. In the proof of Proposition 4.3, where does the assumption that Γ_ϵ is dense in an open interval, that is, the negation of conclusion, actually get used?

4.3.4. Show that if G is any set with finite outer content,

$$G_1 \subseteq G_2 \subseteq G_3 \subseteq \cdots, \quad \text{and} \quad \bigcup_{k=1}^{\infty} G_k = G,$$

then

$$\lim_{k \to \infty} c_e(G_k) \leq c_e(G).$$

4.3.5. Show that if F is any open set,

$$F_1 \supseteq F_2 \supseteq F_3 \supseteq \cdots, \quad \bigcap_{k=1}^{\infty} F_k = F,$$

and F_1 has finite outer content, then

$$\lim_{k \to \infty} c_e(F_k) = c_e(F).$$

4.3.6. Give an example of a sequence of sets $F_1 \supseteq F_2 \supseteq F_3 \supseteq \cdots$, where F_1 is bounded and for which

$$\lim_{k \to \infty} c_e(F_k) \neq c_e \left(\bigcap_k F_k \right).$$

4.3.7. Show that if $F_1 \supseteq F_2 \supseteq F_3 \supseteq \cdots$, where F_1 is bounded, then

$$\lim_{k \to \infty} c_e(F_k) \geq c_e \left(\bigcap_k F_k \right).$$

4.3.8. Show that x is *not* a Γ-point if and only if for each $\alpha > 0$, there is a neighborhood of x, $N_\epsilon(x)$, and a positive integer N such that that for all $y \in N_\epsilon(x)$ and all $n \geq N$, $|f_n(y)| < \alpha$.

4.3.9. Consider the sequence of functions (f_n) defined on $[-1, 1]$ by

$$f_n(x) = \begin{cases} 0, & \text{if } x = 0 \text{ or } |x| > 2/n, \\ \dfrac{1}{n} \sin(\pi/x), & 0 < |x| < 1/n, \\ \sin(\pi/x), & 1/n \leq |x| \leq 2/n. \end{cases}$$

Show that this sequence converges to 0 for all $x \in (-1, 1)$. Show that this convergence is not uniform in any ϵ-neighborhood of 0. For what values of $\alpha > 0$ is 0 in Γ_α?

4.3.10. Consider the sequence of functions, $(g_n)_{n=1}^\infty$, defined on $(-1, 1)$ by $g_n(0) = 0$ and for $x \neq 0$,

$$g_n(x) = \frac{1}{m-1} \left[\left(|x| - \frac{1}{m} \right) m(m-1) \right]^n, \quad \frac{1}{m} \leq |x| < \frac{1}{m-1}, \quad m \geq 2.$$

Show that this sequence converges to 0 for all $x \in [-1, 1]$. Show that this convergence is not uniform in any ϵ-neighborhood of 0. Show that 0 is an accumulation point of Γ-points, but it is not a Γ-point. This demonstrates that while each set Γ_α is closed, the union of these sets needs not be closed.

4.4 The Baire Category Theorem

Osgood had shown that if a sequence of continuous functions converges to 0 and is uniformly bounded, then the integral of the limit is the limit of the integrals. This implies that if (f_n) is a sequence of continuous functions, $f_n \to f$, $|f_n - f|$ is bounded, and f is continuous, then $f_n - f$ is a uniformly bounded sequence of continuous functions converging to 0, and so

$$\lim_{n \to \infty} \int_a^b \left(f_n(x) - f(x) \right) dx = 0 \quad \implies \quad \lim_{n \to \infty} \int_a^b f_n(x) \, dx = \int_a^b f(x) \, dx.$$

Not every uniformly bounded sequence of continuous functions converges to a continuous function. Fourier series are one of the best examples. But for Fourier series, it *is* true that the integral of the limit equals the limit of the integrals. What more, beyond the restriction that $|f_n - f|$ is uniformly bounded, do we need before we can conclude that term-by-term integration is legitimate?

Leopold Kronecker (1823–1891) obtained his doctorate at Berlin University in 1845, working on a problem in number theory under the direction of Dirichlet. A wealthy man involved in his family's banking business, he would not hold an academic appointment until 1883 when he was appointed chair at Berlin University, but he was active in research and began lecturing at Berlin University in 1862, following his election to the Berlin Academy. He was suspicious of the direction in which Heine, Cantor, and others were leading the study of analysis and tried to convince them not to publish the key papers we have described. Nevertheless, he did make one important contribution to our story in a paper published in 1879.

Theorem 4.6 (Kronecker's Theorem). *Let $(f_n)_{n=1}^{\infty}$ be a sequence of integrable functions on $[a, b]$ that converge to the integrable function f. If $|f_n - f|$ is bounded on $[a, b]$ and if, for all $\sigma > 0$,*

$$\lim_{n \to \infty} c_e\left(E(n, \sigma)\right) = 0, \qquad (4.13)$$

where

$$E(n, \sigma) = \left\{ x \in [a, b] \,\middle|\, |f_n(x) - f(x)| \geq \sigma \right\},$$

then

$$\int_a^b f(x)\,dx = \lim_{n \to \infty} \int_a^b f_n(x)\,dx. \qquad (4.14)$$

Proof. Let B be a bound on $|f_n - f|$. Given $\epsilon > 0$, we need to find an N so that $n \geq N$ implies that

$$\left| \int_a^b f_n(x)\,dx - \int_a^b f(x)\,dx \right| < \epsilon.$$

Choose $\sigma < \epsilon / \left(2(b - a) \right)$, an N so that $n \geq N$ implies that

$$c_e\left(E(n, \sigma)\right) < \epsilon/3B,$$

and a finite union of open intervals, U, that contains $E(n, \sigma)$ and whose outer content is less than $\epsilon/2B$. We then have that

$$\left| \int_a^b f_n(x)\,dx - \int_a^b f(x)\,dx \right| \leq \int_a^b \left| f_n(x) - f(x) \right| dx$$

$$= \int_U \left| f_n(x) - f(x) \right| dx$$

$$+ \int_{[a,b]-U} \left| f_n(x) - f(x) \right| dx$$

$$< B \cdot \frac{\epsilon}{2B} + \sigma \cdot (b-a) < \epsilon.$$

$\square$

In 1885, Cesare Arzelà proved that if (f_n) is a sequence of integrable functions that converges to an integrable function and if $|f_n - f|$ is uniformly bounded, then $\lim_{n \to \infty} c_e(E(n, \sigma)) = 0$ (equation (4.13)). This raised the question whether one could have a uniformly bounded sequence of integrable functions for which

$$\lim_{n \to \infty} c_e(E(n, \sigma)) \neq 0.$$

The first explicit example of such a sequence was given by René Baire in his doctoral dissertation of 1899, (Example 4.5 on p. 100).

René-Louis Baire (1874–1932) entered the École Normale Supérieure as an undergraduate in 1892 and earned his doctorate there in 1899. He went on to teach at the University of Montpellier in 1902 and Dijon in 1905. Baire suffered from both physical and psychological disorders that became progressively debilitating. By 1914, they completely prevented him from teaching or continuing his mathematics. He spent his last years in bitter solitude.

Baire's dissertation had a profound effect on Henri Lebesgue and the further development of integration. Baire's thesis, *Sur les fonctions de variables réelles* (On functions of real variables), clarified the intimate connection between the structure of the real numbers and properties of functions. In the process, he made it very clear that outer content is a fundamentally flawed way of measuring the size of a set. We begin with the central result of his thesis.

Theorem 4.7 (Baire Category Theorem). *An open interval cannot be expressed as the countable union of nowhere dense sets.*

We have seen that nowhere dense sets can be quite large in the sense that their outer content can be as close as desired to the length of the interval in which they lie. Baire realized that not even a countable union of them could fill that interval. This is called the Baire *category* theorem because of Baire's definition.

> **Definition: Category**
>
> A set is of **first category** if it is a countable union of nowhere dense sets. A set
> that is not of first category is said to be of **second category**.

Thus, the Baire category theorem is more succinctly put as: "Every open interval
is of second category."

Proof. We lose no generality if we assume that our interval is $(0, 1)$. Let $(S_n)_{n=1}^{\infty}$ be
a sequence of nowhere dense subsets of $(0, 1)$. We must show that there is at least
one $x \in (0, 1)$ that is not in the union of the S_n.

Since S_1 is nowhere dense, we can find an open subinterval of $(0, 1)$ that contains
no points of S_1. If necessary, we come in slightly from each endpoint to find a closed
interval $[a_1, b_1] \subseteq (0, 1)$, $a_1 < b_1$, that contains no points of S_1. Since S_2 is nowhere
dense, we can find a subinterval $[a_2, b_2] \subseteq (a_1, b_1)$, $a_2 < b_2$, that contains no points
of S_2. In general, once we have defined $[a_{n-1}, b_{n-1}]$, $a_{n-1} < b_{n-1}$, we choose a
subinterval $[a_n, b_n] \subseteq (a_{n-1}, b_{n-1})$, $a_n < b_n$, that contains no points of S_n. By the
nested interval principle, the intersection $\bigcap_{n=1}^{\infty} [a_n, b_n]$ contains at least one point,
and this point is not in any of the S_n.   □

This is not a hard proof. Baire's genius lay in recognizing how important this
simple observation can be.

Applications of Baire's Theorem

Notice that any countable set is of first category – any set with a single element is
nowhere dense. We immediately get Cantor's theorem that the set of elements in
$(0, 1)$ is not countable. Another easy consequence is given by the next corollary.

Corollary 4.8 (Complement of First Category). *The complement of any first-
category set in $\mathbb{R}$ is dense in $\mathbb{R}$. In fact, the complement of any first-category set
has an uncountable intersection with every open interval.*

Proof. Let S be of first category. Any subset of a first-category set is again of first
category (Exercise 4.4.7), so the intersection of S with any open interval is of first
category. It follows that every open interval contains a point of S^C, and thus S^C is
dense in $\mathbb{R}$. It is left for Exercise 4.4.8 to show that the intersection of S^C with any
open interval cannot be countable.   □

Now we get to the heart of what interested Baire, the characterization of discontinuous functions. Recall Hankel's distinction (p. 45) between totally discontinuous
functions, such as Dirichet's function, Example 1.1, which is discontinuous at every point, and pointwise discontinuous functions, such as Riemann's function,

Example 2.1, which is discontinuous at every rational point with even denominator but is still continuous at all other points.

It will be convenient to follow Baire and consider the continuous functions as a subset of the pointwise discontinuous functions. A continuous function is simply a pointwise discontinuous function for which the set of points of discontinuity has shrunk all the way down to the empty set.

Corollary 4.9 (Characterization of Pointwise Discontinuous). *A function on* [a, b] *is pointwise discontinuous if and only if the set of points at which it is discontinuous is of first category.*

Proof. One direction is easy and follows from Corollary 4.8. If the set of discontinuities is of first category, then its complement, the set of points at which the function is continuous, is dense.

For the other direction, we assume that f is pointwise discontinuous and show this implies that the set of discontinuities is of first category. We begin by recalling (Proposition 2.4) that a function f is continuous at c if and only if the oscillation of f at c, $\omega(f;c)$, is zero. Let

$$P_k = \left\{ x \in [a, b] \,\middle|\, \omega(f;x) \geq 1/k \right\}.$$

The set of points at which f is discontinuous is the countable union $\bigcup_{k=1}^{\infty} P_k$. If we can show that each P_k is nowhere dense, then we are done. We need to show that every interval in [a, b] contains a subinterval with no points of P_k.

Pick any interval $(\alpha, \beta) \subseteq [a, b]$. Since f is pointwise discontinuous, we can find a point of continuity, say c, in (α, β). Continuity implies that we can control the change in f by staying close enough to c, so we should be able to find a neighborhood of c in which every point has oscillation strictly less than $1/k$. We now show how to find this neighborhood.

First find a δ response so that $(c - \delta, c + \delta) \subseteq (\alpha, \beta)$ and

$$|x - c| < \delta \quad \Longrightarrow \quad |f(x) - f(c)| < \frac{1}{4(k + 1)}.$$

Now consider the interval $(c - \delta/2, c + \delta/2)$. If x is in this interval and $|x - y| < \delta/2$, then $|y - c| < \delta$ and

$$|f(x) - f(y)| \leq |f(x) - f(c)| + |f(c) - f(y)|$$
$$< \frac{1}{4(k + 1)} + \frac{1}{4(k + 1)} = \frac{1}{2(k + 1)}.$$

This implies that the oscillation of f at x must be less than or equal to $1/(k + 1) < 1/k$. We have shown that P_k is nowhere dense, and therefore the set of points of discontinuity is of first category. $\square$

The function g of Exercise 1.1.15 (p. 15) is an example of a function that is discontinuous at every rational number but continuous at every irrational. Volterra showed that we cannot have a function that is continuous at the rational numbers and discontinuous at the irrationals. In fact, he showed that we cannot have two pointwise discontinuous functions for which the points of continuity of one are the points of discontinuity of the other, and vice versa. This result is an easy corollary of Baire's result. We can even strengthen it.

Corollary 4.10 (Volterra's Theorem Strengthened). *Let f_1, f_2, ... be any countable collection of pointwise discontinuous functions on (a, b). There are uncountably many points in (a, b) at which all of these functions are continuous.*

Proof. By Corollary 4.9, the set of points of discontinuity for each function is a set of first category. A countable union of sets of first category is a countable union of countable unions of nowhere dense sets, so it is also of first category. By Corollary 4.8, there are uncountably many points of (a, b) not in this union.   □

Baire's Big Theorem

One of the points of this work was to be able to say something meaningful about limits of continuous functions. How discontinuous can a Fourier series be? Baire's result is impressively strong. We first need to explain what is meant by continuity relative to a set. See the definition given below.

For example, Dirichlet's characteristic function of the rationals, $\chi_{\mathbb{Q}}$, is continuous relative to the set of irrationals. It is also continuous relative to the set of rationals. It is not continuous relative to any set in which both the rationals and irrationals are dense.

Theorem 4.11 (Limit of Cont Fcns $\implies$ Ptwise Discont). *If a function f is the limit of continuous functions, then it is pointwise discontinuous. In fact, if we restrict f to any closed set, S, then the points of continuity of f relative to S form a dense subset of S.*

Recall that continuous functions are considered to be a special case of pointwise discontinuous functions. The limit of continuous functions can be continuous. It

---

Definition: Continuity relative to a set

The function f is continuous at $c \in S$, relative to the set S, if given any $\epsilon > 0$ there is a response $\delta > 0$ so that $x \in S$ and $|x - c| < \delta$ implies that $\left| f(x) - f(c) \right| < \epsilon$.

> **Definition: Class**
>
> Continuous functions constitute **class 0**. Pointwise discontinuous functions that are not continuous constitute **class 1**. Inductively, if f is the limit of functions in class n but it is not in any class $k \leq n$, then we say that f is in **class $n + 1$**.

cannot be as discontinuous as Dirichlet's function, which is discontinuous at every value. It cannot even be as discontinuous as the characteristic function of the set of points of the Cantor ternary set, SVC(3), that are not endpoints of the deleted intervals (points of SVC(3) that are not of the form an integer divided by a power of 3). If we take this characteristic function and restrict it to the Cantor ternary set, a closed set, we still do not have any points of continuity relative to SVC(3).

I shall discuss the implications of Theorem 4.11 before sketching a proof. With this theorem in mind, Baire defined classes of functions. Essentially, the larger the class number, the more discontinuous the function.

Dirichlet's function is not in class 1, but it is the limit of class 1 functions. To see this, choose positive integers k and n. The function $\left(\cos\left(k!x\pi\right)\right)^{2n}$ is continuous. The limit as n approaches ∞ is

$$f_k(x) = \lim_{n \to \infty} \left(\cos\left(k!\, x\, \pi\right)\right)^{2n} = \begin{cases} 1, & \text{if } k!\, x \in \mathbb{N}, \\ & i.e.,\ x \in \mathbb{Q} \text{ and} \\ & \qquad \text{denominator divides } k!, \\ 0, & \text{otherwise.} \end{cases}$$

This function is not continuous, but it is the limit of continuous functions, so it is in class 1. We now take the limit

$$f(x) = \lim_{k \to \infty} f_k(x) = \begin{cases} 1, & x \in \mathbb{Q}, \\ 0, & \text{otherwise.} \end{cases}$$

This is Dirichlet's function which we know is not in class 1. It must be in class 2. Are there functions in class 3, 4, ... up to any finite number? Are there functions so discontinuous that they are not in *any* finite class? In 1905, Henri Lebesgue would show that the answer to both questions is "yes."

Lebesgue's Proof of Theorem 4.11

The previous year, 1904, Lebesgue had published a greatly simplified proof of Theorem 4.11. It is still more complicated than I want to pursue in this book, but an outline of his proof is instructive, for it both demonstrates the role of the Baire category theorem and illustrates a clever idea that is the hallmark of Lebesgue's approach and that will dominate the next several chapters. *Lebesgue partitioned the range of the function.*

Let f be the limit of continuous functions, $f_n \to f$, on $[a, b]$. As before, we let P_k denote the set of points at which the oscillation of f is greater than or equal to $1/k$. If we can show that each P_k is nowhere dense, then f is pointwise discontinuous. We take any open interval $(\alpha, \beta) \subseteq [a, b]$ and partition the entire y-axis from $-\infty$ to ∞ using points $\cdots < m_{-1} < m_0 < m_1 < m_2 < \cdots$ for which $m_{i+1} - m_i < 1/2k$. Consider the set

$$E_i = \left\{ x \in (\alpha, \beta) \,\middle|\, m_i < f(x) < m_{i+2} \right\}.$$

Notice that if $f(x) = m_i$, then $x \in E_{i-1}$. If $m_i < f(x) < m_{i+1}$, then $x \in E_{i-1} \cap E_i$. We have that

$$(\alpha, \beta) = \bigcup_{i=-\infty}^{\infty} E_i \text{ and } x_1, x_2 \in E_i \implies \left| f(x_1) - f(x_2) \right| < m_{i+2} - m_i < \frac{1}{k}.$$

The oscillation of f on E_i is less than $1/k$.

Lebesgue begins by using the fact that f is the limit of continuous functions to prove that each E_i is a countable union of closed sets. In fact, he does more than this. He proves that f is a limit of continuous functions if and only if, for each $k \in \mathbb{N}$, the domain can be represented as a countable union of closed sets so that the oscillation of f on each set is strictly less than $1/k$.

He next proves that given any set E that is a countable union of closed sets, we can construct a function for which the points of discontinuity are precisely the points of E. Let ϕ_i be a function on (α, β) for which the points of discontinuity are precisely the points in E_i. Could all of the functions ϕ_i, $-\infty < i < \infty$, be pointwise discontinuous? If they were, then by Corollary 4.10 there would be a point in (α, β), call it c, where all of them are continuous. But $f(c) \in E_j$ for some j, and that means that ϕ_j is not continuous at c, a contradiction. At least one of the ϕ_i must be totally discontinuous.

If ϕ_j is totally discontinuous on (α, β), then there is an open subinterval of (α, β) for which ϕ_j is discontinuous at every point of this subinterval. By the way we defined ϕ_j, the set E_j contains an open subinterval of (α, β). From the definition of E_j, the oscillation is less than $1/k$ at every point in this subinterval. We have shown that P_k is nowhere dense, and therefore, f is pointwise discontinuous.

Discontinuities of Derivatives

We conclude this section with a corollary of Theorem 4.11. It was Darboux who first observed that if a derivative is discontinuous, then its discontinuities must be like those in the derivative of Volterra's function. Even though $\lim_{x \to c} f'(x)$

does not exist, f' must still satisfy the intermediate value property.[1] Baire showed that Volterra's example illustrates the worst possible case in terms of the size and density of the set of discontinuities of f'.

Corollary 4.12 (Derivative $\Longrightarrow$ Dense Set of Continuities). *Every function that is a derivative is of class 0 or 1.*

Proof. Let f be differentiable on (a, b), and let f' be its derivative. Since f is differentiable, it is continuous on (a, b). For each $k \geq 1$, the function defined by

$$f_k(x) = \frac{f(x + 1/k) - f(x)}{1/k}$$

is also continuous on $(a, b - 1/k)$. Choose a positive integer K. Since f is differentiable, $\lim_{k \to \infty} f_k(x)$, $k \geq K$, exists and equals $f'(x)$ for all $x \in (a, b - 1/K)$. By Theorem 4.11, f' is pointwise discontinuous on $(a, b - 1/K)$. Its points of discontinuity form a set of first category. Since this is true for every $K \geq 1$, the set of points of discontinuity of f' on

$$(a, b) = \bigcup_{K=1}^{\infty} (a, b - 1/K)$$

is also of first category. Therefore, f' is pointwise discontinuous, which implies that it is of class either 0 or 1. $\qquad\square$

Exercises

4.4.1. Consider the sequence of functions (f_n) defined in Example 4.5 on p. 100. Describe the sets

$$E(n, \sigma) = \left\{ x \in [0, 1] \,\middle|\, |f_n(x) - f(x)| \geq \sigma \right\}.$$

Find the value of $\lim_{n \to \infty} c_e\left(E(n, \sigma)\right)$.

4.4.2. Give an example of a sequence of integrable functions on $[0, 1]$ that converge to an integrable function and such that

$$\lim_{n \to \infty} c_e\left(E(n, \sigma)\right) = 0,$$

for all $\sigma > 0$, but such that

$$\int_a^b f(x)\, dx \neq \lim_{n \to \infty} \int_a^b f_n(x)\, dx.$$

Thus, we really do need the hypothesis that $|f_n - f|$ is bounded.

---

[1] See Theorem 1.7.

4.4.3. Prove that any finite union of nowhere dense sets is nowhere dense.

4.4.4. Let C_1 denote the Cantor ternary set, $C_1 = \text{SVC}(3)$. Let C_2 be the subset of $[0, 1]$ formed by putting a copy of C_1 inside every open interval in $[0, 1] - C_1$. Let C_3 be the subset of $[0, 1]$ formed by putting a copy of C_1 inside every open interval in $[0, 1] - (C_1 \cup C_2)$. In general, let C_n be the subset of $[0, 1]$ formed by putting a copy of C_1 inside every open interval in $[0, 1] - (C_1 \cup C_2 \cup \cdots \cup C_{n-1})$. Show that $C = \bigcup_{n=1}^{\infty} C_n$ is first category.

4.4.5. For the set C defined in Exercise 4.4.4, find a description of the elements of C in terms of their representation in base 3.

4.4.6. Let V_1 denote Volterra's set, $V_1 = \text{SVC}(4)$. Exactly as in Exercise 4.4.4, construct a sequence of nowhere dense sets $V_1, V_2, V_3, \ldots$ such that V_n is the subset of $[0, 1]$ formed by putting a copy of V_1 inside every open interval in $[0, 1] - (V_1 \cup V_2 \cup \cdots \cup V_{n-1})$. Show that $V = \bigcup_{n=1}^{\infty} V_n$ is first-category and its outer content is 1.

4.4.7. Prove that any subset of a first-category set is of first category.

4.4.8. Show that if S is of first category and I is an open interval, then $S^C \cap I$ cannot be a countable set.

4.4.9. Show that if f is totally discontinuous on some subinterval of $(-\pi, \pi)$, then it cannot be represented by a trigonometric series,

$$a_0 + \sum_{k=1}^{\infty} \left(a_k \cos(kx) + b_k \sin(kx) \right).$$

4.4.10. We have seen (Corollary 4.12) that any derivative is continuous on a dense set of points. Can a derivative also be discontinuous on a dense set of points? To see that the answer to this question is "yes," let $(r_1, r_2, \ldots)$ be an ordering of the rational numbers in $[0, 1]$. Let

$$f(x) = x^2 \sin(1/x), \ x \neq 0, \quad f(0) = 0,$$

and define

$$g(x) = \sum_{n=1}^{\infty} \frac{1}{n^2} f(x - r_n).$$

Show that f is differentiable at every point in $[0, 1]$ and that its derivative is discontinuous at each r_n.

4.4.11. Prove that any countable union of first-category sets is first category.

4.4.12. Define

$$f(x) = \begin{cases} q, & \text{if } x = p/q \in \mathbb{Q} \text{ where } q \geq 1 \text{ and } \gcd(p, q) = 1, \\ 0, & \text{otherwise} \end{cases}$$

Prove that f is of class 2.

4.4.13. Let $(r_n)_{n=1}^{\infty}$ be a sequence of real numbers chosen so that no two differ by a rational number, $r_i - r_j \notin \mathbb{Q}$ if $i \neq j$. Define

$$Q_n = r_n + \mathbb{Q} = \{r_n + q \mid q \in \mathbb{Q}\}, \quad Q = \bigcup_{n=1}^{\infty} Q_n.$$

Define the characteristic function of Q, $\chi_Q(x) = 1$ if $x \in Q$, $= 0$ if $x \notin Q$. Prove that this function is at most of class 3.

5

The Development of Measure Theory

Through the 1880s and 1890s, the Riemann integral piled up a list of inconveniences, including the following:

1. It is defined only for bounded functions. While improper integrals had been introduced to deal with unbounded functions, this fix appears *ad hoc*. Furthermore, recourse to improper integrals can work only if the set of points with unbounded oscillation has outer content zero.
2. It is possible to have an integrable function with positive oscillation on a dense set of points, and therefore the integral is not differentiable at any of the points in this dense set (Riemann's function, example 2.1). This violates the antidifferentiation part of the fundamental theorem of calculus on this dense set.
3. It is possible to have a bounded derivative that cannot be integrated (Volterra's function, Example 4.4). This violates the evaluation part of the fundamental theorem of calculus.
4. The limit of a bounded sequence of integrable functions is not necessarily integrable (Baire's sequence, Example 4.5).
5. The question of finding necessary and sufficient conditions under which term-by-term integration is valid was turning out to be extremely difficult.

Despite these inconveniences, few mathematicians were dissatisfied with the Riemann integral. One of the few was Weierstrass. In a letter written to Paul du Bois-Reymond in 1885, he expressed his unhappiness with the need to consider the values of the integrand at points where the oscillation is positive. Hankel had proven that if a function is Riemann integrable, then it is at worst pointwise discontinuous. That is to say, the points where the function is continuous must be dense. While it is true that any Riemann integrable function is at worst pointwise discontinuous (see Exercise 5.1.1), recall from Section 2.3 that Hankel went further and also asserted

120

that every pointwise discontinuous function is Riemann integrable. This is false (see Exercise 5.1.2).

Weierstrass suggested modifying Riemann's definition so that in the Riemann sum,

$$\sum_{i=1}^{n} f(t_i)(x_i - x_{i-1}),$$

the t_i, $x_{i-1} \leq t_i \leq x_i$, are restricted to be points of continuity. It appears that he was hoping to expand the class of integrable functions so that, for example, the derivative of Volterra's function would now be integrable.

As du Bois-Reymond pointed out in his response, this does not save us. Even under Weierstrass's definition, the derivative of Volterra's function fails to be integrable. Any interval that contains a point of the set we have called SVC(4) will contain points of continuity at which the value of f is 1, and points of continuity at which the value of f is -1.

A year later, in his Berlin lectures, Weierstrass took a different approach. He returned to the idea of integral as area. Given a nonnegative function f, $f(x) \geq 0$ for all $x \in [a, b]$, we consider the set of points in the plane

$$S_f = \{(x, y) \mid a \leq x \leq b, \ 0 \leq y \leq f(x)\}. \tag{5.1}$$

The integral $\int_a^b f(x)\,dx$ should be the area of S_f. We can extend this to any function. Define

$$f^+(x) = \max\{f(x), 0\}, \qquad f^-(x) = \max\{-f(x), 0\}.$$

We can then define the integral as

$$\int_a^b f(x)\,dx = \text{area}(S_{f^+}) - \text{area}(S_{f^-}).$$

The only question is "what do we mean by the area of a set of points in the plane?"

We can extend the idea of outer content to the plane. The area of any rectangle is its length times its width. In exact analogy with the definition of outer content on the real number line, given any set S, we let $\mathcal{C}$ denote the set of all coverings C of S by a finite number of rectangles, let area(C) be the sum of the areas of the rectangles in C, and define

$$c_e(S) = \inf_{C \in \mathcal{C}} \text{area}(C).$$

This gives us the Weierstrass integral,

$$(W)\!\int_a^b f(x)\,dx = c_e(S_{f^+}) - c_e(S_{f^-}).$$

> **Definition: Characteristic function**
>
> The **characteristic function** of a set S, χ_S, is defined to be 1 if $x \in S$, 0 if $x \notin S$.

The Weierstrass integral has the advantage that every bounded function is integrable. It yields the desired value for the integral of the derivative of Volterra's function.

It does, however, have a noticeable drawback. Consider the characteristic function of a set.

If S and T are disjoint sets that are both dense in $[0, 1]$ (e.g, S could be the rationals and T the irrationals), then

$$(W)\int_0^1 \left(\chi_S(x) + \chi_T(x) \right) dx = 1 \neq (W)\int_0^1 \chi_S(x)\, dx + (W)\int_0^1 \chi_T(x)\, dx = 2.$$

It is important that integration should be additive, $\int (f + g) = \int f + \int g$. Weierstrass's integral would not be the solution, but it was pushing in the right direction. The key to integration would come from a better understanding of area.

5.1 Peano, Jordan, and Borel

Giuseppe Peano (1858–1932) studied and then taught at the university of Turin (Torino). He began teaching there in 1880. Peano is best known for his construction in 1890 of a space-filling curve. This is a curve that passes through every point in the two-dimensional region $0 \leq x \leq 1$, $0 \leq y \leq 1$. He is also known for his axioms, published in 1889, that define the natural numbers in terms of sets, creating the foundations for later work in logic.

Peano's work on area came early in his career. In 1883, he showed how to use the upper and lower Darboux integrals that had been invented by Volterra to provide simplified proofs of many results for Riemann integrals. In 1887, as a further elaboration of the ideas in the 1883 paper, Peano published *Applicazione geometriche del calcolo infinitesimale* (Geometric applications of infinitesimal calculus) in which he became one of the first to provide precise definitions of the interior and the boundary of a set. He distinguished inner and outer content. The **inner content** of a set, $c_i(S)$, is obtained by considering all finite unions of disjoint intervals contained in S. The inner content is defined as the supremum, taken over all such unions, of the sum of the lengths of the intervals in the union. Any set that does not contain any open intervals has inner content zero.

Inner and outer content are easily extended to sets in $\mathbb{R}^2$ or higher dimensions. Instead of working with intervals, we work with rectangles or rectangular blocks whose areas or volumes are defined to be the product of the lengths in each dimension. Peano defined a set as having **area** if and only if the inner content is

---

Definition: Content

Let $\mathcal{C}_S^e$ be the set of all finite coverings of the set $S \subseteq \mathbb{R}^n$ using n-dimensional rectangular boxes, and let $\mathcal{C}_S^i$ be the set of all pairwise disjoint finite collections of open n-dimensional rectangular boxes for which the union is contained in S. The volume of a rectangular box is the product of the lengths of the sides, and the volume of $C \in \mathcal{C}_S^e$ or $\in \mathcal{C}_S^i$, denoted vol(C), is the sum of the volumes of the boxes. The **inner** and **outer content** of a bounded set S are defined, respectively, as

$$c_i(S) = \sup_{C \in \mathcal{C}_S^i} \text{vol}(C), \qquad c_e(S) = \inf_{C \in \mathcal{C}_S^e} \text{vol}(C).$$

If $c_i(S) = c_e(S)$, then we say that S has **content**

$$c(S) = c_i(S) = c_e(S).$$

---

equal to the outer content, in which case we can denote this area as simply the **content** of the set.

If S is not bounded, let $N_k(0)$ be the neighborhood of the origin with radius k and define

$$c_i(S) = \lim_{k \to \infty} c_i\left(S \cap N_k(0)\right), \qquad c_e(S) = \lim_{k \to \infty} c_e\left(S \cap N_k(0)\right).$$

Content corresponds to the usual concept of length in $\mathbb{R}$, area in $\mathbb{R}^2$, and volume in $\mathbb{R}^3$. Inner and outer contents differ only for sparse sets. For example, $\mathbb{Q} \cap [0, 1]$ has inner content 0 and outer content 1. The set SVC(4) has inner content 0 – because it contains no intervals – and outer content $1/2$. Its complement in $[0, 1]$, $[0, 1] - \text{SVC}(4)$, has inner content $1/2$ and outer content 1.

Peano recognized the relationship between inner and outer content given in the following proposition. The concept was both popularized and made rigorous by Camille Jordan in the first volume of *Cours d'analyse*, published in 1893. Recall that the boundary of S, denoted ∂S, consists of all points for which every neighborhood contains at least one point of S and at least one point not in S.

Proposition 5.1 (Inner versus Outer Content). *Let S be a bounded set in $\mathbb{R}^n$. We have that*

$$c_e(S) = c_i(S) + c_e(\partial S). \tag{5.2}$$

*As a consequence, the set S has a well-defined area, called the **content** of the set, if and only if $c_e(\partial S) = 0$.*

Proof. We shall prove this theorem for two-dimensional sets. The same idea works in any number of dimensions. We subdivide $\mathbb{R}^2$ into squares, 2^{-m} by 2^{-m}, and restrict our attention to those squares that have nonempty intersection with S. Let $S_{e,m}$ be the union of the squares that contain at least one point of S. This is a cover of S. As m increases, the area of this cover decreases and approaches the outer content of S,

$$\lim_{m \to \infty} \text{area}(S_{e,m}) = c_e(S).$$

The squares in $S_{e,m}$ are of two types: those that contain boundary points of S and those that do not. Let $S_{\partial,m}$ be the union of the squares that contain boundary points, and $S_{i,m}$ the union of the squares that do not and, therefore, are completely contained within S,

$$S_{e,m} = S_{i,m} \cup S_{\partial,m}, \qquad \text{area}(S_{e,m}) = \text{area}(S_{i,m}) + \text{area}(S_{\partial,m}).$$

As m increases, the area of $S_{i,m}$ also increases and approaches the inner content of S,

$$\lim_{m \to \infty} \text{area}(S_{i,m}) = c_i(S).$$

As m increases, the area of $S_{\partial,m}$ decreases and approaches the outer content of the boundary of S,

$$\lim_{m \to \infty} \text{area}(S_{\partial,m}) = c_i(\partial S).$$

Therefore,

$$\begin{aligned} c_e(S) &= \lim_{m \to \infty} \text{area}(S_{e,m}) \\ &= \lim_{m \to \infty} \text{area}(S_{i,m}) + \lim_{m \to \infty} \text{area}(S_{\partial,m}) \\ &= c_i(S) + c_e(\partial S). \end{aligned} \qquad \square$$

Peano observed that if f is a nonnegative function defined on $[a, b]$ and if S_f is the set of points under f as defined in equation (5.1), then

$$\underline{\int_a^b} f(x)\, dx = c_i(S_f), \qquad \overline{\int_a^b} f(x)\, dx = c_e(S_f). \tag{5.3}$$

It follows that f is Riemann integrable if and only if S_f has area in the sense that its inner and outer contents are equal.

Jordan Measure

Camille Jordan (1838–1922) earned his doctorate in 1861 but worked as an engineer until 1876 when he took a position as professor of analysis at the École

---

Definition: Jordan measure

A set S is **Jordan measurable** if and only if the inner and outer contents are the same, $c_i(S) = c_e(S)$. The **Jordan measure** of S is its content, given by either the inner or outer content, $c(S) = c_i(S) = c_e(S)$.

Polytechnique in Paris. His interests ranged widely, and he is known today for his contributions to group theory, topology, and number theory as well as analysis. His three-volume analysis textbook, *Cours d'analyse de l'École Polytechnique* (Course in analysis for the Polytechnical Institute), published 1893–1896, established Peano's content as the basis for calculus.

The problem that forced Jordan to focus on content was the issue of multidimensional integrals. In particular, he had to explain how to integrate real-valued functions in two real variables for which the domain might be a very irregular region. As long as the inner and outer contents of the domain were equal, one could make sense of the integral. A critical piece of this is the fact that if we have a finite collection of pairwise disjoint sets, then the content of the union is the sum of the contents.

Proposition 5.2 (Finite Additivity of Content). *Let $S_1, S_2, \ldots, S_n$ be a finite set of pairwise disjoint Jordan measurable sets. The content of their union is equal to the sum of their contents,*

$$c\left(S_1 \cup S_2 \cup \cdots \cup S_n\right) = c(S_1) + c(S_2) + \cdots + c(S_n). \qquad (5.4)$$

Proof. From the definition of inner and outer content, we have that

$$\sum_{k=1}^{n} c_i(S_k) \leq c_i\left(S_1 \cup S_2 \cup \cdots \cup S_n\right) \leq c_e\left(S_1 \cup S_2 \cup \cdots \cup S_n\right) \leq \sum_{k=1}^{n} c_e(S_k).$$

Since each set is Jordan measurable,

$$\sum_{k=1}^{n} c_i(S_k) = \sum_{k=1}^{n} c_e(S_k). \qquad \square$$

Jordan's *Cours d'analyse* was very influential. Henri Lebesgue studied it while an undergraduate at the École Normale Supérieure. Lebesgue later recounted how it had prepared the way for his own approach to integration. But using content to define area had one major flaw; for too many important sets, inner and outer content are not equal. Volterra's example of a nonintegrable derivative relies on the fact that the inner and outer content of SVC(4) are different. This suggested to several people that a more all-encompassing definition of area might get around the difficulties of Volterra's function. As we saw with Weierstrass's integral, outer content alone would not do it. Every bounded set has a well-fined outer content, but the Weierstrass integral is not additive because outer content is not additive. It is possible to have two disjoint sets, $S \cap T = \emptyset$, for which

$$c_e(S \cup T) \neq c_e(S) + c_e(T).$$

Borel Measure

Émile Borel (1871–1956) was only 22 when he became chair of mathematics at the University of Lille. He returned to Paris in 1896 to teach at the École Normale Supérieure where Lebesgue was then an undergraduate. In 1909 he became a professor at the Sorbonne. We have already encountered some of his work in the Heine–Borel theorem. We now turn to his study of area published in 1898 in *Leçons sur la théorie des fonctions* (Lectures on the theory of functions).

In Section 3.3, we discussed Borel's paper of 1895 and how his study of the convergence of certain infinite series led to the discovery of the Heine–Borel theorem. It did more than that. Borel was interested in the size of his set of points on which the series must converge,

$$E_c = \left\{ x \in [0, 1] \,\middle|\, |x - a_n| \geq cA_n^{1/2} \text{ for all } n \geq 1 \right\},$$

where $\{a_n\}$ is dense in $[0, 1]$, $A_n > 0$ for all $n \geq 1$, and $\sum A_n^{1/2} = A < \infty$. If $c < 1/(2A)$, then the complement of E_c cannot be a Jordan measurable set. Its inner content is clearly less than or equal to $\sum 2cA_n^{1/2} < 1$. Since it contains $\{a_n\}$, a dense set of points, its outer content is 1. But there is a very natural definition of the size of E_c. Its complement in $[0, 1]$ can be expressed as a countable union of pairwise disjoint intervals. The size of the complement should be the sum of the lengths of those intervals. The size of E_c should be 1 minus the size of the complement.

A similar set for which the same argument should apply is Volterra's set, SVC(4). Its complement in $[0, 1]$ is a union of pairwise disjoint intervals whose combined length is 1/2. We know that the outer content of SVC(4) is 1/2, but since this set contains no open intervals, its inner content is 0. It is not Jordan measurable. Borel believed that measure should be redefined so that SVC(4) has a well-defined measure equal to 1/2.

The problem with Jordan measure is that it is only finitely additive. Borel realized that he needed a measure that is countably additive. This means that if we are given an infinite sequence of pairwise disjoint sets, $(S_1, S_2, S_3, \ldots)$, $S_i \cap S_j = \emptyset$ for $i \neq j$, then we want the measure of the union to equal the sum of the measures,

$$m\left(\bigcup_{k=1}^{\infty} S_k\right) = \sum_{k=1}^{\infty} m(S_k).$$

As an example, such a countably additive measure would imply that the set of rational numbers in $[0, 1]$, a countable union of single points, must have measure 0.

Borel begins with three assumptions that uniquely define Borel measure:

1. The measure of a bounded interval is the length of that interval (whether open, closed, or half open).
2. The measure of a countable union of pairwise disjoint measurable sets is the sum of their measures.
3. If R and S are measurable sets, $R \subseteq S$, then so is $S - R$. Furthermore, $m(S - R) = m(S) - m(R)$.

As an example, each set $SVC(n), n \geq 3$, is the complement in $[0, 1]$ of a countable union of open intervals. It is measurable in Borel's sense, and its measure is $(n - 3)/(n - 2)$. In fact, any open set is a countable union of open intervals, so open sets and closed sets are measurable in Borel's sense.

Borel Sets

Borel came very close to our modern concept of measure, but in all of his discussions of the application of his measure, he restricted himself to sets that could be constructed from intervals using countable unions and complements. Today, we call the sets that can be built in this way **Borel sets**.

It is left for you to show (see Exercises 5.1.13–5.1.15) that under this definition, all open intervals and all half-open intervals are Borel sets, that any countable intersection of Borel sets is also a Borel set, and if A, B are Borel sets, $A \supseteq B$, then $A - B$ is a Borel set.

Definition: σ-algebra

A σ-**algebra**, $\mathcal{A}$, is a collection of sets with the property that

1. $\emptyset \in \mathcal{A}$,
2. if $\{A_1, A_2, \ldots\}$ is any countable (finite or infinite) collection of sets in $\mathcal{A}$, then their union is also in $\mathcal{A}$ (note that we do not need them to be pairwise disjoint),
3. if $A \in \mathcal{A}$, then $A^C \in \mathcal{A}$.

Definition: Borel sets

The collection, $\mathcal{B}$, of **Borel sets** in $\mathbb{R}$ is the smallest σ-algebra in $\mathbb{R}$ that contains all closed intervals.

In the next section, we shall see that any Borel set is measurable in Borel's sense. This will take some work. If we have a countable union of pairwise disjoint sets for which the Borel measure is defined, then the Borel measure of the union is the sum of the Borel measures. But we need to define the Borel measure of *any* countable union of sets with well-defined Borel measure.

The Limitations of Borel Measure

Borel never tried to apply his idea of measure to the problem of integration. In fact, he went so far as to state,

> It will be fruitful to compare the definitions that we have given with the more general definitions that M. Jordan gives in his *Cours d'analyse*. The problem we investigate here is, moreover, totally different from the one resolved by M. Jordan.[1]

Borel measure actually applies to a much smaller collection of sets than Jordan measure. This may seem a strange comment in view of our examples of sets that are measurable in Borel's sense, but not when we try to use Jordan's content. But, as we shall see, the cardinality of $\mathcal{B}$, the collection of all Borel sets, is only $\mathfrak{c}$, the cardinality of $[0, 1]$. On the other hand, any subset of SVC(3) (the Cantor ternary set with content 0) will also have Jordan measure zero. As we saw in Section 4.1, SVC(3) has cardinality $\mathfrak{c}$. The collection of its subsets has cardinality $2^{\mathfrak{c}}$. By Cantor's theorem (Theorem 3.9), this is a larger cardinality than $\mathfrak{c}$.

Proposition 5.3 (Borel Does Not Contain Jordan). *The cardinality of $\mathcal{B}$ is $2^{\aleph_0} = \mathfrak{c}$, the cardinality of $\mathbb{R}$. The cardinality of Jordan measurable sets in $\mathbb{R}$ is $2^{\mathfrak{c}}$. Therefore, there exist Jordan measurable sets that are not Borel measurable.*

In view of the fact that there are so many more Jordan measurable sets than Borel sets, one might expect that it is fairly easy to give an explicit example of a set that is Jordan measurable and not Borel. In fact, finding such explicit sets is difficult (but see Exercises 5.4.8–5.4.11 for an example).

Because it would take us far afield, a discussion of the proof that the cardinality of $\mathcal{B}$ is $\mathfrak{c}$ has been put in Appendix A.1, but there is a simple heuristic argument why this might be the case. The cardinality of the set of all intervals cannot

---

[1] Borel, (1950, p. 46n).

exceed the cardinality of the set of pairs of real numbers, and that is c. Taking complements only doubles the number of elements in the set, so the cardinality is still c. Taking differences, unions of countable collections, and intersections of countable collections of the Borel sets we have already constructed still does not get us beyond cardinality $c^{\aleph_0} = c$. We proceed by induction.

The problem with this approach is that the induction needs to go beyond all finite positive integers. One has to be much more careful than this when arguing by induction with transfinite numbers. But it gives an indication of why we might believe that the cardinality is only c.

Borel knew that it would make sense to assign measure 0 to the subsets of the Cantor ternary set. In general, if $A \subseteq S \subseteq B$, where A and B are Borel sets with the same measure, then Borel recognized that we should assign that value as the measure of S. But he never worked out the implications of this insight. And he never recognized that this would provide the key to the problems of integration. That revelation would come to the young graduate student, Henri Lebesgue.

Exercises

5.1.1. Prove that if f is Riemann integrable, then every open interval contains at least one point at which f is continuous.

5.1.2. Give an example of a pointwise discontinuous function that is not Riemann integrable.

5.1.3. Show that a bounded set S is Jordan measurable if and only if, given $\epsilon > 0$, we can find two finite unions of intervals, E_1 and E_2, such that

$$E_1 \subseteq S \subseteq E_2, \quad \text{and} \quad c(E_2) - c(E_1) < \epsilon.$$

5.1.4. Prove Proposition 5.2 with the weaker hypothesis that instead of having $S_1, S_2, \ldots, S_n$ be pairwise disjoint, we only insist that their interiors are pairwise disjoint.

5.1.5. Show that if S and T are Jordan measurable, then so are $S \cup T$ and $S \cap T$, and

$$c(S \cup T) + c(S \cap T) = c(S) + c(T).$$

5.1.6. Show that if S is Jordan measurable, then so is the interior of S and the closure of S, and all three sets have the same content.

5.1.7. Is the set

$$\left\{ m + \frac{1}{n+1} \mid m, n \in \mathbb{N} \right\}$$

Jordan measurable? Justify your answer.

5.1.8. Let $(r_k)_{k=1}^{\infty}$ be the sequence of rationals in $[0, 1]$. Let I_k be the open interval of length $1/2^{k+1}$ centered at r_k. Show that $\bigcup_{k=1}^{\infty} I_k$ is not Jordan measurable.

5.1.9. Given an example of a bounded open set that is not Jordan measurable.

5.1.10. Give an example of a bounded closed set that is not Jordan measurable.

5.1.11. Give an example of a Borel measurable set that is not Jordan measurable.

5.1.12. Find the smallest σ-algebra that contains all closed intervals with rational endpoints.

5.1.13. Show that under the definition of Borel sets, every open interval and every half-open interval is a Borel set.

5.1.14. Show that under the definition of Borel sets, every countable intersection (finite or infinite) of Borel sets is again a Borel set.

5.1.15. Show that under the definition of Borel sets, if A and B are Borel sets, $A \supseteq B$, then $A - B$ is a Borel set.

5.1.16. Find an example of a Borel set that is neither the countable union of intervals (open, closed, half open, or even a single point), nor is it the complement of such a union.

5.1.17. Define f for $x \geq 0$ by

$$f(x) = x^{-1/4}, \ x > 0; \quad f(0) = 0.$$

This function is not bounded and so not Riemann integrable on $[0, 1]$, but its improper integral does exist. Find the value of the improper integral $\int_0^1 f(x)\,dx$. Consider the function g defined on $[0, 1]$ by $g(x) = 0$ for $x \in \mathrm{SVC}(4)$, and, if (a, b) is one of the disjoint open intervals whose union equals $[0, 1] - \mathrm{SVC}(4)$, then g on (a, b) is given by

$$g(x) = \begin{cases} (x - a)^{-1/4}, & a < x \leq (a + b)/2, \\ (b - x)^{-1/4}, & (a + b)/2 < x < b. \end{cases}$$

Show that even the improper Riemann integral of g over $[0, 1]$ does not exist. Find the integral of g over each of the open intervals of length 4^{-n} on which g is nonzero. Sum these values (recalling that there are 2^{n-1} intervals of length 4^{-n}) to find a value that would be reasonable to assign to this improper integral.

5.1.18. Consider the set C defined in Exercise 4.4.4. Find the Borel measure of C. Justify your answer.

5.1.19. Consider the set V defined in Exercise 4.4.6. Find the Borel measure of V. Justify your answer.

5.1.20. Let f be any real-valued function defined on $\mathbb{R}$. Show that the set of points of continuity of f is a Borel set.

5.1.21. Let (f_n) be a sequence of continuous functions defined on $\mathbb{R}$. Show that the set of points at which this sequence converges is a Borel set.

5.1.22. A real number x is **simply normal** to base 10 if each digit appears with the same asymptotic frequency. Specifically, let $N(x, d, n)$ be the number of occurrences of the digit d among the first n digits in the decimal expansion of x, then $\lim_{n\to\infty} N(x, d, n)/n = 1/10$. A real number x is **normal** to base 10 if each block of k digits appears with the same asymptotic frequency. Let $N(x, B, n)$ be the number of occurrences of the block B (including overlapping occurrences) among the first n digits of the decimal expansion of x, the $\lim_{n\to\infty} N(x, B, n)/n = 10^{-k}$.

1. Show that for each digit d, $\{x \,|\, \lim_{n\to\infty} N(x, d, n)/n\}$ is a Borel set.
2. Show that the set of real number that simply normal to base 10 is a Borel set.
3. Show that for any base $b \geq 2$, the set of real numbers that are simply normal to base b is a Borel set.
4. Show that for any base $b \geq 2$, the set of real numbers that are normal to base b is a Borel set.
5. Show that the set of numbers that are normal to every base $b \geq 2$ is a Borel set.

5.2 Lebesgue Measure

Lebesgue was born in 1875. He entered college at the École Normale Superieur in Paris in 1894. It was there that he studied Jordan's *Cours d'analyse* and met Émile Borel. He graduated in 1897, worked in the library for two years, and then took a high school teaching position in Nancy, all while working on his doctoral dissertation in which he undertook nothing less than a revolutionary approach to the problem of integration. Over the period 1899–1901, he published the results of his studies. The dissertation was formally accepted at the Sorbonne in 1902, and in 1902–1903 he gave the prestigious *Cours Peccot* at the Collège de France, in which he explained his results in *Leçons sur l'intégration et la recherche des fonctions primitives* (Lectures on integration and the search for antiderivatives).

After receiving his doctorate, Lebesgue held professorships in Rennes and Poitier and then at the Sorbonne beginning 1910. In 1921 he was appointed professor at the Collège de France. He was elected to membership in the Académie des Sciences in 1922. Lebesgue was a prolific mathematician, making important contributions in topology, set theory, and partial differential equations. In his later years, he focused on pedagogy and the history of mathematics. Lebesgue died in 1941.

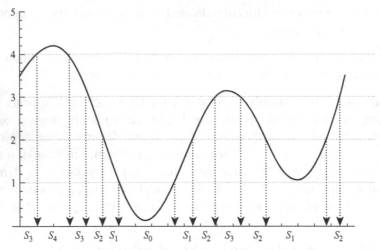

Figure 5.1. Lebesgue's horizontal partition.

His *Lectures on integration*, still in print, is an excellent introduction to the subject of Lebesgue measure and integration. The first third explains the Riemann integral, discusses its strengths and flaws, and goes over much of the history we have presented in the earlier chapters of this book. Lebesgue ends this section with Jordan measure and the theorem that for nonnegative functions the Riemann integral is simply the Jordan measure of S_f, the set of points bounded above by the graph of f and below by the x-axis.

In Chapter 7, he reveals his new idea. To define $\int_a^b f(x)\,dx$, he does not follow Newton, Leibniz, Cauchy, and Riemann who partitioned the x-axis between a and b. Instead, he lets l be the infimum of the values of f, L the supremum, *and then partitions the y-axis between l and L*. That is, instead of cutting the area using a finite number of vertical cuts, he takes a finite number of horizontal cuts (see Figure 5.1).

Consider the partition of the y-axis: $l = l_0 < l_1 < l_2 < \cdots < l_n = L$. For each horizontal strip, $l_i \le y < l_{i+1}$, we consider all points in the domain for which $f(x)$ lies in this strip:

$$S_i = \{x \mid l_i \le f(x) < l_{i+1}\}.$$

Let χ_{S_i} be the characteristic function of this set. It is 1 if x is in the set, 0 if it is not. We can squeeze our function between two sums of characteristic functions:

$$\sum_{i=0}^{n-1} l_i \chi_{S_i}(x) \le f(x) < \sum_{i=0}^{n-1} l_{i+1} \chi_{S_i}(x) \quad \text{for all } x \in [a, b].$$

We are working with finite summations, so the integral of each of these sums should be the sums of the integrals, and integration should preserve the inequalities:

$$\sum_{i=0}^{n-1} l_i \int_a^b \chi_{S_i}(x)\,dx \le \int_a^b f(x)\,dx \le \sum_{i=0}^{n-1} l_{i+1} \int_a^b \chi_{S_i}(x)\,dx.$$

The integral of the characteristic function of a set should be the measure of that set. Our sets S_i are, by the way they have been defined, pairwise disjoint. If they are always Jordan measurable, then the sum of the measures of the S_i is the measure of their union, which is $b - a$. If the lengths of the intervals on the y-axis, $l_{i+1} - l_i$, are all less than ϵ, then the upper and lower limits on the value for our integral differ by at most

$$\sum_{i=0}^{n-1} (l_{i+1} - l_i) \int_a^b \chi_{S_i}(x)\,dx \le \epsilon \sum_{i=0}^{n-1} \int_a^b \chi_{S_i}(x)\,dx = \epsilon(b - a).$$

We can *always* force the upper and lower bounds as close as we wish by taking the partition of the y-axis sufficiently fine.

If we restrict ourselves to Jordan measure, then we are right back at the Riemann integral. But Lebesgue saw that he could use Borel's idea of measure.

Consider V', the derivative of Volterra's function. Our inability to integrate this function comes from the fact that in any neighborhood of a point in SVC(4), this derivative takes on both the values $+1$ and -1. If we slice our function horizontally and look at where the function lies between, say, 0.7 and 0.8, this is a fairly nice set. It is a countable union of disjoint intervals (see Figure 5.2). The set of values of x for which $V'(x)$ lies between 0.7 and 0.8 is not measurable in Jordan's sense, but it is a Borel set. If we use Borel measure, then the derivative of Volterra's function is integrable. The fundamental theorem of calculus (evaluation part) appears to be salvageable.

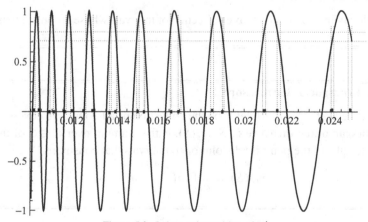

Figure 5.2. Lebesgue's partition of V'.

Improving on Borel

Lebesgue realized that he could not simply substitute Borel measure for Jordan measure. As we have seen, that severely reduces the number of sets that are measurable. What Lebesgue needed was a concept of measure that would encompass all Jordan measurable sets *and* all Borel measurable sets. He laid out three conditions that his measure would have to possess:

1. It is translation invariant: adding the same number to each element of a measurable set does not change its measure.
2. The measure of a countable union of pairwise disjoint measurable sets is equal to the sum of the measures of the individual sets.
3. The measure of the interval $(0, 1)$ is 1.

Unlike Borel, Lebesgue sought to find the most general possible collection of sets for which such a measure could be defined. Lebesgue measure is built on the concept of the **countable cover**.

Any countable union of intervals can be expressed as a countable union of pairwise disjoint intervals. If C is a countable union of pairwise disjoint intervals, then Lebesgue's three conditions imply that the Lebesgue measure of C, denoted $m(C)$, must equal the sum of the lengths of the intervals of C. We use this as our starting point for the general definition of Lebesgue measure.

Lebesgue outer measure satisfies Lebesgue's three conditions, but it still misses one critical property that is present in Jordan and Borel measure: if S and T are Jordan or, respectively, Borel measurable sets, then $S - T$ is also respectively Jordan or Borel measurable, and the respective measure of $S - T$ is the measure of S minus the measure of $S \cap T$. In the case of Borel measure, this is built into

---

Definition: Countable cover

A **countable cover** of S is a countable collection of intervals whose union contains S.

Definition: Lebesgue outer measure

Given a bounded set $S \subseteq [a, b]$, let $\mathcal{C}$ be the collection of all countable covers of S. The **Lebesgue outer measure of** S, $m_e(S)$, is the infimum over $C \in \mathcal{C}$ of the sum of the lengths of the pairwise disjoint open intervals that constitute C,

$$m_e(S) = \inf_{C \in \mathcal{C}} m(C).$$

the definition since measurability is preserved under complementation. For Jordan measure, the justification is a bit more subtle.

Jordan measure is defined in terms of the measure of bounded sets. A bounded set is Jordan measurable when its inner and outer content are the same, and the inner content is defined by the supremum of the sum of the lengths of a finite number of disjoint intervals whose union is contained within the set. The complement of any finite union of disjoint intervals is a finite union of disjoint intervals (allowing an interval to be a single point). If $S \subseteq [a, b]$, then each C, a finite union of disjoint intervals contained in S, corresponds to exactly one K, a finite union of disjoint intervals that contains $[a, b] - S$, and vice versa. It follows that

$$c_i(S) = \sup_{C \in \mathcal{C}_S^i} c(C)$$

$$= \sup_{K \in \mathcal{C}_{[a,b]-S}^e} \left((b - a) - c(K) \right)$$

$$= (b - a) - \inf_{K \in \mathcal{C}_{[a,b]-S}^e} c(K)$$

$$= (b - a) - c_e \left([a, b] - S \right).$$

The condition needed for Jordan measurability, $c_i(S) = c_e(S)$, is precisely the condition we need in order to guarantee that $c_e \left([a, b] - S \right) = (b - a) - c_e(S)$.

Lebesgue outer measure is more complicated because the complement of a countable union of disjoint intervals is no longer necessarily a countable union of disjoint intervals – witness the SVC sets. It might seem that the natural definition of Lebesgue inner measure would be to take the supremum over all countable unions of disjoint intervals contained in S of the sum of the lengths of these intervals. That turns out to be a useless notion because it just recreates the inner content (see Exercise 5.2.11). The right definition of Lebesgue inner measure parallels the

---

Definition: Lebesgue inner measure

Given a bounded set $S \subseteq [a, b]$, the **Lebesgue inner measure of** S, $m_i(S)$, is $b - a$ minus the Lebesgue outer measure of the complement of S in $[a, b]$:

$$m_i(S) = (b - a) - m_e \left([a, b] - S \right).$$

Definition: Lebesgue measure

Given a bounded set $S \subseteq [a, b]$, if $m_i(S) = m_e(S)$, then we say that S is **Lebesgue measurable**, and its measure is defined to be this common value:

$$m(S) = m_e(S) = m_i(S).$$

alternate definition of inner content, the one that shows how to compute the content of a complement.

It may seem that we have stopped short of the full complementarity that we need: If S and T are measurable, then so is $S - T$, and $m(S - T) = m(S) - m(S \cap T)$. As we shall see in the next section, this more general statement of complementarity is a consequence of the statement that for $S \subseteq [a, b]$,

$$m([a, b] - S) = (b - a) - m(S). \tag{5.5}$$

From now on, the terms **outer measure**, **inner measure**, and **measure** will refer to Lebesgue measures. The fact that this definition satisfies the first and third conditions for our measure is easy to check and is left for the exercises. The second condition, countable additivity, will be proven in the next section.

As Lebesgue observed, outer measure is always **subadditive**,

Theorem 5.4 (Subadditivity of Outer Measure). *Lebesgue outer measure is subadditive. That is to say, for any countable collection of sets, $(S_1, S_2, \ldots)$,*

$$m_e \left(\bigcup_{i=1}^{\infty} S_i \right) \leq \sum_{i=1}^{\infty} m_e(S_i). \tag{5.6}$$

Proof. Choose any $\epsilon > 0$ and for each S_i choose a countable open cover $(I_{i1}, I_{i2}, \ldots)$ for which

$$\sum_{n=1}^{\infty} \text{length}(I_{in}) < m_e(S_i) + \epsilon/2^i.$$

We create a countable open cover of $\bigcup S_i$ by taking all of the open intervals in all of the chosen open covers. This is a countable collection of countable collections, so it is still countable. The outer measure of $\bigcup S_i$ is bounded by the sum of the lengths of all of these intervals. This sum is bounded by

$$\sum_{i=1}^{\infty} \left(m_e(S_i) + \epsilon/2^i \right) = \left(\sum_{i=1}^{\infty} m_e(S_i) \right) + \epsilon.$$

Since this upper limit holds for all $\epsilon > 0$, it follows that

$$m_e \left(\bigcup_{i=1}^{\infty} S_i \right) \leq \left(\sum_{i=1}^{\infty} m_e(S_i) \right). \qquad \square$$

It follows that

$$m_e(S) + m_e([a, b] - S) \geq m_e([a, b]) = b - a. \tag{5.7}$$

This can be restated as

$$m_e(S) \geq (b - a) - m_e\left([a, b] - S\right) = m_i(S), \tag{5.8}$$

with exact equality if and only if S is measurable. Note that $S \subseteq [a, b]$ is measurable if and only if $[a, b] - S$ is measurable (see Exercise 5.2.5).

As the next theorem shows, subadditivity allows us to collect many examples of measurable sets.

Theorem 5.5 (Examples of Measurable Sets). *If the set S is bounded, then any of the following conditions implies that S is measurable:*

1. *The outer measure of S is zero,*
2. *S is countable, or*
3. *S is an interval (open, closed, or half open).*

Proof.

1. Because of inequality (5.7), we only need to prove that

$$m_e(S) + m_e\left([a, b] - S\right) \leq b - a.$$

   The outer measure of $[a, b] - S$ is less than or equal to $b - a$, and therefore

$$m_e(S) + m_e\left([a, b] - S\right) \leq 0 + (b - a).$$

2. Any countable set has outer measure zero because given any $\epsilon > 0$, we can put the nth point inside an open interval of length less than $\epsilon/2^n$ and so obtain a countable open cover for which the sum of the lengths of the intervals is less than ϵ.
3. If S is an interval, we can choose $[a, b]$ to be the closure of S (see Exercise 5.2.4). The set $[a, b] - S$ consists of at most two points, so its outer measure is zero. $\qquad\square$

Alternate Definition of Lebesgue Measure

Lebesgue defined his outer measure in terms of countable covers which pushed him into a somewhat awkward definition of the inner measure. It was later realized that there is a simpler formulation of these definitions. As we have shown, every open set is a countable union of pairwise disjoint open intervals. We shall now assume what will be proven in the next section, that any countable union of pairwise disjoint intervals is measurable, and its measure is the sum of the lengths of these intervals. It follows that every open set is measurable. Therefore, if S is any bounded set and U is any open set that contains S, then U is a countable cover of S, and U has a

> **Definition: Lebesgue measure**
>
> Given a bounded set S, the **Lebesgue outer measure** of S is the infimum of $m(U)$ taken over all open sets U that contain S. The **Lebesgue inner measure** of S is the supremum of $m(F)$ taken over all closed sets F contained in S. The set S is measurable if and only if the inner and outer measures are equal.

well-defined measure. Furthermore, given any $\epsilon > 0$ and any countable cover C of S, we can always expand the length of the ith interval to $\epsilon/2^i$ and turn C into an open set U for which $m(U) \leq m(C) + \epsilon$. In other words, we lose nothing if we restrict our countable covers to consist of open sets.

If $S \subseteq [a, b]$ and U is an open set that contains $[a, b] - S$, then $[a, b] - U = [a, b] \cap U^C$ is a closed set contained in S. If we define the measure of $F = [a, b] - U$ to be $(b - a) - m(U)$, then the inner measure of S is $b - a$ minus the outer measure of $[a, b] - S$, which is the supremum over all closed sets F contained in S of the measure of F (see Exercise 5.2.16). We have established the equivalent definition of Lebesgue inner and outer measure given above.

Exercises

5.2.1. Prove that if A and B are measurable sets and $A \subseteq B$ then $m(A) \leq m(B)$.

5.2.2. Let x be a real number, $x + S = \{x + s \mid x + s, \ s \in S\}$. Prove that $m_e(x + S) = m_e(S)$. Show that this implies that if S is measurable, then $m(x + S) = M(S)$.

5.2.3. Prove that $(0, 1)$ is measurable and its measure is equal to 1.

5.2.4. Show that the definition of the inner measure does not depend on the choice of the interval $[a, b]$. Let $\alpha = \inf S$ and $\beta = \sup S$. Show that

$$m_e \left(S^C \cap [a, b] \right) = (b - a) - (\beta - \alpha) + m_e \left(S^C \cap [\alpha, \beta] \right).$$

5.2.5. Prove that if $S \subseteq [a, b]$ then S is measurable if and only if $S^C \cap [a, b]$ is measurable.

5.2.6. Prove that if $m_e(S) = 0$ then $m_e(S \cup T) = m_e(T)$.

5.2.7. Prove that for any bounded set S and any $\epsilon > 0$, we can always find an open set $U \supseteq S$ such that

$$m_e(U) < m_e(S) + \epsilon.$$

5.2.8. Prove that for any bounded set S, we can always find a set T that is a countable intersection of open sets (and, thus, a Borel set) for which $S \subseteq T$ and

$$m_e(S) = m_e(T).$$

5.2.9. Prove that for any bounded set S and any $\epsilon > 0$, we can always find a closed set $K \subseteq S$ such that

$$m_e(K) > m_e(S) - \epsilon.$$

5.2.10. Prove that for any bounded set S, we can always find a set L that is a countable union of closed sets (and, thus, a Borel set) for which $S \supseteq L$ and

$$m_e(S) = m_e(L).$$

5.2.11. Given a set S, let $\mathcal{K}_S$ be the collection of all countable unions of pairwise disjoint intervals contained in S. Prove that

$$\sup_{K \in \mathcal{K}_S} m(K) = c_i(S).$$

5.2.12. Let S and T each be a countable union of pairwise disjoint intervals, $S = \bigcup_{n=1}^{\infty} I_n$ and $T = \bigcup_{n=1}^{\infty} K_n$. Show that

$$m\left(S \cup T\right) + m\left(S \cap T\right) = m(S) + m(T).$$

5.2.13. Show that if S and T are bounded sets, then

$$m_e\left(S \cup T\right) + m_e\left(S \cap T\right) \leq m_e(S) + m_e(T).$$

5.2.14. Let S and T be bounded sets such that

$$\inf\left\{|x - y| \mid x \in S,\ y \in T\right\} > 0.$$

Show that

$$m_e(S \cup T) = m_e(S) + m_e(T).$$

5.2.15. Show that if S is a bounded, measurable set and $T \supseteq S$, then

$$m_e(T - S) = m_e(T) - m(S).$$

5.2.16. Assuming that all open sets and all closed sets are measurable, show that if $S \subseteq [a, b]$, then the infimum over all open sets U that contain $[a, b] - S$ of $m(U)$ is equal to $b - a$ minus the supremum over all closed sets $F \subseteq S$ of $m(F)$.

5.2.17. Prove that if $S \subset (a, b)$ has measure 0, then $(a, b) - S$ is dense in (a, b) and is uncountable.

5.2.18. Using Exercise 5.2.17 and the fact that any uncountable closed subset of $\mathbb{R}$ has cardinality $\mathfrak{c}$ (see p. 96), prove that if $S \subset (a, b)$ has measure 0, then $(a, b) - S$ has cardinality $\mathfrak{c}$.

5.3 Carathéodory's Condition

Constantin Carathéodory (1873–1950) came from a family of the Greek urban elite of Constantinople (modern Istanbul). His father served as Ottoman ambassador to Brussels from 1875 until 1900. Carathéodory grew up in Belgium where he studied engineering. From 1897 until 1900 he worked on the construction of the Assiut dam in Egypt, studying Jordan's *Course d'analyse* in his spare time. He then went to the University of Berlin where he earned his doctorate in mathematics under the direction of Hermann Minkowski. For most of his career, he taught in German universities.

In 1914, Carathéodory offered an alternate definition of measurability. It arose from his extension of the notion of measure to k-dimensional subsets of $\mathbb{R}^q$, $k < q$, an extension that, in 1919, would lead Felix Hausdorff to define and explore sets with noninteger dimension. We want measurable sets to have the property found in Borel sets that if X and S are measurable, then so is $X - S$. Furthermore, $m(X - S) = m(X) - m(X \cap S)$. Carathéodory showed that this follows from Lebesgue's definition of measure and that something much stronger is true.

What Carathéodory's condition says is that we can take any measurable set and use it to cut any other set. The outer measure of the set that is being cut – in this case, X – will equal the sum of the outer measures of the two pieces. The first part of this section will be devoted to proving that Carathéodory's condition is a consequence of Lebesgue's definition of measurability. The heart of this section will use Carathéodory's condition to prove that any countable union or intersection of measurable sets is again measurable. This establishes the fact that Lebesgue measurable sets form a σ-algebra, and thus include all Borel sets.

Note also that Lebesgue's original definition of measure was restricted to bounded sets. Carathéodory's condition enables us to determine when an un-bounded set is measurable.

If a bounded set satisfies Carathéodory's condition, then it satisfies this equality when $X = [a, b] \supseteq S$, and so it satisfies Lebesgue's definition of a measurable set. It will take some work to show that any set that satisfies Lebesgue's condition also satisfies Carathéodory's.

Definition: Lebesgue measure, Carathéodory condition

A set S is **measurable** if and only if for every set X with finite outer measure,

$$m_e(X - S) = m_e(X) - m_e(X \cap S). \tag{5.9}$$

If S is measurable, then the measure of S is defined to be $m_e(S)$.

Theorem 5.6 (Lebesgue $\Longrightarrow$ Carathéodory). *If a bounded set S satisfies*

$$m_e\left([a, b] - S\right) = (b - a) - m_e(S) \tag{5.10}$$

for any interval $[a, b] \supseteq S$, then it satisfies Carathéodory's condition.

Before we prove this theorem, we need a lemma.

Lemma 5.7 (Local Additivity). *Let S be any bounded set and $(I_1, I_2, \ldots)$ any countable collection of pairwise disjoint intervals, then*

$$m_e\left(S \cap \bigcup_i I_i\right) = \sum_i m_e(S \cap I_i). \tag{5.11}$$

Proof. Given any $\epsilon > 0$, choose a countable open cover, $(J_1, J_2, \ldots)$, of $S \cap \bigcup_i I_i$ such that $\sum_j m(J_j) < m_e(S \cap \bigcup_i I_i) + \epsilon$. Because the intervals $I_1, I_2, \ldots$ are pairwise disjoint, we have that

$$\sum_i m(J_j \cap I_i) \leq m(J_j).$$

By the subadditivity of the outer measure, Theorem 5.4, and the fact that

$$S \cap \bigcup_i I_i = \bigcup_i (S \cap I_i),$$

we see that

$$
\begin{aligned}
m_e\left(S \cap \bigcup_i I_i\right) &\leq \sum_i m_e(S \cap I_i) \\
&\leq \sum_i m_e\left(\bigcup_j (J_j \cap I_i)\right) \\
&\leq \sum_{i,j} m_e\left(J_j \cap I_i\right) \\
&\leq \sum_j m(J_j) \\
&< m_e\left(S \cap \bigcup_i I_i\right) + \epsilon.
\end{aligned}
\tag{5.12}
$$

Since this is true for all $\epsilon > 0$, the first inequality must be an equality. $\square$

Proof. (**Theorem 5.6**) We assume that S satisfies equation (5.10). Our first step is to show that S satisfies equation (5.9) when X is a bounded interval.

We find two intervals, Y immediately to the left of X and Z immediately to the right, so that Y, X, and Z are pairwise disjoint and $Y \cup X \cup Z$ is a single closed interval that contains S. From equation (5.10), we see that

$$m_e(S) + m_e \left(S^C \cap (Y \cup X \cup Z) \right) = m(Y \cup X \cup Z) = m(Y) + m(X) + m(Z).$$
(5.13)

By Lemma 5.7, we know

$$m_e(S) = m_e(S \cap Y) + m_e(S \cap X) + m_e(S \cap Z), \quad (5.14)$$

$$m_e \left(S^C \cap (Y \cup X \cup Z) \right) = m_e(S^C \cap Y) + m_e(S^C \cap X) + m_e(S^C \cap Z).$$
(5.15)

Combining equations (5.13)–(5.15), we see that

$$\begin{aligned}
m(Y) + m(X) + m(Z) &= m_e(S) + m_e \left(S^C \cap (Y \cup X \cup Z) \right) \\
&= \left(m_e(S \cap Y) + m_e(S^C \cap Y) \right) \\
&\quad + \left(m_e(S \cap X) + m_e(S^C \cap X) \right) \\
&\quad + \left(m_e(S \cap Z) + m_e(S^C \cap Z) \right).
\end{aligned}$$
(5.16)

Subadditivity implies that for each interval I, we have

$$m_e(S \cap I) + m_e(S^C \cap I) \ge m(I).$$

The only way we can get equality (5.16) is if we have $m_e(S \cap I) + m_e(S^C \cap I) = m(I)$ for each of the three intervals. Therefore,

$$m(X) = m_e(S \cap X) + m_e(S^C \cap X).$$

We now let X be any set with finite outer measure. Given any $\epsilon > 0$, we choose a countable open cover, $(I_1, I_2, \ldots)$, of X so that

$$\sum_i m(I_i) < m_e(X) + \epsilon.$$

We use subadditivity and the first part of our proof:

$$\begin{aligned}
m_e(X) &\le m_e(S \cap X) + m_e(S^C \cap X) \\
&\le m_e \left(S \cap \bigcup_i I_i \right) + m_e \left(S^C \cap \bigcup_i I_i \right) \\
&\le \sum_i \left(m_e(S \cap I_i) + m_e(S^C \cap I_i) \right) \\
&= \sum_i m(I_i) \\
&< m_e(X) + \epsilon.
\end{aligned}$$
(5.17)

Again, since this is true for all $\epsilon > 0$, the first inequality must be equality. $\qquad \square$

We now take the first step toward proving that any countable union of measurable sets is measurable. This theorem, in addition to moving us toward that result, is very important in its own right.

Theorem 5.8 (Countable Additivity). *If $(S_1, S_2, \ldots)$ are pairwise disjoint measurable sets whose union has finite outer measure, then*

$$m_e\left(\bigcup_{i=1}^{\infty} S_i\right) = \sum_{i=1}^{\infty} m(S_i). \tag{5.18}$$

Proof. We start with two disjoint measurable sets, S_1 and S_2, and invoke the Carathéodory condition with $X = S_1 \cup S_2$. Since $S_1^C \cap X = S_2$, the Carathéodory condition gives us exactly what we need,

$$m(S_1) + m(S_2) = m(S_1) + m(S_1^C \cap X) = m_e(X) = m_e(S_1 \cup S_2).$$

We now proceed by induction. Assume that additivity holds if we have $n-1$ pairwise disjoint measurable sets, but that we are faced with n pairwise disjoint measurable sets. Let $X = S_1 \cup S_2 \cup \cdots \cup S_n$. Then $S_1^C \cap X = S_2 \cup \cdots \cup S_n$. From the Carathédory condition and our induction hypothesis, we see that

$$m_e(X) = m(S_1) + m_e(S_2 \cup \cdots \cup S_n) = m(S_1) + m(S_2) + \cdots + m(S_n).$$

Finally, we consider a countably infinite collection of pairwise disjoint measurable sets. For any finite value of n,

$$m_e\left(\bigcup_{i=1}^{\infty} S_i\right) \geq m_e\left(\bigcup_{i=1}^{n} S_i\right) = \sum_{i=1}^{n} m(S_i).$$

Since this upper bound holds for all n, the summation converges as n approaches infinity and

$$m_e\left(\bigcup_{i=1}^{\infty} S_i\right) \geq \sum_{i=1}^{\infty} m(S_i).$$

The inequality in the other direction follows from subadditivity. $\square$

We would like to be able to say that these unions are also measurable. In fact, we would like to be able to say that any countable union of measurable sets is measurable. The first step is to consider finite unions and intersections.

Theorem 5.9 (Finite Unions and Intersections). *Any finite union or intersection of measurable sets is measurable.*

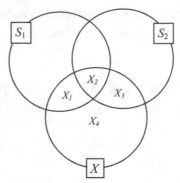

Figure 5.3. The set X cut by S_1 and S_2.

Proof. It is enough to show that the union of two measurable sets is measurable. The intersection of two sets is the complement of the union of their complements,

$$S_1 \cap S_2 = (S_1^C \cup S_2^C)^C,$$

and by induction we can then conclude that any finite union or intersection of measurable sets is measurable.

Let X be the arbitrary set to be cut by $S_1 \cup S_2$. We divide X into four disjoint subsets (see Figure 5.3):

$$X_1 = X \cap (S_1 \cap S_2^C), \qquad X_2 = X \cap (S_1 \cap S_2),$$
$$X_3 = X \cap (S_1^C \cap S_2), \qquad X_4 = X \cap (S_1^C \cap S_2^C). \tag{5.19}$$

To show that $S_1 \cup S_2$ is measurable, we need to show that

$$m_e(X_1 \cup X_2 \cup X_3) + m_e(X_4) = m_e(X). \tag{5.20}$$

We first cut $X_1 \cup X_2 \cup X_3$ by S_1:

$$m_e(X_1 \cup X_2 \cup X_3) + m_e(X_4) = m_e(X_1 \cup X_2) + m_e(X_3) + m_e(X_4). \tag{5.21}$$

We next use S_2 to glue together X_3 and X_4:

$$m_e(X_1 \cup X_2) + m_e(X_3) + m_e(X_4) = m_e(X_1 \cup X_2) + m_e(X_3 \cup X_4). \tag{5.22}$$

Finally, we use S_1 to glue together $X_1 \cup X_2$ and $X_3 \cup X_4$:

$$m_e(X_1 \cup X_2) + m_e(X_3 \cup X_4) = m_e(X). \tag{5.23}$$

$\square$

Theorem 5.10 (Countable Unions and Intersections). *Any countable union or intersection of measurable sets is measurable.*

Proof. Again, it is enough to prove this theorem for countable unions. Let $T_n = S_1 \cup S_2 \cup \cdots \cup S_n$, $T = \bigcup_{i=1}^{\infty} S_i$. We shall also use the sets U_n, where $U_1 = T_1$ and $U_n = T_n - T_{n-1}$ for $n > 1$. In other words, U_n consists of all elements of S_n that are not in $S_1, S_2, \ldots,$ or S_{n-1}. By their construction, the sets T_n and U_n are measurable, and the U_n are pairwise disjoint. The union of U_1 through U_n is T_n. Since T_n is measurable, we know that for any set X with finite outer measure,

$$m_e(X) = m_e(X \cap T_n) + m_e\left(X \cap T_n^C\right). \tag{5.24}$$

Since U_n is measurable,

$$m_e(X \cap T_n) = m_e(X \cap T_n \cap U_n) + m_e\left(X \cap T_n \cap U_n^C\right)$$
$$= m_e(X \cap U_n) + m_e(X \cap T_{n-1}). \tag{5.25}$$

By induction,

$$m_e(X \cap T_n) = \sum_{k=1}^{n} m_e(X \cap U_k). \tag{5.26}$$

Since $T \supseteq T_n$, we know that $T^C \subseteq T_n^C$, and therefore $m_e(X \cap T_n^C) \geq m_e(X \cap T^C)$. We can rewrite equation (5.24) as an inequality,

$$m_e(X) \geq \sum_{k=1}^{n} m_e(X \cap U_k) + m_e(X \cap T^C). \tag{5.27}$$

We use the same trick we used in the proof of Theorem 5.8. Our summation has an upper bound independent of n, so the infinite summation must converge and

$$m_e(X) \geq \sum_{k=1}^{\infty} m_e(X \cap U_k) + m_e(X \cap T^C).$$

By Theorem 5.8,

$$\sum_{k=1}^{\infty} m_e(X \cap U_k) \geq m_e\left(\bigcup_{k=1}^{\infty}(X \cap U_k)\right) = m_e\left(X \cap \bigcup_{k=1}^{\infty} U_k\right) = m_e(X \cap T).$$

Therefore,

$$m_e(X) \geq m_e(X \cap T) + m_e(X \cap T^C).$$

Subadditivity gives us the inequality in the other direction. $\square$

An important consequence of this result is the next theorem that shows that Lebesgue measure is, in a real sense, not very far removed from Jordan content.

> **Definition: Symmetric difference**
>
> The **symmetric difference** of two sets is the set of points that are in exactly one of these sets. The symmetric difference of S and T is written $S \triangle T$,
>
> $$S \triangle T = (S - T) \cup (T - S) = (S \cap T^C) \cup (T \cap S^C).$$

For example, the symmetric difference of the overlapping intervals $[0, 2]$ and $[1, 3]$ is

$$[0, 2] \triangle [1, 3] = [0, 1) \cup (2, 3].$$

Theorem 5.11 (Approximation by Finite Number of Open Intervals). *If $S \subseteq [a, b]$ is a measurable set, then for any $\epsilon > 0$ we can find a finite union of open intervals, U, such that*

$$m(S \triangle U) < \epsilon. \tag{5.28}$$

Borel called this the "Second Fundamental Theorem of Measure Theory."[2] We shall use it several times over the next few chapters.

Proof. By the definition of outer measure, we can find a countable union of open intervals that contains S, call it V, such that

$$m(S) \leq m(V) < m(S) + \epsilon/2.$$

While S and V are subsets of $[a, b]$, it may be helpful to think of them as subsets of the plane (see Figure 5.4). Since both S and V are measurable, $m(V - S) < \epsilon/2$. Let W denote a countable union of open intervals that contains $S^C \cap [a, b]$ and such that $m_e\left(W - (S^C \cap [a, b])\right) < \epsilon/2$. Since W is open, W^C is closed. We know that

$$W - \left(S^C \cap [a, b]\right) \supseteq W \cap S = S - W^C. \tag{5.29}$$

Therefore,

$$m\left(S - W^C\right) \leq m\left(W - \left(S^C \cap [a, b]\right)\right) < \epsilon/2.$$

Since $W \supseteq S^C \cap [a, b]$, S contains $W^C \cap [a, b]$, which is closed. We have now sandwiched our set S between a closed set and an open set,

$$W^C \cap [a, b] \subseteq S \subseteq V.$$

---

[2] Borel, *Leçons sur la Théorie des Fonctions*, 4th ed. 1950. The first fundamental theorem is the Heine–Borel theorem.

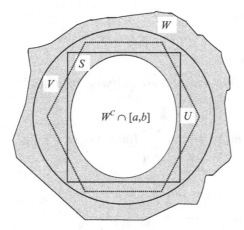

Figure 5.4. $V \supseteq S \supseteq W^C \cap [a, b]$. The shaded region is W. U is the region inside the dotted hexagon.

Since $W^C \cap [a, b]$ is closed and bounded, it is compact. Since V is a countable union of open intervals that contains $W^C \cap [a, b]$, the Heine–Borel theorem promises us a finite subcollection of open intervals, $U = \bigcup_{i=1}^{n} I_i$, such that

$$W^C \cap [a, b] \subseteq U \subseteq V.$$

By subadditivity,

$$m_e(S \Delta U) \leq m_e(S \cap U^C) + m_e(S^C \cap U) \leq m_e(S - W^C) + m_e(V - S) < \epsilon.$$

$\square$

We finish with a corollary that stands in stark contrast to Lemma 4.4 on page 105 which was both much more complicated and much more restricted in its assumptions.

Corollary 5.12 (Limit of Measure). *If $S_1, S_2, \ldots$ are measurable sets such that*

$$S_1 \subseteq S_2 \subseteq S_3 \subseteq \cdots, \qquad \bigcup_{i=1}^{\infty} S_i = S,$$

then S is measurable and

$$m(S) = \lim_{i \to \infty} m(S_i). \tag{5.30}$$

Similarly, if $T_1 \supseteq T_2 \supseteq T_3 \supseteq \cdots$ are measurable, T_1 has finite measure, and

$$\bigcap_{i=1}^{\infty} T_i = T,$$

then T is measurable and

$$m(T) = \lim_{i \to \infty} m(T_i). \tag{5.31}$$

Proof. The sets $S_2 - S_1$, $S_3 - S_2$, ... are pairwise disjoint. If we define $S_0 = \emptyset$, then we can write

$$\lim_{i \to \infty} m(S_i) = \lim_{i \to \infty} \sum_{j=1}^{i} \left(m(S_j) - m(S_{j-1}) \right)$$

$$= \lim_{i \to \infty} \sum_{j=1}^{i} m(S_j - S_{j-1})$$

$$= m \left(\bigcup_{j=1}^{\infty} S_j - S_{j-1} \right) = m(S).$$

To prove equation (5.31), we set $S_i = T_1 - T_i$ and apply equation (5.30) (see Exercise 5.3.5). $\qquad \square$

All intervals are measurable. With Theorem 5.10, we see that all Borel sets are measurable. I leave it as an exercise (Exercise 5.3.4) to verify that for any bounded set S,

$$c_i(S) \le m_i(S) \le m_e(S) \le c_e(S),$$

and therefore any Jordan measurable set is measurable in Lebesgue's sense. Are there *any* sets that are not measurable? That is an important question with a very surprising answer that will be revealed in the next section.

Exercises

5.3.1. Show that both $\mathbb{R}$ and $\emptyset$ satisfy Carathéodory's condition.

5.3.2. Let S be an unbounded set with bounded outer measure such that for every $k \in \mathbb{N}$, $S \cap [-k, k]$ satisfies Lebesgue's condition for measurability,

$$m_e \left([-k, k] - S \right) = 2k - m_e \left(S \cap [-k, k] \right).$$

Show that S satisfies the Carathéodory condition and that $m(S) = \lim_{k \to \infty} m \left(S \cap [-k, k] \right)$.

5.3.3. Show directly that the intersection of two measurable sets is measurable by proving that

$$m_e(X_2) + m_e(X_1 \cup X_3 \cup X_4) = m_e(X). \tag{5.32}$$

for the sets defined in equation (5.19) on p. 144.

5.3.4. Show that if the set S is bounded, then

$$c_i(S) \leq m_i(S) \leq m_e(S) \leq c_e(S).$$

5.3.5. Use equation (5.30) to prove equation (5.31).

5.3.6. Justify the set containment

$$W - \left(S^C \cap [a, b]\right) \supseteq W \cap S$$

given in (5.29).

5.3.7. Let U be an open set such that $S \subseteq U$ and $U \cap T = \emptyset$. Show that

$$m_e\left(S \cup T\right) = m_e(S) + m_e(T).$$

5.3.8. Show that if S and T are measurable, then

$$m(S \cup T) + m(S \cap T) = m(S) + m(T).$$

5.3.9. Show that if $m_e(S) < \infty$ and there is a measurable subset $T \subseteq S$ such that $m(T) = m_e(S)$, then S is measurable.

5.3.10. Let S be any subset of $\mathbb{R}$. Show that for any $\epsilon > 0$ there is an open set $U \supseteq S$ such that $m_e(S) < m(U) + \epsilon$. Show that there is a countable intersection of open sets, G, such that $G \supseteq S$ and $m_e(S) = m(G)$.

5.3.11. Show that for any bounded set S, the following statements are equivalent:

1. S is measurable.
2. Given any $\epsilon > 0$, there is an open set $U \supseteq S$ such that $m_e(U - S) < \epsilon$.
3. There is a countable intersection of open sets $G \supseteq S$ such that $m_e(G - S) = 0$.
4. Given any $\epsilon > 0$, there is a closed set $C \supseteq S$ such that $m_e(S - C) < \epsilon$.
5. There is a countable union of closed sets $F \subseteq S$ such that $m_e(S - F) = 0$.

5.3.12. Show that the statements of Exercise 5.3.11 are also equivalent when S is *any* subset of $\mathbb{R}$.

5.3.13. Let S and T be sets with finite outer measure. Show that

$$m_e(S \cup T) = m_e(S) + m_e(T)$$

if and only if there are measurable sets S_1 and T_1 such that $S \subseteq S_1$, $T \subseteq T_1$, and $m(S_1 \cap T_1) = 0$.

5.3.14. Let S and T be sets with finite outer measure. Show that if $m_e(S \cup T) = m_e(S) + m_e(T)$, then $m_i(S \cup T) = m_i(S) + m_i(T)$.

5.3.15. For a sequence $(S_n)_{n=1}^{\infty}$ of sets in $\mathbb{R}$, we define the supremum and the infimum[3] as

$$\overline{\lim_{n \to \infty}} \, S_n = \bigcap_{k=1}^{\infty} \bigcup_{n=k}^{\infty} S_n \quad \text{and} \quad \underline{\lim_{n \to \infty}} \, S_n = \bigcup_{k=1}^{\infty} \bigcap_{n=k}^{\infty} S_n.$$

1. Show that if each S_n is measurable, then

$$m \left(\underline{\lim_{n \to \infty}} \, S_n \right) \le \underline{\lim_{n \to \infty}} \, m(S_n).$$

2. Show that if, in addition, $m(S_n \cup S_{n+1} \cup \cdots) < \infty$ for at least one $n \ge 1$, then

$$m \left(\overline{\lim_{n \to \infty}} \, S_n \right) \ge \overline{\lim_{n \to \infty}} \, m(S_n).$$

5.3.16. We say that a sequence of sets, $(S_n)_{n=1}^{\infty}$, converges if $\overline{\lim}_{n \to \infty} S_n = \underline{\lim}_{n \to \infty} S_n$. We denote this common value by $\lim_{n \to \infty} S_n$.

1. Show that any monotonic sequence of sets converges.
2. Show that if (S_n) is a convergent sequence of measurable sets, $S_n \subseteq T$ for all $n \ge 1$, and $m_e(T) < \infty$, then

$$m \left(\lim_{n \to \infty} S_n \right) = \lim_{n \to \infty} m(S_n).$$

5.3.17. Let S be the set of points in $[0, 1]$ that do not require the use of the digit 7 in their decimal expansion. Show that S is measurable and $m(S) = 0$.

5.3.18. Find the Lebesgue measure of the set of points in $[0, 1]$ for which there is a decimal expansion that uses all of the digits 1 through 9.

5.4 Nonmeasurable Sets

In 1905, Giuseppe Vitali (1875–1932) published an example of a nonmeasurable set. Vitali graduated from the Scuola Normale Superiore in Pisa in 1899. He worked with Dini for two years before taking a job as a high school teacher and then entering politics, representing the Socialist Party on the city council in Genoa. With the rise of the Fascists in 1922 and the dissolution of the Socialist Party, he returned to mathematics. He suffered a stroke in 1926 that left half of his body paralyzed, but he continued to make important contributions to analysis until his death in 1932 from a heart attack.

The idea behind Vitali's construction is as follows. We separate $\mathbb{R}$ into **equivalence classes**. Two numbers are in the same equivalence class if they differ by a rational number. All of the rational numbers constitute one equivalence class. All numbers of the form $\sqrt{2} - a$ where $a \in \mathbb{Q}$ is another equivalence class. All

---

[3] First introduced by Borel in 1905.

numbers of the form $\pi - b$ where $b \in \mathbb{Q}$ is a third. From each equivalence class, we select one number that lies within $(0, 1)$ and call the resulting set $\mathcal{N}$.

Theorem 5.13 (Existence of Nonmeasurable Set). *The set $\mathcal{N}$ is not measurable.*

Proof. Let q be any rational number and define the translation $\mathcal{N} + q$ to be $\{a + q \mid a \in \mathcal{N}\}$. We have added a rational number to each element of $\mathcal{N}$, so $\mathcal{N} + q$ also consists of exactly one element from each equivalence class. If q_1 and q_2 are distinct rational numbers, then $\mathcal{N} + q_1$ is disjoint from $\mathcal{N} + q_2$.

Every real number in $(0, 1)$ is contained in $\mathcal{N} + q$ for exactly one rational value of q, and this value of q lies strictly between -1 and 1. To see this, take any real number $\alpha \in (0, 1)$ and find the equivalent number $\beta \in \mathcal{N}$. By definition of the equivalence, $\beta - \alpha \in \mathbb{Q}$ and $-1 < \beta - \alpha < 1$. We can bound the union of the pairwise disjoint sets $\mathcal{N} + q$ for rational q between -1 and 1:

$$(0, 1) \subseteq \bigcup_{q \in \mathbb{Q} \cap (-1,1)} (\mathcal{N} + q) \subseteq (-1, 2).$$

The outer measure is translation invariant (because it is based on interval lengths, and interval lengths are translation invariant), so $m_e(\mathcal{N} + q) = m_e(\mathcal{N})$. By subadditivity,

$$1 \le m_e \left(\bigcup_{q \in \mathbb{Q} \cap (-1,1)} (\mathcal{N} + q) \right) \le \sum_{q \in \mathbb{Q} \cap (-1,1)} m_e(\mathcal{N} + q) = \sum_{q \in \mathbb{Q} \cap (-1,1)} m_e(\mathcal{N}).$$

This tells us that $m_e(\mathcal{N}) > 0$.

If $\mathcal{N}$ is measurable, then all of the sets $\mathcal{N} + q$ must be measurable. This implies that $\bigcup_{q \in \mathbb{Q} \cap (-1,1)} (\mathcal{N} + q)$ is a countable union of measurable, pairwise disjoint sets, and so

$$3 \ge m \left(\bigcup_{q \in \mathbb{Q} \cap (-1,1)} (\mathcal{N} + q) \right) = \sum_{q \in \mathbb{Q} \cap (-1,1)} m(\mathcal{N}) = \infty.$$

The set $\mathcal{N}$ cannot be measurable. $\qquad \square$

Difficulties

This would seem to settle the matter. There are nonmeasurable sets. But Vitali's paper landed in the very center of a raging controversy among mathematicians. The construction of $\mathcal{N}$ requires selecting one number from each equivalence class. We have uncountably many equivalence classes.

Is it possible to have a set whose definition requires uncountably many choices?

Just a year earlier, many prominent mathematicians, Lebesgue among them, asserted that this should not be allowed. To appreciate what Vitali stepped into, we need to back up and investigate some of the issues created by Georg Cantor's work on transfinite numbers.

When Cantor first introduced his different sizes of infinity, $\aleph_0$, $\mathfrak{c} = 2^{\aleph_0}$, $2^{\mathfrak{c}}$, he thought of them only as **cardinal numbers**, descriptions of the relative size of a set. Finite cardinal numbers can also be thought of as **ordinal numbers**, describing a position in the ordered sequence of numbers. Thus "5" describes the size of the set $\{A, B, C, D, E\}$. It is also the integer that comes after 4 and before 6. Is there a similar ordering of the transfinite numbers? Can we write $\aleph_0 < \mathfrak{c} = 2^{\aleph_0} < 2^{\mathfrak{c}}$?

It may seem that the answer is obviously "yes," but there is a subtlety here. What enables us to say "$5 < 6$" is the fact that in any set of cardinality 6, we can always find a subset of cardinality 5. In every set of cardinality $\mathfrak{c}$, is there always a subset of cardinality $\aleph_0$? In general, given any two transfinite cardinals, α and β, is it always true that $\alpha < \beta$, $\alpha = \beta$, or $\beta < \alpha$? This became known as the **trichotomy property**. If two sets are not in one-to-one correspondence, is one of the sets always in one-to-one correspondence with a subset of the other? For infinite sets, the answer is not clear. "Always" is a very big word. Mathematicians had learned by now not to trust intuition when working with infinite sets.

In 1895, Cantor asserted that transfinite numbers possess the trichotomy property. In a letter to Richard Dedekind written in 1899 he claimed that this would follow from a property of transfinite sets that he had begun investigating in 1883, the notion of a **well-ordered** set.

Cantor was correct. In 1904 Ernst Zermelo proved that the trichotomy property follows if all sets are well ordered. Friedrich Hartogs would prove in 1915 that these properties are equivalent: If the trichotomy property always holds, then every set can be well ordered.

Definition: Total order

A **total order** on a set is a relation, call it $\prec$, such that (1) if $a \prec b$ and $b \prec c$, then $a \prec c$, and (2) for any two elements a and b, exactly one of the following is true: $a \prec b$, $a = b$, or $b \prec a$.

Definition: Well-ordered set

We say that a set S is **well ordered** if we can put a total order on the set so that every nonempty subset has a smallest element. This implies that S has a first element, and, given any element of S, there is a well-defined next element.

Zermelo (1871–1953) earned his doctorate in 1894 at the University of Berlin, working on the calculus of variations. After moving to Göttingen in 1897, and at the urging of David Hilbert, he turned his attention to the problems of set theory. He taught for several years at the University of Zurich before retiring to the Black Forest of Germany because of poor health. He was awarded an honorary chair at the University of Freiburg in 1926, a position he resigned in 1935 in protest against Hitler's government.

To illustrate what is meant by a well-ordered set, we begin with the rational numbers between 0 and 1. The rational numbers are not well ordered if we rely on the usual order according to position on the real number line. With this order, the set of rational numbers between 0 and 1 does not have a smallest element. If we use an order that puts these rational numbers into one-to-one correspondence with the natural numbers:

$$\frac{1}{2} \prec \frac{1}{3} \prec \frac{2}{3} \prec \frac{1}{4} \prec \frac{3}{4} \prec \frac{1}{5} \prec \frac{2}{5} \prec \cdots,$$

this is a well ordering. Every subset has a first element when we use this order.

Is it possible to well order the real numbers between 0 and 1? Cantor thought that it should be possible, but he could not find such an order. The problem hung unanswered for 17 years, occasionally prodded by those few individuals truly dedicated to set theory, but ignored by most mathematicians. Then in 1900, David Hilbert, probably the most influential mathematician of the age, delivered an address at the International Congress of Mathematicians in which he described the 23 most pressing and important unsolved problems in mathematics. Problem number 1 on his list was to settle the continuum hypothesis (see p. 75). The specific question he asked is whether or not there exists an infinite subset of [0, 1] whose cardinality is neither $\aleph_0$ nor c.

In his explanation of problem 1, Hilbert raised the question whether it is possible to well order the real numbers. Suddenly, this became an important problem. Opinion was divided. In 1904, Julius König announced a proof that such a well ordering could not exist. Flaws in his proof were quickly discovered. The same year, Ernst Zermelo published his proof that it could be done, not just for the real numbers but for any set.

What Zermelo actually accomplished was to show that every set can be well ordered if and only if the axiom of choice always holds.

Definition: Axiom of choice

The axiom of choice says that given any set S, there is a mapping that assigns to each nonempty subset of S one of the elements of that subset.

In other words, given any collection of subsets, even an uncountable collection, we can always choose one element from each subset. Vitali used this axiom to define $\mathcal{N}$ by selecting one element from each equivalence class.

If the set S is well ordered, we can always make the assignment by using this well ordering, assigning the least element in the subset. Thus, the axiom of choice follows from the well ordering principle. Zermelo's accomplishment was to show that the axiom of choice also implies the well ordering principle.

This was the first time that the axiom of choice was stated explicitly. The principle had been used implicitly many times previously. No one had raised serious objections. But now that it was shown to be equivalent to well ordering, there was doubt. It seemed too much like sleight-of-hand to claim that now we can well order the real numbers.

Battle lines were drawn. Hilbert considered the problem solved. Hadamard agreed. In opposition stood Borel, Jourdain, Bernstein, Schönflies, Baire, and Lebesgue. Vitali's nonmeasurable set, appearing less than a year later, was greeted by Lebesgue and many others as an empty exercise. They wanted an example of a nonmeasurable set whose construction would not depend on the axiom of choice.

Pursuing the Axiom of Choice

Other mathematicians would find additional examples of nonmeasurable sets, Van Vleck in 1908 and Bernstein the same year, but they all were dependent on the axiom of choice. Also in 1908, Zermelo published his axioms of set theory. Sets were recognized as forming the foundation for all of mathematics, and Zermelo attempted to clarify the assumptions that enable us to construct and work with sets. His axioms were modified by Fraenkel and Skolem in 1922, creating what today are known as the Zermelo–Fraenkel (**ZF**) axioms. If we add the axiom of choice, the system is known as **ZFC**. The axioms of **ZF** are the assumptions we need if we are to build the mathematics that we know. In particular, they enable us to construct the real number system. One of the great problems now facing those working on the foundations of mathematics was whether the axiom of choice could be shown to be a consequence of the **ZF** axioms, in contradiction to those axioms, or independent of them.

At the same time, other mathematicians were discovering results that began to cast doubt on the axiom of choice. In 1914, Felix Hausdorff used the axiom of choice to create a most peculiar decomposition of the surface of a sphere. He first removed a particular set of points, a set that is easily shown to be countable, and then took the remainder of the sphere and decomposed it into three pairwise disjoint pieces, call them A, B, and C, with the following properties:

1.   It is possible to rotate A so that it exactly matches up with B.

2. It is possible to rotate A so that it exactly matches up with C. Note that so far, there is nothing particularly remarkable about this decomposition.
3. It is possible to rotate A so that it exactly matches up with $B \cup C$.

In view of the first two properties, the last one *is* remarkable. The pieces are, of course, nonmeasurable sets determined by the axiom of choice.

Then, in 1924, Stefan Banach and Alfred Tarski took Hausdorff's construction and showed how it could be used to take a solid ball, cut it into five pieces, and then reassemble those five pieces using only rigid motions (rotations and translations) into two solid balls, each of the same size as the original. With just a little more work, it was possible to modify the argument so that one could begin with any solid object, dissect it into finitely many pieces, and reassemble those pieces using rigid motions into any other solid object.

This result is sometimes referred to as "the pea and the sun theorem." One can take a pea, cut it into finitely many (nonmeasurable) pieces, and reassemble those pieces into a sphere the size of the sun.[4] To those accepting the axiom of choice, this simply illustrated how meaningless are our intuitive understandings of area or volume once we begin working with nonmeasurable sets. To others, it confirmed the implausibility of this axiom.

The next big step forward came from Kurt Gödel in 1938. He proved that the Zermelo–Fraenkel axioms **ZF** are consistent with the continuum hypothesis, and they are consistent with the axiom of choice. Chalk one up for those supporting the axiom of choice.

In 1964, Paul Cohen published his proof that the axiom of choice could be false without contradicting **ZF**. He also proved that the continuum hypothesis could be false without contradicting **ZFC** (Zermelo–Fraenkel plus the axiom of choice). In other words, the assumptions needed to define the real number line are consistent with or without the axiom of choice. The axiom of choice is our choice to make.

What does this say about the existence of nonmeasurable sets? In 1970, Robert Solovay proved that, while the existence of nonmeasurable sets is not enough to imply the axiom of choice, it is an assumption that goes beyond **ZF**. We do not violate any of our assumptions about the real number line if we assume that all sets are measurable.

Do Nonmeasurable Sets Exist?

Does Theorem 5.13 say anything meaningful? Does the set $\mathcal{N}$ actually exist? As we have seen, it exists if we accept the axiom of choice but its existence is a choice we get to make. Life may seem much simpler if we choose to reject the existence of nonmeasurable sets, but this is not the road taken by the mathematical community.

---

[4] For a delightful and very accessible proof of this result, see *The Pea and the Sun* by Leonard P. Wapner.

We have encountered three decision points in our construction of the real number line. The first was whether or not to include infinitesimals. The judgment to reject them was made by Cauchy and his contemporaries in the early nineteenth century. As we have seen, calculus could have been placed on a firm foundation with their acceptance, but this was not fully realized until the work of Abraham Robinson in the 1960s. It requires paying the price of greatly complicating the structure of the real numbers and violating the intuitive principle of commensurability. Robinson's nonstandard analysis has many supporters who believe that it should be the standard approach to analysis, but they consitute a minority of mathematicians.

The second decision point involves the continuum hypothesis. We are free to decide that there either are or are not infinite subsets of $\mathbb{R}$ with cardinality other than $\aleph_0$ or c. Given this choice, most mathematicians would probably opt for no other cardinalities. It keeps life simpler. Beyond those who work directly in set theory, no one worries much about this. This preference does not impact other branches of mathematics.

The axiom of choice is far more problematic because it does affect many other branches of mathematics. There are results whose proofs are greatly simplified by appeal to the axiom of choice, others that are possible only because of this axiom. In 1918, Wacław Sierpiński published a list of such results. In 1929, Krull used the axiom of choice to prove that in a commutative ring, every proper ideal can be extended to a maximal prime ideal. In 1932, Hausdorff used the well-ordering principle to prove that every vector space has a basis. In 1936, Teichmüller extended this proof to show that every Hilbert space has an orthonormal basis. It is not necessary to know what these statements mean to recognize that much modern mathematics presupposes the axiom of choice. We certainly could live without it or with a weaker form that creates fewer apparent paradoxes, but that would create complications that most mathematicians would prefer to live without.

Sierpiński was unhappy calling this the "axiom of choice" since nothing is chosen. Rather, this axiom asserts the existence of something that we can never explicitly construct. In a letter to Émile Borel written in 1905, Jacques Hadamard described this debate as centering on the distinction "between what is *determined* and what can be *described*." Nonmeasurable sets can be determined in the sense that they can be prescribed; they cannot be described. Hadamard goes on to compare this debate to "the one which arose between Riemann and his predecessors over the notion of function. The *rule* that Lebesgue demands appears to me to resemble closely the analytic expression on which Riemann's adversaries insisted so strongly." Here Hadamard adds a footnote:

I believe it necessary to reiterate this point, which, if I were to express myself fully, apppears to form the essence of the debate. From the invention of the infinitesimal calculus to the present, it seems to me, the essential progress in mathematics has resulted from successively annexing notions which, for the Greeks or the Renaissance geometers or the predecessor of Riemann, were "outside mathematics" because it was impossible to describe them.[5]

The power of mathematical thinking is manifested precisely when we are willing to explore promising avenues even when they lead us outside preexisting expectations. To me, the proper response to the Banach–Tarski paradox is fascination and delight that the axiom of choice can lead us to such surprising conclusions.

Exercises

5.4.1. Prove that if $q_1 \neq q_2$ are rational numbers, then

$$(\mathcal{N} + q_1) \cap (\mathcal{N} + q_2) = \emptyset.$$

Exercises 5.4.2–5.4.7 establish the fact that if we accept the existence of a nonmeasurable set, $\mathcal{N}$, then every set of positive outer measure contains a nonmeasurable set.

5.4.2. Show that any set of positive outer measure contains a bounded set of positive outer measure.

5.4.3. Show that any set of positive outer measure contains a closed set of positive measure. With Exercise 5.4.2, we can conclude that any set of positive outer measure contains a closed and bounded set of positive measure.

5.4.4. Let S be any closed, bounded set and let U be any open set that contains S. Using the fact S and U^C are disjoint, closed sets, show that there is $\delta > 0$ such for any $x \in \mathbb{R}$ with $|x| < \delta$, the set $x + S = \{x + s \mid s \in S\}$ is contained in U.

5.4.5. Show that if S is measurable, closed, and bounded and has positive measure, then we can always choose the open set U that contains S to have measure strictly less than $2m(S)$. Use this fact to show that if $\delta > 0$ is chosen so that $|x| < \delta$ implies that $x + S \subseteq U$, then $(x + S) \cap S \neq \emptyset$.

5.4.6. Define $S \ominus S = \{s - t \mid s, t \in S\}$. Show that if $(x + S) \cap S \neq \emptyset$ for all $|x| < \delta$, then

$$(-\delta, \delta) \subseteq S \ominus S.$$

Putting this result with Exercises 5.4.2–5.4.5, show that if S is measurable and has positive measure, then $S \ominus S$ contains an open interval.

5.4.7. Let S be a set with positive outer measure and define $S_q = S \cap (q + \mathcal{N})$, $q \in \mathbb{Q}$. If S_q is measurable with positive measure, then by Exercise 5.4.6, $S_q \ominus S_q$

---

[5] Translation due to G. H. Moore, *Zermelo's Axiom of Choice*, pp. 317–318.

contains an open interval, and therefore so does $(q + \mathcal{N}) \ominus (q + \mathcal{N}) = \mathcal{N} \ominus \mathcal{N}$. Explain why this cannot happen. Explain why S_q cannot have measure 0 for all $q \in \mathbb{Q}$. Complete the proof that every set with positive measure contains a nonmeasurable subset.

Exercises 5.4.8–5.4.11 show that there exists a Jordan measurable set (and thus a Lebesgue measurable set) that is not a Borel set. Recall that for any set S in the domain of f, $f(S) = \{f(s) \mid s \in S\}$.

5.4.8. Show that if ψ is a continuous, strictly increasing function, then so is ψ^{-1}, and any set U in the domain of ψ is open if and only $\psi(U)$ is open.

5.4.9. Show that if ψ is a continuous, strictly increasing function, then any set S in the domain of ψ is a Borel set if and only if $\psi(S)$ is a Borel set.

5.4.10. Define $\psi(x) = x + DS(x)$, where DS is the Devil's staircase, Example 4.1 on p. 86. Show that ψ is a continuous and strictly increasing function from $[0, 1]$ onto $[0, 2]$. Let $C = SVC(3)$ be the Cantor ternary set on $[0, 1]$. Show that $m\left(\psi(C)\right) = 1$, and thus $\psi(C)$ contains a nonmeasurable set, $\mathcal{M}$.

5.4.11. Let $\mathcal{M}$ be a nonmeasurable set contained in $\psi(C)$. Show that $\psi^{-1}\left(\mathcal{M}\right)$ is Jordan measurable, but it cannot be a Borel set.

6

The Lebesgue Integral

In Section 5.2, we saw that the idea behind the Lebesgue integral of f is to partition the y-axis, $l = l_0 < l_1 < l_2 < \cdots < l_n = L$, define $S_i = \{x \mid l_i \leq f(x) < l_{i+1}\}$, and then bound the integral by the summations

$$\sum_{i=0}^{n-1} l_i \, m(S_i) \leq \int_a^b f(x) \, dx \leq \sum_{i=0}^{n-1} l_{i+1} \, m(S_i).$$

As the partition of the y-axis gets finer, these sums will approach each other and so approach a value for the integral. The only catch is that these sets, the S_i, must be measurable.

In view of the difficulty involved in finding a nonmeasurable set, we should expect that for reasonable functions, the S_i are measurable. But there is something to prove here. We shall call such functions **measurable functions**. The most important result of the first section is that every Riemann integrable function is measurable. We will lose nothing (and gain a great deal) by switching from the Riemann integral to the Lebesgue integral.

Our greatest gain will be Lebesgue's dominated convergence theorem, stated and proven in Section 6.3. Here at last we shall see a broadly applicable sufficient condition allowing for term-by-term convergence. In Section 6.4, we shall explore the connection between measurability and uniform convergence. This will lead into a discussion of some of the varied ways in which sequences can converge, a theme that will be picked up and developed much further in Chapter 8.

6.1 Measurable Functions

We begin with the formal definition of a measurable function. This may not appear to agree with the definition of the S_i, but that is taken care of with the next proposition.

Definition: Measurable function

The function f is **measurable** on the interval $[a, b]$ if for all $c \in \mathbb{R}$, the set $\left\{ x \in [a, b] \mid f(x) > c \right\}$ is measurable.

Proposition 6.1 (Equivalent Definitions of Measurability). *The following statements are equivalent:*

1. *for all $c \in \mathbb{R}$, $\{x \in [a, b] \mid f(x) > c\}$ is measurable,*
2. *for all $c \in \mathbb{R}$, $\{x \in [a, b] \mid f(x) \geq c\}$ is measurable,*
3. *for all $c \in \mathbb{R}$, $\{x \in [a, b] \mid f(x) < c\}$ is measurable, and*
4. *for all $c \in \mathbb{R}$, $\{x \in [a, b] \mid f(x) \leq c\}$ is measurable.*

Proof. Since complements and countable intersections of measurable sets are measurable and

$$\left\{ x \in [a, b] \mid f(x) \geq c \right\} = \bigcap_{n=1}^{\infty} \left\{ x \in [a, b] \mid f(x) > c - 1/n \right\}, \qquad (6.1)$$

$$\left\{ x \in [a, b] \mid f(x) < c \right\} = [a, b] - \left\{ x \in [a, b] \mid f(x) \geq c \right\}, \qquad (6.2)$$

$$\left\{ x \in [a, b] \mid f(x) \leq c \right\} = \bigcap_{n=1}^{\infty} \left\{ x \in [a, b] \mid f(x) < c + 1/n \right\}, \text{ and} \quad (6.3)$$

$$\left\{ x \in [a, b] \mid f(x) > c \right\} = [a, b] - \left\{ x \in [a, b] \mid f(x) \leq c \right\}, \qquad (6.4)$$

it follows that statement 1 implies 2 which implies 3 which implies 4 which implies 1. $\qquad \square$

Corollary 6.2 (Lebesgue Sets S_i Are Measurable). *If f is measurable on $[a, b]$, then $\left\{ x \in [a, b] \mid c \leq f(x) < d \right\}$ is measurable.*

Proof. The intersection of measurable sets is measurable, so

$$\begin{aligned}
&\left\{ x \in [a, b] \mid c \leq f(x) < d \right\} \\
&= \left\{ x \in [a, b] \mid c \leq f(x) \right\} \cap \left\{ x \in [a, b] \mid f(x) < d \right\}.
\end{aligned}$$
$\qquad \square$

The next proposition shows us that simple combinations of measurable functions are also measurable.

Proposition 6.3 (Measurable Functions Closed under $+, \times, | \cdot |$). *If f and g are measurable functions on $[a, b]$ and if k is any constant, then kf, f^2, $f + g$, fg, and $|f|$ are also measurable on $[a, b]$.*

Proof. If $k = 0$, then $kf = 0$, and any constant function is measurable. If $k > 0$, then

$$\{x \in [a, b] \mid kf(x) > c\} = \{x \in [a, b] \mid f(x) > c/k\}.$$

If $k < 0$, then the second set is $\{x \in [a, b] \mid f(x) < c/k\}$. If $c < 0$, then $\{x \in [a, b] \mid f^2(x) > c\} = [a, b]$. If $c \geq 0$, then

$$\{x \in [a, b] \mid f^2(x) > c\} = \{x \in [a, b] \mid f(x) < -\sqrt{c}\}$$
$$\cup \{x \in [a, b] \mid f(x) > \sqrt{c}\}.$$

If q is any rational number then

$$S_q = \{x \in [a, b] \mid f(x) > q\} \cap \{x \in [a, b] \mid g(x) > c - q\}$$

is measurable. I leave it as Exercise 6.1.2 to verify that

$$\{x \in [a, b] \mid f(x) + g(x) > c\} = \bigcup_{q \in \mathbb{Q}} S_q. \qquad (6.5)$$

Since

$$fg = \frac{1}{4}\left[(f + g)^2 - (f - g)^2\right],$$

fg is measurable. If $c < 0$, then $\{x \in [a, b] \mid |f(x)| > c\} = [a, b]$. If $c \geq 0$, then

$$\{x \in [a, b] \mid |f(x)| > c\} = \{x \in [a, b] \mid f(x) < -c\} \cup \{x \in [a, b] \mid f(x) > c\}.$$

$\square$

Limits of Measurable Functions

For the Riemann integral, the limit of a sequence of integrable functions is not necessarily integrable (see example (4.5) on p. 100). This creates serious complications when we try to find conditions that allow term-by-term integration. As we shall now see, limits of measurable functions *are* measurable.

Proposition 6.4 (Inf and Sup of Measurable Functions). *If $(f_n)_{n=1}^{\infty}$ is a sequence of measurable functions on $[a, b]$, then the functions defined by $\inf_{n \geq 1} f_n(x)$, $\sup_{n \geq 1} f_n(x)$, $\underline{\lim}_{n \to \infty} f_n(x)$, $\overline{\lim}_{n \to \infty} f_n(x)$, are measurable functions.*

Definition: Almost everywhere

When we say that something happens **almost everywhere**, we mean that it happens for all x in our domain except for a set of measure zero (which could be the empty set).

Proof. The measurability of $\inf_{n\geq 1} f_n(x)$ and of $\sup_{n\geq 1} f_n(x)$ follows from the equalities

$$\left\{ x \in [a,b] \,\middle|\, \left(\inf_{n\geq 1} f_n(x)\right) \geq c \right\} = \bigcap_{n=1}^{\infty} \left\{ x \in [a,b] \,\middle|\, f_n(x) \geq c \right\}, \qquad (6.6)$$

$$\left\{ x \in [a,b] \,\middle|\, \left(\sup_{n\geq 1} f_n(x)\right) > c \right\} = \bigcup_{n=1}^{\infty} \left\{ x \in [a,b] \,\middle|\, f_n(x) > c \right\}. \qquad (6.7)$$

The measurability of $\underline{\lim}_{n\to\infty} f_n(x)$ and of $\overline{\lim}_{n\to\infty} f_n(x)$ now follows from the definition:

$$\overline{\lim_{n\to\infty}} \, f_n(x) = \inf_{n\geq 1} \left(\sup_{m\geq n} f_m(x) \right), \qquad (6.8)$$

$$\underline{\lim_{n\to\infty}} \, f_n(x) = \sup_{n\geq 1} \left(\inf_{m\geq n} f_m(x) \right). \qquad (6.9)$$

$\square$

It follows immediately that if $\lim_{n\to\infty} f_n(x)$ converges to $f(x)$, then f is a measurable function. Something even stronger is true. We might have a few values of x at which the sequence fails to converge or converges to something other than $f(x)$ and still be able to conclude that f is measurable. Let S be the set of points for which $\lim_{n\to\infty} f_n(x) \neq f(x)$. If the measure of S is zero, then f is measurable. This idea that something happens except at points in a set of measure zero will become so common that we give it a formal name, **almost everywhere**.

For example, for $x \in [a,b]$

$$\lim_{n\to\infty} f_n(x) = f(x) \text{ almost everywhere}$$

means that this equality holds except possibly for values of $x \in [a,b]$ in a set of measure zero. On this set of measure zero, the equality could fail to hold because $\lim_{n\to\infty} f_n(x)$ does not exist, because $f(x)$ is not well defined, or because both exist but are not equal.

Theorem 6.5 (Limit of Measurable Functions). *If (f_n) is a sequence of measurable functions and $\lim_{n\to\infty} f_n(x) = f(x)$ almost everywhere in $[a,b]$, then f is a measurable function.*

Proof. Recall that any set of outer measure zero is measurable (Theorem 5.5), and thus any subset of a set of measure zero also has measure zero. There is a set S of measure 0 such that

$$\varliminf_{n \to \infty} f_n(x) = \varlimsup_{n \to \infty} f_n(x) = f(x)$$

for all $x \in [a, b] - S$, where $m(S) = 0$. If we choose some $c \in \mathbb{R}$, then the sets

$$F_1 = \{x \mid f(x) > c\} \quad \text{and} \quad F_2 = \{x \mid \varliminf f_n(x) > c\}$$

are not necessarily identical, but any element that is in one but not in the other must be in S.

Let

$$S_1 = F_1 - F_2 \quad \text{and} \quad S_2 = F_2 - F_1.$$

From Proposition 6.4, we know that F_2 is measurable. Since S_1 and S_2 are subsets of S, they are also measurable. I leave it for you (Exercise 1.2.14) to show that

$$F_1 = (F_2 \cup S_1) \cap S_2^C.$$

Therefore, F_1 is measurable. $\qquad\qquad\square$

We now focus on the kind of function we want to use in our approximation to the Lebesgue integral.

A function is simple if and only if it can be written as a finite linear combination of characteristic functions of measurable functions. Simple functions admit many different representations. For example,

$$\chi_{[0,2]} + \chi_{[1,3]} = \chi_{[0,1)} + 2\chi_{[1,2]} + \chi_{(2,3]}$$
$$= 0 \cdot \chi_{(-\infty,0)\cup(3,\infty)} + \chi_{[0,1)\cup(2,3]} + 2\chi_{[1,2]}.$$

However, there is always a unique representation in the form

$$\phi(x) = \sum_{i=1}^{n} l_i \, \chi_{S_i}(x),$$

where $l_i \in \mathbb{R}$ are distinct, χ_{S_i} is the characteristic function of S_i, and the sets S_i are measurable, pairwise disjoint, and their union is the domain of ϕ. In dealing with a generic simple function, we shall assume that the representation we are using is this unique representation.

> **Definition: Simple function**
>
> A function ϕ is called **simple** if its image consists of a finite number of values and the set of points that map to each of these values is measurable.

Theorem 6.6 (Measurable Functions as Limits). *A function f is measurable on $[a, b]$ if and only if it is the limit almost everywhere of simple functions on $[a, b]$. If f is measurable and bounded below, then it is the limit of a monotonically increasing sequence of simple functions.*

Proof. I leave it for Exercise 6.1.5 to prove that simple functions are measurable. It follows from Theorem 6.5 that the limit almost everywhere of simple functions, if it exists, is a measurable function.

In the other direction, let f be measurable. For each positive integer n and for $-n2^n < k < n2^n$, define

$$E_{n,k} = \left\{ x \in [a, b] \ \middle| \ \frac{k}{2^n} \le f(x) < \frac{k+1}{2^n} \right\}.$$

Define

$$E_{n,-n2^n} = \left\{ x \in [a, b] \ \middle| \ f(x) < -n + 2^{-n} \right\}.$$
$$E_{n,n2^n} = \left\{ x \in [a, b] \ \middle| \ n \le f(x) \right\}.$$

Since all of the sets $E_{n,k}$, $E_{n,-n2^n}$, and $E_{n,n2^n}$ are measurable, we can define the sequence of simple functions

$$f_n(x) = \sum_{k=-n\cdot 2^n}^{n\cdot 2^n} \frac{k}{2^n}\, \chi_{E_{n,k}}(x).$$

For $n \ge |f(x)|$, we know that

$$\left| f_n(x) - f(x) \right| < 2^{-n},$$

and so f is the limit of these simple functions. If f is bounded below, $f(x) \ge A$ for all $x \in [a, b]$, then for all $n > |A|$, this sequence is monotonically increasing. $\square$

Corollary 6.7 (Nonnegative Measurable Functions as Monotonic Limits). *A nonnegative measurable function on $[a, b]$ is the limit of a monotonically increasing sequence of simple functions.*

Note that f does not need to be a bounded function. For the Riemann integral, we had to twist ourselves in knots to handle unbounded functions. For the Lebesgue integral, unbounded functions present no special problems.

Farewell to the Riemann Integral

We now prove that every Riemann integrable function is measurable. In the next section, we shall see that every bounded measurable function is Lebesgue integrable.

Lebesgue integration completely subsumes all integrals that can be defined using Riemann's definition and, fortunately, when they both exist, the values of the Riemann and Lebesgue integrals are the same.

Theorem 6.8 (Riemann Integrable $\Longrightarrow$ Measurable). *Every Riemann integrable function on $[a, b]$ is measurable on $[a, b]$.*

Proof. Let f be a Riemann integrable function on $[a, b]$. For each positive integer n, we define P_n as the partition of $[a, b]$ into 2^n equal intervals of length $(b - a)/2^n$. Let $I_{n,k}$ be the kth interval of this partition,

$$I_{n,k} = \left[a + (a - b)\frac{k - 1}{2^n}, a + (a - b)\frac{k}{2^n} \right) \text{ for } 1 \leq k < 2^n;$$

$$I_{n,2^n} = \left[b - 2^{-n}, b \right],$$

and let $m_{n,k} = \inf_{x \in I_{n,k}} f(x)$. We define the simple function

$$\phi_n = \sum_{k=1}^{2^n} m_{n,k}\, \chi_{I_{n,k}}.$$

Each of our partitions is a refinement of the previous partition, $P_n \supseteq P_{n-1}$, and therefore $\phi_n(x) \geq \phi_{n-1}(x)$. For each $x \in [a, b]$, $\phi_n(x)$ forms an increasing sequence. Since f is Riemann integrable, it must be bounded, and therefore this sequence converges,

$$\lim_{n \to \infty} \phi_n(x) = \phi(x) \leq f(x).$$

All that remains is to prove that $\phi(x) = f(x)$ almost everywhere. Choose any $x \in [a, b]$. For each n, choose k so that $x \in I_{n,k}$. The oscillation of f over $I_{n,k}$ is

$$\omega(f; I_{n,k}) \geq f(x) - \phi_n(x) \geq f(x) - \phi(x).$$

Every open interval that contains x will contain $I_{n,k}$ for some n, k, and therefore the oscillation of f at x is at least $f(x) - \phi(x)$. Therefore, the set of points for which $\lim_{n \to \infty} \phi_n(x) \neq f(x)$ is contained in the set of points at which f is not continuous.

To complete the proof, we shall show that if f is Riemann integrable, then the set of points at which it is not continuous has measure zero. In other words, if a function is Riemann integrable, then it is continuous almost everywhere.

In Theorem 2.5, we saw that a function is Riemann integrable if and only if for each $\sigma > 0$, the set S_σ, the set of points at which the oscillation is at least σ, has outer content zero. Since outer measure is always less than or equal to outer content, the measure of S_σ is zero. The set of points at which the oscillation is positive is the union

$$\bigcup_{k=1}^{\infty} S_{1/k}.$$

This is a countable union of sets of measure zero, so it also has measure zero.

We have shown that $\lim_{n \to \infty} \phi_n(x) = f(x)$ almost everywhere. Since each ϕ_n is measurable, Theorem 6.5 implies that f is measurable. $\qquad\square$

In proving that every Riemannn integrable function is measurable, we discovered that every Riemann integrable function is continuous almost everywhere. Lebesgue realized that this is also true in the other direction, yielding a simple characterization of Riemann integrable functions.

Theorem 6.9 (Lebesgue's Characterization of Riemann Integrability). *A bounded function defined on a closed and bounded interval is Riemann integrable if and only if it is continuous almost everywhere.*

Proof. We confirmed one direction in the previous proof. All that remains is to show that if f is bounded and continuous almost everywhere, then it is Riemann integrable. Again let S_σ be the set of points at which the oscillation is greater than or equal to $\sigma > 0$. By Theorem 2.5, we need to show that this set has outer content zero. Since it is a subset of the set of points at which f is discontinuous, we know that S_σ has measure zero. The problem is that $m(S_\sigma) \leq c_e(S_\sigma)$, and we need to show that $c_e(S_\sigma) = 0$. We shall need to be clever.   •

The fact that $m(S_\sigma) = 0$ means that for any $\epsilon > 0$, we can find a countable open cover of S_σ for which the sum of the lengths of the intervals is less than ϵ. If we can show that S_σ is closed, then we can use the Heine–Borel theorem to conclude that there is finite subcover of S_σ for which the sum of the lengths of the intervals is less than ϵ. This is exactly what we need to conclude that $c_e(S_\sigma) = 0$.

Our proof has come down to showing that S_σ is closed. We will show that its complement is open. If $c \in S_\sigma^C$, then the oscillation at c is strictly less than σ,

$$\omega(f; c) = \varlimsup_{x \to c} f(x) - \varliminf_{x \to c} f(x) < \sigma.$$

Let $\delta = (\sigma - \omega(f; c))/3$. By the definition of $\varlimsup$ and $\varliminf$, we can find an open neighborhood of c in which

$$\varliminf_{x \to c} f(x) - \delta < f(x) < \varlimsup_{x \to c} f(x) + \delta.$$

The distance between these upper and lower bounds is

$$\left(\overline{\lim_{x \to c}} \, f(x) + \delta \right) - \left(\underline{\lim_{x \to c}} \, f(x) - \delta \right) = \omega(f;c) + 2\delta = \frac{1}{3}\omega(f;c) + \frac{2}{3}\sigma < \sigma.$$

This open neighborhood of c is entirely contained in S_σ^C. Since we can find such a neighborhood for any element of S_σ^C, this set is open, and S_σ is closed. $\qquad \square$

There is a delicious irony here. Riemann introduced his definition of the integral for the purpose of understanding how discontinuous a function could be and still be integrable. It appeared that it could be very discontinuous, having discontinuities at all rational numbers. In fact, there are Riemann integrable functions with discontinuities at the points in a set with cardinality $\mathfrak{c}$. Now that we are finally putting the Riemann integral behind us, we get the answer that Riemann was seeking. A Riemann integrable function is always a very continuous function. It is continuous almost everywhere. A function that is discontinuous only at the rational numbers is not very discontinuous.

The Lebesgue integral enables us to handle truly discontinuous functions. Dirichlet's function, the characteristic function of the rationals, was created to show that a function could be so discontinuous that it would make no sense to talk about its integral. This was considered a function beyond the pale. Yet, as we shall see, the Lebesgue integral of this function has a simple and natural meaning. Over any interval, the integral of this function is the measure of the set of rationals in that interval, which is zero.

Exercises

6.1.1. Using the definition of a measurable function, show that any constant function is measurable.

6.1.2. Prove equation (6.5).

6.1.3. Show that

$$\chi_{S \cap T} = \chi_S \cdot \chi_T,$$
$$\chi_{S \cup T} = \chi_S + \chi_T - \chi_S \cdot \chi_T,$$
$$\chi_{S^C} = 1 - \chi_S.$$

6.1.4. Let $\left(S_n\right)_{n=1}^{\infty}$ be a sequence of sets. Prove the equality of the characteristic function of the infimum of these sets (defined in Exercise 5.3.15) and the lim inf of the characteristic functions of these sets,

$$\chi_{\underline{\lim}_{n \to \infty} S_n} = \underline{\lim_{n \to \infty}} \, \chi_{S_n}.$$

6.1.5. Prove that every simple function is measurable.

6.1.6. Show that if ϕ is a simple function, then it is possible to find a representation,

$$\phi(x) = \sum_{i=1}^{n} l_i \chi_{S_i}(x),$$

for which the sets S_i are both measurable and pairwise disjoint.

6.1.7. Prove that any sum or product of finitely many simple functions is a simple function.

6.1.8. If $|f|$ is measurable, does it necessarily follow that f is measurable?

6.1.9. Let f be a real-valued function defined on $\mathbb{R}$. Show that the condition

$$\{x \in [a, b] \mid f(x) = c\} \text{ is measurable for all } c \in \mathbb{R}$$

is *not* enough to guarantee that f is measurable on $[a, b]$.

6.1.10. Let S be a dense subset of $\mathbb{R}$. Show that f is measurable on the interval $[a, b]$ if and only if $\{x \in [a, b] \mid f(x) \geq c\}$ is measurable for every $c \in S$.

6.1.11. Show that a real-valued function f defined on $[a, b]$ is measurable if and only $f^{-1}(U)$ is measurable for every open set $U \subseteq \mathbb{R}$.

6.1.12. Show that if a real-valued function f defined on $\mathbb{R}$ is measurable, then $f^{-1}(B)$ is measurable for every Borel set $B \subseteq \mathbb{R}$.

6.1.13. Prove that any continuous function defined on $[a, b]$ is measurable.

6.1.14. Prove that if f is measurable and $f = g$ almost everywhere, then g is measurable.

6.1.15. Assume that f is continuous on $[a, b]$. Show that f satisfies the condition

$$S \subseteq [a, b] \quad \text{and} \quad m(S) = 0 \quad \text{implies} \quad m\left(f(S)\right) = 0$$

if and only if

for any measurable set $M \subseteq [a, b]$, its image, $f(M)$, is measurable.

6.1.16. Show that if g is measurable on $I = [a, b]$ and f is continuous on $g(I)$, then $f \circ g$ is measurable on I.

6.1.17. Suppose that g is continuous on $I = [a, b]$ and h is measurable on $g(I)$. Does it necessarily follow that $h \circ g$ is measurable on I?

6.1.18. Suppose that g is measurable and f satisfies the condition that for every open set U, the inverse image $f^{-1}(U)$ is a Borel set. Show that $f \circ g$ is measurable.

6.1.19. Give an example of measurable function whose inverse is not measurable.

6.1.20. Let f be differentiable on $[a, b]$, Show that its derivative, f', is measurable on $[a, b]$.

6.2 Integration

We begin with the definition of the Lebesgue integral of a simple function (see below) and prove a few of the properties we would expect of an integral.

Proposition 6.10 (Properties of Lebesgue Integral, Simple Functions). *Let ϕ and ψ be simple functions, $c \in \mathbb{R}$, and $E = E_1 \cup E_2$, where E_1 and E_2 are disjoint measurable sets. The following properties hold:*

1. $\int_E c\phi(x)\,dx = c \int_E \phi(x)\,dx,$
2. $\int_E \left(\phi(x) + \psi(x)\right) dx = \int_E \phi(x)\,dx + \int_E \psi(x)\,dx,$
3. *if $\phi(x) \le \psi(x)$ for all $x \in E$, then $\int_E \phi(x)\,dx \le \int_E \psi(x)\,dx$, and*
4. $\int_E \phi(x)\,dx = \int_{E_1} \phi(x)\,dx + \int_{E_2} \phi(x)\,dx.$

The last of these statements may look a little unusual. It is simply a generalization of the identity

$$\int_a^c f(x)\,dx = \int_a^b f(x)\,dx + \int_b^c f(x)\,dx.$$

Proof. We begin by setting

$$\phi = \sum_{i=1}^m k_i\, \chi_{S_i}, \qquad \psi = \sum_{j=1}^n l_j\, \chi_{T_j},$$

where the k_i are distinct, the l_j are distinct, the S_i are pairwise disjoint, the T_j are pairwise disjoint, and $E = \bigcup_{i=1}^m S_i = \bigcup_{j=1}^n T_j$.

1.

$$\int_E c\,\phi(x)\,dx = \sum_{i=1}^m c\,k_i\, m(S_i \cap E) = c \sum_{i=1}^m k_i\, m(S_i \cap E) = c \int_E \phi(x)\,dx.$$

Definition: Lebesgue integral of simple function

Given a simple function

$$\phi = \sum_{i=1}^n l_i\, \chi_{S_i},$$

where the l_i are distinct, the S_i are pairwise disjoint, and $\bigcup_{i=1}^n S_i$ is the domain of ϕ, we define its Lebesgue integral over the measurable set E to be

$$\int_E \phi(x)\,dx = \sum_{i=1}^n l_i\, m(S_i \cap E).$$

---

2. The sum $\phi + \psi$ is a simple function given by

$$\phi + \psi = \sum_{i,j} (k_i + l_j) \chi_{(S_i \cap T_j)}.$$

It follows that

$$
\begin{aligned}
\int_E (\phi(x) + \psi(x)) \, dx &= \sum_{i,j} (k_i + l_j) m \left(S_i \cap T_j \cap E \right) \\
&= \sum_{i=1}^{m} k_i \sum_{j=1}^{n} m \left(S_i \cap T_j \cap E \right) \\
&\quad + \sum_{j=1}^{n} l_j \sum_{i=1}^{m} m \left(S_i \cap T_j \cap E \right) \\
&= \sum_{i=1}^{m} k_i \, m(S_i \cap E) + \sum_{j=1}^{n} l_j \, m(T_j \cap E) \\
&= \int_E \phi(x) \, dx + \int_E \psi(x) \, dx.
\end{aligned}
$$

3. The assumption $\phi \leq \psi$ for all $x \in E$ implies that we can rewrite our functions as

$$\phi = \sum_{i,j} k_i \chi_{(S_i \cap T_j)}, \qquad \psi = \sum_{i,j} l_j \chi_{(S_i \cap T_j)},$$

and for each pair i, j for which $S_i \cap T_j \cap E \neq \emptyset$, we have $k_i \leq l_j$. It follows that

$$\int_E \phi(x) \, dx = \sum_{i,j} k_i \, m(S_i \cap T_j \cap E) \leq \sum_{i,j} l_j \, m(S_i \cap T_j \cap E) = \int_E \psi(x) \, dx.$$

4.

$$
\begin{aligned}
\int_E \phi(x) \, dx &= \sum_{i=1}^{m} k_i \left(m(S_i \cap E_1) + m(S_i \cap E_2) \right) \\
&= \sum_{i=1}^{m} k_i \, m(S_i \cap E_1) + \sum_{i=1}^{m} k_i \, m(S_i \cap E_2) \\
&= \int_{E_1} \phi(x) \, dx + \int_{E_2} \phi(x) \, dx.
\end{aligned}
$$

$\square$

We have one more result to prove about integrals of simple functions. Like many of our results, it applies to monotonic sequences.

Proposition 6.11 (Monotone Convergence, Simple Functions). *Let* (ϕ_n) *be a monotonically increasing sequence of nonnegative simple functions. If there is a finite A for which* $\int_E \phi_n(x)\,dx < A$ *for all n, then* ϕ_n *converges to a finite-valued function f almost everywhere.*

Proof. Since every bounded increasing sequence converges, the conclusion of this proposition is equivalent to the statement that U, the set of $x \in E$ for which $(\phi_n(x))_{n=1}^{\infty}$ is unbounded, is a set of measure zero. Choose any $\epsilon > 0$, and define

$$E_n = \left\{ x \in E \mid \phi_n(x) > A/\epsilon \right\}.$$

Since ϕ_n is nonnegative,

$$\frac{A}{\epsilon} m(E_n) \leq \int_{E_n} \phi_n(x)\,dx \leq \int_E \phi_n(x)\,dx < A.$$

Therefore, $m(E_n) < \epsilon$. Since (ϕ_n) is monotonically increasing, we see that $E_1 \subseteq E_2 \subseteq \cdots$, and U is contained in the union of the E_n (see exercise 6.2.11). By Corollary 5.12,

$$m(U) \leq m \left(\bigcup_{n=1}^{\infty} E_n \right) = \lim_{n \to \infty} m(E_n) \leq \epsilon.$$

Since this is true for every $\epsilon > 0$, $m(U) = 0$. $\qquad\square$

Integration of Measurable Functions

We are almost ready to define the integral of a measurable function. The definition that we give will need to apply to unbounded functions. If a function is bounded below and unbounded above, we can do this easily. If it is unbounded both below and above, we run into potential problems of offsetting infinities. To avoid such problems, we restrict our attention to nonnegative functions, $f(x) \geq 0$. We shall first define the integral of a nonnegative function. If f is measurable but takes on both positive and negative values, we can write f as a difference of two nonnegative functions:

$$f = f^+ - f^-, \quad \text{where} \tag{6.10}$$
$$f^+(x) = \max\{0, f(x)\}, \tag{6.11}$$
$$f^-(x) = \max\{0, -f(x)\}. \tag{6.12}$$

We saw in Theorem 6.6 that any nonnegative function is the limit of a monotonically increasing sequence of simple functions. We define the integral of f in terms of the integrals of simple functions.

Definition: Lebesgue integral of measurable function

Given a nonnegative measurable function f and a measurable set E, we define

$$\int_E f(x)\,dx = \sup_{\phi \leq f} \int_E \phi(x)\,dx, \qquad (6.13)$$

where the supremum is taken over all simple functions ϕ for which $\phi(x) \leq f(x)$ for all $x \in E$. For all other measurable functions, we say that f is **integrable** if it is measurable and both $\int_E f^+(x)\,dx$ and $\int_E f^-(x)\,dx$ are finite.

If either $\int_E f^+(x)\,dx$ or $\int_E f^-(x)\,dx$ is finite, then we define

$$\int_E f(x)\,dx = \int_E f^+(x)\,dx - \int_E f^-(x)\,dx. \qquad (6.14)$$

If they both are infinite, then $\int_E f(x)\,dx$ does not exist.

Note that an integral might have the value $+\infty$ or $-\infty$ even though the function is not integrable. This is analogous to the situation of a sequence that diverges to infinity. Such a sequence does not converge, but to say that this sequence approaches $+\infty$ or that it approaches $-\infty$ still says something meaningful.

The following proposition follows immediately from the definition of the Lebesgue integral.

Proposition 6.12 (Integral over Set of Measure Zero). *If f is any measurable function and $m(E) = 0$, then*

$$\int_E f(x)\,dx = 0. \qquad (6.15)$$

If $E = E_1 \cup E_2$, where E_1 and E_2 are disjoint measurable sets, then

$$\int_E f(x)\,dx = \int_{E_1} f(x)\,dx + \int_{E_2} f(x)\,dx. \qquad (6.16)$$

Combining these results, we see that if $f = g$ almost everywhere, then

$$\int_E f(x)\,dx = \int_E g(x)\,dx. \qquad (6.17)$$

Because of this proposition, if we wish to integrate $f(x) = \lim_{n \to \infty} f_n(x)$ over a set E and discover that the limit does not exist on a subset of measure zero, we can safely ignore those values of x at which the limit does not exist. No matter how we choose to define f at the points where the limit does not exist, it will not change the value of the integral of f.

Proposition 6.13 (Null Integral $\Longleftrightarrow$ Zero AE). *Let f be a nonnegative measurable function on the measurable set E. Then $f = 0$ almost everywhere if and only if*

$$\int_E f(x)\, dx = 0.$$

Proof. We first assume that the integral is zero. We set

$$E_n = \{x \in E \mid f(x) > 1/n\}.$$

Since f is nonnegative, we see that

$$0 = \int_E f(x)\, dx = \int_{E_n} f(x)\, dx \geq \frac{1}{n} m(E_n),$$

and therefore $m(E_n) = 0$. Since

$$\{x \in E \mid f(x) > 0\} = \bigcup_{n=1}^{\infty} E_n, \quad \text{and} \quad E_1 \subseteq E_2 \subseteq \cdots,$$

we can invoke Corollary 5.12 to conclude that

$$m\left(\{x \in E \mid f(x) > 0\}\right) = \lim_{n \to \infty} m(E_n) = 0.$$

The other direction follows from Proposition 6.12. $\qquad \square$

The Monotone Convergence Theorem

We now get our first result that enables us to interchange integration and a limit. This is still a good deal weaker than the Arzelà–Osgood theorem, but it is an important first step. In the years following Lebesgue's publication of his new integral, many mathematicians studied it, discovering new properties and better proofs of the fundamental relationships. One of these was Beppo Levi (1875–1961) who published five papers on the Lebesgue integral in 1906. Levi was born in Torino (Turin) where he also studied. His first professorship came in 1906 at the University of Cagliari on the island of Sardinia. He went on the University of Parma in 1910 and then to the University of Bologna in 1928. He was fired in 1938 because of his Jewish heritage and took the position of director of the newly created Universidad del Litoral in Rosario, Argentina, where he remained until his death. Levi's primary work was in algebraic geometry, but he made important contributions to our understanding of the Lebesgue integral. The following theorem was first proven by Levi in 1906. Its proof is fairly complicated, but it has many immediate corollaries and provides a very efficient route to Lebesgue's dominated convergence theorem.

$$
\begin{array}{cccccccccccc}
\phi_{1,1} & \le & \phi_{1,2} & \le & \phi_{1,3} & \le & \cdots & \le & \phi_{1,n-1} & \le & \phi_{1,n} & \nearrow & f_1 \\
\phi_{2,1} & \le & \phi_{2,2} & \le & \phi_{2,3} & \le & \cdots & \le & \phi_{2,n-1} & \le & \phi_{2,n} & \nearrow & f_2 \\
\phi_{3,1} & \le & \phi_{3,2} & \le & \phi_{3,3} & \le & \cdots & \le & \phi_{3,n-1} & \le & \phi_{3,n} & \nearrow & f_3 \\
\vdots & & \vdots & & \vdots & & & & \vdots & & \vdots & & \vdots \\
\phi_{n-1,1} & \le & \phi_{n-1,2} & \le & \phi_{n-1,3} & \le & \cdots & \le & \phi_{n-1,n-1} & \le & \phi_{n-1,n} & \nearrow & f_{n-1} \\
\phi_{n,1} & \le & \phi_{n,2} & \le & \phi_{n,3} & \le & \cdots & \le & \phi_{n,n-1} & \le & \phi_{n,n} & \nearrow & f_n \\
\vdots & & \vdots & & \vdots & & & & \vdots & & \vdots & & \vdots
\end{array}
$$

Figure 6.1. Monotonically increasing sequences of simple functions.

Theorem 6.14 (Monotone Convergence). *Let* $(f_n)_{n=1}^{\infty}$ *be a monotonically increasing sequence of nonnegative measurable functions. If there is a finite A for which* $\int_E f_n(x)\,dx < A$ *for all n, then* f_n *converges to a finite-valued function f almost everywhere, f is integrable, and*

$$
\int_E f(x)\,dx = \lim_{n \to \infty} \int_E f_n(x)\,dx. \tag{6.18}
$$

Proof. We need to establish that $(f_n)_{n=1}^{\infty}$ is bounded almost everywhere. If, for a given x, $(f_n(x))_{n=1}^{\infty}$ is bounded, then the sequence converges. We know from Theorem 6.5 that on the set of x where we have convergence, $f = \lim f_n$ is a nonnegative measurable function. Since f is nonnegative and $\int_E f_n(x)\,dx$ is bounded, equation (6.18) is all that remains to be shown before we can conclude that f is integrable. There are three parts to our proof. First, we show that U, the set of $x \in E$ for which $(f_n(x))_{n=1}^{\infty}$ is unbounded, is a set of measure zero. Next, we demonstrate the fairly easy inequality $\lim_{n \to \infty} \int_E f_n(x)\,dx \le \int_E f(x)\,dx$. Finally, we tackle the more difficult inequality, $\lim_{n \to \infty} \int_E f_n(x)\,dx \ge \int_E f(x)\,dx$.

For the first part of the proof, we want to be able to use Proposition 6.11. We know from Theorem 6.5 that each f_n is the limit of a monotonically increasing sequence of simple functions, $\phi_{n,k} \nearrow f_n$ (see Figure 6.1). We define a new sequence of simple functions, (ϕ_n), by

$$
\phi_n(x) = \max_{1 \le j,k \le n} \phi_{j,k}(x).
$$

By Proposition 6.4, ϕ_n is measurable. (The maximum is just the supremum taken over a finite set.) We recognize that

$$
\phi_n(x) = \max_{1 \le j,k \le n} \phi_{j,k}(x) \ge \max_{1 \le j,k \le n-1} \phi_{j,k}(x) = \phi_{n-1}(x),
$$

so this is a monotonically increasing sequence. Since $\phi_n \le f_n$, the integrals $\int_E \phi_n(x)\,dx$ are bounded.

If x is a point at which $(f_n(x))$ is unbounded, then for any M, we can find an n for which $f_n(x) > M + 1$. Since $\phi_{n,k} \nearrow f_n$, we can find a k for which $\phi_{n,k}(x) > M$. This implies that $\phi_{\max\{n,k\}}(x) > M$, and so the sequence $(\phi_n(x))$ is also unbounded. By Proposition 6.11, U, the set of x on which $(f_n(x))$ is unbounded, has measure zero. By Proposition 6.12, we can define f however we wish on U. We choose to define $f(x) = 0$ for $x \in U$. This concludes the first part of the proof.

For the second part, we observe that if $g, h \in \mathcal{M}^+$ and $g(x) \le h(x)$ for all $x \in E$, then the set of simple functions that are less than or equal to g is contained in the set of simple functions that are less than or equal to h. Therefore,

$$\int_E g(x)\,dx = \sup_{\phi \le g} \int_E \phi(x)\,dx \le \sup_{\phi \le h} \int_E \phi(x)\,dx = \int_E h(x)\,dx. \qquad (6.19)$$

From Proposition 6.12, if $g(x) \le h(x)$ except possibly for $x \in U$, where $m(U) = 0$, then

$$\int_E g(x)\,dx = \int_{E-U} g(x)\,dx \le \int_{E-U} h(x)\,dx = \int_E h(x)\,dx.$$

Therefore, functional inequalities that hold almost everywhere imply integral inequalities that hold on all of E. It follows that for all n, $\int_E f_n(x)\,dx \le \int_E f(x)\,dx$, and therefore

$$\lim_{n \to \infty} \int_E f_n(x)\,dx \le \int_E f(x)\,dx. \qquad (6.20)$$

For the third part of the proof, we begin with any simple function less than or equal to f, $0 \le \phi = \sum_{i=1}^k l_i \, \chi_{S_i} \le f$, where the S_i are pairwise disjoint and $\bigcup_{i=1}^k S_i = E$. Choose an α between 0 and 1, and define the sets

$$A_n = \left\{ x \in E \mid f_n(x) \ge \alpha \phi(x) \right\}.$$

Each A_n is a measurable set (explain why in Exercise 6.2.5) and

$$A_1 \subseteq A_2 \subseteq \cdots, \qquad \bigcup_{n=1}^{\infty} A_n = E.$$

By Corollary 5.12,

$$\lim_{n \to \infty} m(S_i \cap A_n) = m(S_i \cap E).$$

Choose an $\epsilon > 0$. For each set S_i, we can find an N_i so that $n \ge N_i$ implies that $m(S_i \cap A_n) > (1 - \epsilon/(\alpha l_i))m(S_i \cap E)$. Let $N = \max\{N_1, N_2, \ldots, N_k\}$.

We now see that for $n \geq N$,

$$\int_E f_n(x)\,dx \geq \int_{A_n} f_n(x)\,dx$$

$$\geq \alpha \int_{A_n} \phi(x)\,dx$$

$$= \alpha \sum_{i=1}^{k} l_i\, m(S_i \cap A_n)$$

$$> \alpha \sum_{i=1}^{k} l_i \left(1 - \frac{\epsilon}{\alpha l_i}\right) m(S_i \cap E)$$

$$= \alpha \sum_{i=1}^{k} l_i\, m(S_i \cap E) - \epsilon \sum_{i=1}^{k} m(S_i \cap E)$$

$$= \alpha \int_E \phi(x)\,dx - \epsilon\, m(E).$$

Since this is true for every $\epsilon > 0$ and for every α between 0 and 1, it follows that

$$\int_E f_n(x)\,dx \geq \int_E \phi(x)\,dx.$$

We have shown that every simple function $\phi \leq f$ has an integral that is dominated by $\int_E f_n(x)\,dx$ for all n sufficiently large. We can conclude that

$$\lim_{n \to \infty} \int_E f_n(x)\,dx \geq \sup_{\phi \leq f} \int_E \phi(x)\,dx = \int_E f(x)\,dx. \qquad (6.21)$$

$\square$

It took some work to prove this theorem, but we are rewarded with four important corollaries.

Corollary 6.15 (Properties of Lebesgue Integral). *Let f, g be integrable functions and c any constant. It follows that $|f|$, cf, and $f + g$ are integrable and*

$$\int_E cf(x)\,dx = c \int_E f(x)\,dx, \qquad (6.22)$$

$$\int_E \big(f(x) + g(x)\big)\,dx = \int_E f(x)\,dx + \int_E g(x)\,dx. \qquad (6.23)$$

Proof. We know that $|f|$, cf, and $f + g$ are measurable. Since the integrals of f^+ and f^- are finite, so is

$$\int_E |f(x)|\,dx = \int_E f^+(x)\,dx + \int_E f^-(x)\,dx.$$

To show that cf and $f + g$ are integrable, we need to establish that the integrals of $(cf)^+$, $(cf)^-$, $(f + g)^+$, and $(f + g)^-$ are finite. We see that

$$(cf)^+ \leq |c|(f^+ + f^-), \quad (cf)^- \leq |c|(f^+ + f^-),$$
$$(f + g)^+ \leq f^+ + g^+, \quad (f + g)^- \leq f^- + g^-.$$

Since these functions are nonnegative and we know that the integrals of the larger functions are all finite, so are the integrals of the smaller functions.

We first establish equations (6.22) and (6.23) for f, g nonnegative, $c \geq 0$. If $c = 0$, then equation (6.22) is trivially true. Let ϕ_n be a monotonically increasing sequence of simple functions that converges to f. It follows that $c\phi_n$ is a monotonically increasing sequence of simple functions that converges to cf. By Theorem 6.14,

$$\int_E cf(x)\,dx = \lim_{n\to\infty} \int_E c\,\phi_n(x)\,dx = \lim_{n\to\infty} c \int_E \phi_n(x)\,dx = c \int_E f(x)\,dx.$$

If ψ_n is a monotonically increasing sequence of simple functions that converges to g, then $\phi_n + \psi_n$ converges to $f + g$. Again by Theorem 6.14,

$$\int_E \left(f(x) + g(x)\right)\,dx = \lim_{n\to\infty} \int_E \left(\phi_n(x) + \psi_n(x)\right)\,dx$$
$$= \lim_{n\to\infty} \int_E \phi_n(x)\,dx + \lim_{n\to\infty} \int_E \psi_n(x)\,dx$$
$$= \int_E f(x)\,dx + \int_E g(x)\,dx.$$

We now write $f = f^+ - f^-$, $g = g^+ - g^-$. If $c \geq 0$, then the positive part of cf is $c \cdot f^+$ and the negative part of cf is $c \cdot f^-$. If $c < 0$, then the positive part of cf is $-c \cdot f^-$ and the negative part is $-c \cdot f^+$. In either case, we have

$$\int_E cf(x)\,dx = \int_E c \cdot f^+(x)\,dx - \int_E c \cdot f^-(x)\,dx$$
$$= c \left(\int_E f^+(x)\,dx - \int_E f^-(x)\,dx \right)$$
$$= c \int_E f(x)\,dx.$$

To conclude the proof of equation (6.23), we begin with the observation that

$$(f + g)^+ - (f + g)^- = f + g = f^+ - f^- + g^+ - g^-,$$

and therefore

$$(f + g)^+ + f^- + g^- = (f + g)^- + f^+ + g^+.$$

We integrate each side of this equality. All of the summands are nonnegative measurable functions, so we can write each integral as a sum of integrals:

$$\int_E (f + g)^+(x)\,dx + \int_E f^-(x)\,dx + \int_E g^-(x)\,dx$$

$$= \int_E (f + g)^-(x)\,dx + \int_E f^+(x)\,dx + \int_E g^+(x)\,dx,$$

$$\int_E (f + g)^+(x)\,dx - \int_E (f + g)^-(x)\,dx$$

$$= \int_E f^+(x)\,dx - \int_E f^-(x)\,dx + \int_E g^+(x)\,dx - \int_E g^-(x)\,dx,$$

$$\int_E (f + g)(x)\,dx = \int_E f(x)\,dx + \int_E g(x)\,dx. \qquad \square$$

In the next corollary, we see that term-by-term integration is correct for series of nonnegative measurable functions provided that the sum of the integrals converges. The proof is left as Exercise 6.2.7.

Corollary 6.16 (Term-by-term Integration, Summands ≥ 0). *If $\sum_{k=1}^{\infty} f_k(x)$ is a series of nonnegative measurable functions and $\sum_{k=1}^{\infty} \left(\int_E f_k(x)\,dx \right)$ converges, then $\sum_{k=1}^{\infty} f_k(x)$ converges almost everywhere and*

$$\int_E \left(\sum_{k=1}^{\infty} f_k(x) \right) dx = \sum_{k=1}^{\infty} \left(\int_E f_k(x)\,dx \right). \qquad (6.24)$$

The next corollary of the monotone convergence theorem gives us an important result for approximating integrable functions by simple functions.

Corollary 6.17 (Approximation by Simple Function). *Let f be an integrable function on $[a, b]$. Given any $\epsilon > 0$, we can find a simple function ϕ such that*

$$\int_a^b \left| f(x) - \phi(x) \right| dx < \epsilon. \qquad (6.25)$$

Proof. We write f as the difference of two nonnegative functions, $f = f^+ - f^-$. By Corollary 6.7, we can find monotonically increasing sequences of simple functions that converge to f^+, and f^-, respectively,

$$\phi_n \nearrow f^+, \quad \psi_n \nearrow f^-.$$

By the monotone convergence theorem, we can find an N so that

$$0 \le \int_a^b f^+(x)\,dx - \int_a^b \phi_N(x)\,dx < \frac{\epsilon}{2},$$

$$0 \le \int_a^b f^-(x)\,dx - \int_a^b \psi_N(x)\,dx < \frac{\epsilon}{2}.$$

Let $\Phi = \phi_N - \psi_N$, which is also a simple function. We have that

$$\int_a^b \left| f(x) - \Phi(x) \right| dx = \int_a^b \left| f^+(x) - \phi(x) - f^-(x) + \psi(x) \right| dx$$

$$\le \int_a^b \left| f^+(x) - \phi(x) \right| dx + \int_a^b \left| f^-(x) - \psi(x) \right| dx$$

$$< \frac{\epsilon}{2} + \frac{\epsilon}{2} = \epsilon.$$

$\square$

Our final corollary will be a very important result that shows that for any integrable function, even an unbounded function, we can force the integral to be as small as we wish by taking a domain with sufficiently small measure.

Corollary 6.18 (Small Domain $\implies$ Small Integral). *Let f be an integrable function on $[a, b]$ and $\epsilon > 0$ any positive bound. There is always a positive response $\delta > 0$ so that for any measurable set $S \subseteq [a, b]$ with $m(S) < \delta$, we have that*

$$\left| \int_S f(x)\,dx \right| < \epsilon. \tag{6.26}$$

Proof. We know from Corollary 6.17 that for any $\epsilon > 0$ we can approximate f by a simple function ϕ so that

$$\int_a^b \left| f(x) - \phi(x) \right| dx < \frac{\epsilon}{2}.$$

Since ϕ is simple, it is bounded, say $|\phi(x)| < B$ for all $x \in [a, b]$. We can choose $\delta = \epsilon/2B$. Given any set $S \subseteq [a, b]$ of measure less than $\epsilon/2B$,

$$\int_S |\phi(x)|\,dx < \frac{\epsilon}{2B} \cdot B = \frac{\epsilon}{2}.$$

It follows that

$$
\begin{aligned}
\left| \int_S f(x)\, dx \right| &\le \int_S |f(x)|\, dx \\
&\le \int_S |f(x) - \phi(x)|\, dx + \int_S |\phi(x)|\, dx \\
&\le \int_a^b |f(x) - \phi(x)|\, dx + \int_S |\phi(x)|\, dx \\
&< \frac{\epsilon}{2} + \frac{\epsilon}{2} = \epsilon.
\end{aligned} \qquad \square
$$

Exercises

6.2.1. Find the Lebesgue integral over $[0, 1]$ of the function f defined by

$$
f(x) = \begin{cases} x^2, & x \in [0, 1] - \mathbb{Q}, \\ 1, & x \in [0, 1] \cap \mathbb{Q}. \end{cases}
$$

Is this function Riemann integrable over $[0, 1]$?

6.2.2. Using the Cantor ternary set, SVC(3), define the function g on $[0, 1]$ by

$$
g(x) = \begin{cases} 0, & x \in \text{SVC}(3), \\ n, & x \text{ is in a removed interval of length } 1/3^n. \end{cases}
$$

Find the value of the integral $\int_0^1 g(x)\, dx$. Is this function Riemann integrable over $[0, 1]$?

6.2.3. Using the Cantor ternary set, SVC(3), define the function h by

$$
h(x) = \begin{cases} \sin(\pi x), & x \in [0, 1/2] - \text{SVC}(3), \\ \cos(\pi x), & x \in [1/2, 1] - \text{SVC}(3), \\ x^2, & x \in \text{SVC}(3). \end{cases}
$$

Find the value of the integral $\int_0^1 h(x)\, dx$. Is this function Riemann integrable over $[0, 1]$?

6.2.4. Show that if ϕ is a simple function given by

$$
\phi(x) = \sum_{i=1}^n k_i\, \chi_{T_i}(x),
$$

where $(k_1, \dots, k_n)$ are any n real numbers, not necessarily distinct, and $(T_1, \dots, T_n)$ are measurable sets, not necessarily pairwise disjoint and whose union is not necessarily the domain of ϕ, but for which $\phi(x) = 0$ for all $x \in \left(\bigcup_{i=1}^{n} T_i\right)^{C}$, then it is still true that for any measurable set E,

$$\int_E \phi(x)\, dx = \sum_{i=1}^{n} k_i\, m\left(T_i \cap E\right).$$

6.2.5. Explain why it is that if f_n and ϕ are measurable functions, then $A_n = \{x \in E \mid f_n(x) \geq \alpha \phi(x)\}$ is a measurable set.

6.2.6. Compare the hypotheses of the monotone convergence theorem, Theorem 6.14 with those of the Arzelà–Osgood theorem, Theorem 4.5 on p. 106. Give an example of a sequence of functions that satisfies the hypotheses of the Arzelà–Osgood theorem but not those of the monotone convergen theorem. Give an example of a sequence of functions that satisfies the hypotheses of the monotone convergence theorem but not those of the Arzelà–Osgood theorem.

6.2.7. Prove Corollary 6.16.

6.2.8. Show that the conclusion of Corollary 6.16 can be false if we do not require that $f_n \geq 0$ for all n, even if we strengthen the bounding condition to $\left| \int_E \left(\sum_{k=1}^{N} f(x) \right) dx \right| < A$.

6.2.9. Let $\sum_{n=1}^{\infty} f_n$ be a series of integrable functions for which

$$\sum_{n=1}^{\infty} \int_a^b \left| f_n(x) \right| dx$$

converges. Show that the series $\sum_{n=1}^{\infty} f_n$ converges almost everywhere on $[a, b]$ and

$$\sum_{n=1}^{\infty} \int_a^b f_n(x)\, dx = \int_a^b \left(\sum_{n=1}^{\infty} f_n(x) \right) dx.$$

6.2.10. Show that the conclusion of Exercise 6.2.9 can be false if the assumption is weakened to

$$\sum_{n=1}^{\infty} \int_a^b f_n(x)\, dx$$

converges.

6.2.11. In the proof of Proposition 6.11, explain why $E_n \subseteq E_{n+1}$. Then show that if $(\phi_n(x))_{n=1}^{\infty}$ is unbounded for some x, then x must be contained in at least one of the E_k.

6.2.12. Show that if f is Lebesgue integrable on E and if

$$S_n = \left\{ x \in E \mid |f(x)| \geq n \right\},$$

then $\lim_{n \to \infty} n \cdot m(S_n) = 0$.

6.2.13. Show that if f and g are integrable over E and $f(x) \le g(x)$ for all $x \in E$, then

$$\int_E f(x)\,dx \le \int_E g(x)\,dx.$$

6.2.14. Show that we can still conclude that $\int_E f(x)\,dx \le \int_E g(x)\,dx$ with the weaker hypothesis that $f(x) \le g(x)$ almost everywhere on E.

6.2.15. Prove the three identities of Proposition 6.12.

6.2.16. Show that if $\int_S f(x)\,dx = 0$ for every measurable subset $S \subseteq E$ and if $m(E) > 0$, then $f = 0$ almost everywhere on E. Show that the conclusion still holds if all that we know is that $\int_S f(x)\,dx = 0$ for every closed subset $S \subseteq E$.

6.2.17. Find an example of an integrable function f on a set of positive measure E so that $\int_S f(x)\,dx = 0$ for every open subset $S \subseteq E$, but f is not zero almost everywhere.

6.2.18. Show that if f is Lebesgue measurable on E and

$$\left| \int_E f(x)\,dx \right| = \int_E |f(x)|\,dx,$$

then either $f \ge 0$ almost everywhere on E or $f \le 0$ almost everywhere on E.

6.2.19. Let f be a nonnegative, measurable function on the set E, $m(E) < \infty$. Prove that f is Lebesgue integrable if and only if $\sum_{k=0}^{\infty} k\, m(E_k)$ converges, where

$$E_k = \left\{ x \in E \mid k \le f(x) < k+1 \right\}.$$

6.2.20. Let f be a nonnegative, measurable function on the set F, $m(F) < \infty$. Prove that f is Lebesgue integrable if and only if $\sum_{k=0}^{\infty} m(F_k)$ converges, where

$$F_k = \left\{ x \in F \mid f(x) \ge k \right\}.$$

6.2.21. Let f be a nonnegative, measurable function on the set G, $m(G) < \infty$. For $\epsilon > 0$, define

$$S(\epsilon) = \sum_{k=0}^{\infty} k\epsilon\, m(G_k), \quad \text{where} \quad G_k = \left\{ x \in G \mid k\epsilon \le f(x) < (k+1)\epsilon \right\}.$$

Prove that

$$\lim_{\epsilon \to 0} S(\epsilon) = \int_G f(x)\,dx.$$

6.3 Lebesgue's Dominated Convergence Theorem

In 1904, Lebesgue published his solution to the problem of term-by-term integration. Lebesgue's solution gives a sufficient condition rather than a necessary condition, but it hews so closely to what is necessary that it gives us a simple yet practical guide to determine when term-by-term integration is allowed. In this section, we state the theorem, and then discuss how it is used and what it means, and finally prove it.

Theorem 6.19 (Dominated Convergence Theorem). *Let* $(f_n)_{n=1}^{\infty}$ *be a sequence of integrable functions that converges almost everywhere to* f *over the measurable set* E. *If there is an integrable function* g *for which*

$$|f_n| \leq g \quad \text{almost everywhere in } E,$$

then f *is integrable and*

$$\int_E f(x)\,dx = \lim_{n \to \infty} \int_E f_n(x)\,dx. \tag{6.27}$$

In terms of infinite series, this says that if the f_n are integrable functions, if the series converges almost everywhere, and if the partial sums are bounded by an integrable function, g,

$$\left| \sum_{n=1}^{N} f_n(x) \right| \leq g(x) \quad \text{almost everywhere in } E,$$

then the integral of the series exists, is finite, and

$$\int_E \left(\sum_{n=1}^{\infty} f_n(x) \right) dx = \sum_{n=1}^{\infty} \left(\int_E f_n(x)\,dx \right). \tag{6.28}$$

Uniform Convergence

Weierstrass had shown that if a sequence of Riemann integrable functions converges uniformly, then the integral of the limit is the limit of the integrals. We shall show that this is a special case of Theorem 6.19, the dominated convergence theorem.

To say that (f_n) converges uniformly to f over $[a, b]$ means that for any $\epsilon > 0$ we can find a response N so that $n \geq N$ implies that

$$\left| f_n(x) - f(x) \right| < \epsilon \quad \text{for all } x \text{ in } [a, b].$$

Since we are working with Riemann integrable functions, each of the f_n as well as f is bounded on $[a, b]$. If we define the function g by

$$g(x) = \max \left\{ \left| f_1(x) \right|, \left| f_2(x) \right|, \ldots, \left| f_N(x) \right|, \left| f(x) \right| + \epsilon \right\},$$

then g will also be measurable and bounded, and therefore, integrable on $[a, b]$. Since

$$\left| f_n(x) \right| \leq g(x) \quad \text{for all } x \in [a, b],$$

the conditions of Theorem 6.19 are satisfied.

Bounded Convergence

Arzelà's generalization of Osgood's theorem says that if a sequence of integrable functions converges to an integrable function and there is a finite bound A such that

$$\left| f_n(x) - f(x) \right| \leq A \qquad \text{for all } n \geq 1 \text{ and for all } x \text{ in } [a, b],$$

then the integral of the limit is the limit of the integrals. Again, this is a special case of Theorem 6.19, the dominated convergence theorem. In this case, we define

$$g(x) = \left| f_1(x) \right| + 2A.$$

The function g is integrable. For every $x \in [a, b]$ we have that

$$\left| f_n(x) \right| \leq \left| f_n(x) - f(x) \right| + \left| f(x) - f_1(x) \right| + \left| f_1(x) \right| \leq g(x).$$

Example 4.7 from Section 4.3

In Section 4.3, we saw how Osgood's theorem explains Example 4.8. This was a sequence of continuous functions that do not converge uniformly, but do have bounded convergence. But the Arzelà and Osgood results did not help us with Example 4.7,

$$B_n(x) = \frac{n^2 x}{1 + n^3 x^2}.$$

These functions converge to zero on $[0, 1]$, but the convergence is not bounded. The maximum value of this function on $[0, 1]$ occurs at $x = n^{-3/2}$ and is equal to $\sqrt{n}/2$, a value that does not stay bounded. Nevertheless, as we saw in equation (4.9) on p. 102, the integral of the limit is equal to the limit of the integrals. This example can be explained by the dominated convergence theorem.

Define $g(0) = 0$ and

$$g(x) = \frac{2^{2/3}}{3} x^{-1/3} \geq \sup_{n \geq 1} \frac{n^2 x}{1 + n^3 x^2}, \quad 0 < x \leq 1$$

(see Exercise 6.3.3). This is an unbounded function, but it is integrable over $[0, 1]$ in the Lebesgue sense.

We can use an improper Riemann integral to verify that g is Lebesgue integrable:

$$\lim_{\alpha \to 0^+} \int_\alpha^1 \frac{2^{2/3}}{3} x^{-1/3} \, dx = \lim_{\alpha \to 0^+} \left(2^{1/3} - 2^{1/3} \alpha^{2/3} \right) = 2^{1/3}.$$

The unproven theorem that we are using is that any strictly nonnegative function for which the improper Riemann integral exists will be Lebesgue integrable (see Exercise 6.3.4).

For a rigorous verification that g is Lebesgue integrable, we show how to express g as a limit of simple functions. For $1 \leq i < m$, let S_i^m be the interval $[(i - 1)/m, i/m)$, which is closed on the left, open on the right, and let $S_m^m = [(m - 1)/m, 1]$. Our function g can be written as a limit of step functions,

$$g(x) = \lim_{m \to \infty} \sum_{i=1}^m \frac{2^{2/3}}{3} \left(\frac{i}{m} \right)^{-1/3} \chi_{S_i^m}(x).$$

We see that

$$\lim_{m \to \infty} \int_0^1 \sum_{i=1}^m \frac{2^{2/3}}{3} \left(\frac{i}{m} \right)^{-1/3} \chi_{S_i^m}(x) \, dx = \lim_{m \to \infty} \sum_{i=1}^m \frac{2^{2/3}}{3} \left(\frac{i}{m} \right)^{-1/3} \frac{1}{m}$$

$$= \lim_{m \to \infty} \frac{2^{2/3}}{3m^{2/3}} \sum_{i=1}^m i^{-1/3}.$$

We can bound this summation by integrals,

$$1 + \int_1^m x^{-1/3} \, dx > \sum_{i=1}^m i^{-1/3} > \int_1^m x^{-1/3} \, dx,$$

$$\frac{3}{2} m^{2/3} - \frac{1}{2} > \sum_{i=1}^m i^{-1/3} > \frac{3}{2} m^{2/3} - \frac{3}{2}.$$

If we divide by $m^{2/3}$ and then take the limit as m approaches ∞, we get

$$\lim_{m \to \infty} \frac{2^{2/3}}{3m^{2/3}} \sum_{i=1}^m i^{-1/3} = \frac{2^{2/3}}{3} \cdot \frac{3}{2} = 2^{1/3}.$$

The integrals of the step functions are bounded, and therefore g is integrable.

However we verify it, g *is* integrable on $[0, 1]$, and therefore

$$\int_0^1 \lim_{n\to\infty} B_n(x)\,dx = \lim_{n\to\infty} \int_0^1 B_n(x)\,dx.$$

Example 4.6 from Section 4.3

What about Example 4.6,

$$A_n(x) = nxe^{-nx^2}?$$

As we saw there,

$$\int_0^1 \left(\lim_{n\to\infty} A_n(x) \right) dx = 0 \quad \text{but} \quad \lim_{n\to\infty} \left(\int_0^1 A_n(x)\,dx \right) = \frac{1}{2}.$$

The conclusion of the dominated convergence theorem is false, so the hypothesis had better be false. Indeed, we see that

$$\sup_{n\geq 1} \left(nxe^{-nx^2} \right) = x^{-1}e^{-1},$$

which is not integrable over $[0, 1]$.

Sufficient but Not Necessary

The dominated convergence theorem says that if the functions f_n are integrable and if $g = \sup_{n\geq 1} |f_n|$, which will always be measurable, has a finite integral, then the integral of the limit equals the limit of the integrals. What if the integral of g is infinite? Does it follow that the integral of the limit does not equal the limit of the integrals. In a word, "no."

Example 6.1. Consider $f_n = n \cdot \chi_{[1/n,(n+1)/n^2]}$, the function that is n for $1/n \leq x \leq 1/n + 1/n^2$ and zero everywhere else.

Each f_n is nonzero on a distinct interval, and therefore the integral of the supremum over $[0, 2]$ is the sum of the integrals of the f_n:

$$\int_0^2 \sup_{n\geq 1} f_n(x)\,dx = \sum_{n=1}^{\infty} n \cdot \frac{1}{n^2}.$$

This sum diverges to infinity. On the other hand, at each x,

$$\lim_{n\to\infty} f_n(x) = 0,$$

and

$$\lim_{n\to\infty} \int_0^2 f_n(x)\,dx = \lim_{n\to\infty} \frac{1}{n} = 0 = \int_0^2 \lim_{n\to\infty} f_n(x)\,dx.$$

This is a sequence that is not dominated by an integrable function, and yet the limit of the integrals *does* equal the integral of the limit. Lebesgue's condition is sufficient but not necessary.

Nevertheless, the dominated convergence theorem is extremely useful. It gives us a very generous condition under which term-by-term integration is always allowed.

Fatou's Lemma

Pierre Fatou (1878–1929) studied as an undergraduate at the École Normale Supérieure, attending from 1898 to 1901. The result that carries his name was part of his doctoral thesis of 1906. He worked as an astronomer at the Paris observatory. Much of his mathematics involved proving the existence of solutions to systems of orbital differential equations. He also studied iterative processes and was the first to investigate what today we call the Mandelbrot set.

If we think back to our examples where the limit of the integrals of a sequence of functions is not equal to the integral of the limit (such as Example 4.6 on p. 100), we see that the integral of the limit was always less than the limit of the integrals. Fatou's lemma says what should be intuitively apparent, that if the functions are nonnegative then we can never get the inequality to go in the other direction.

Theorem 6.20 (Fatou's Lemma). *If* (f_n) *is a sequence of nonnegative, integrable functions, then*

$$\int_E \lim_{n\to\infty} f_n(x)\,dx \leq \varliminf_{n\to\infty} \int_E f_n(x)\,dx. \tag{6.29}$$

If f_n *converges to* f *almost everywhere on* E *and if* $\int_E f_n(x)\,dx$ *is bounded by a constant independent of* n, *then* f *is integrable and*

$$\int_E f(x)\,dx \leq \varliminf_{n\to\infty} \int_E f_n(x)\,dx. \tag{6.30}$$

Proof. Define the sequence (g_m) by $g_m(x) = \inf_{n \geq m} f_n(x)$. It follows that for all $n \geq m$, we have

$$\int_E g_m(x)\,dx \leq \int_E f_n(x)\,dx,$$

and therefore

$$\int_E g_m(x)\,dx \le \varliminf_{n\to\infty} \int_E f_n(x)\,dx.$$

By definition, the sequence (g_m) is monotonically increasing and converges to $\varliminf f_n$. By the monotone convergence theorem (Theorem 6.14), we have that

$$\int_E \varliminf_{n\to\infty} f_n(x)\,dx = \lim_{m\to\infty} \int_E g_m(x)\,dx \le \varliminf_{n\to\infty} \int_E f_n(x)\,dx.$$

If f_n converges to f almost everywhere, then $g_m \nearrow f$ almost everywhere. If $\int_E f_n(x)\,dx$ is bounded by a constant independent of n, then the monotone convergence theorem tells us that f is integrable.   □

Proof of the Dominated Convergence Theorem

Finally, we prove Lebesgue's theorem.

Proof. **(Dominated Convergence Theorem, Theorem 6.19)** By Theorem 6.5, f is measurable. We also know that

$$|f| = f^+ + f^- \le g \quad \text{almost everywhere.}$$

Since f^+ and f^- are nonnegative, they are each bounded above almost everywhere by g. By Proposition 6.12, if we change the value of a function on a set of measure zero, it does not change the value of the integral. Therefore, the integrals of f^+ and of f^- are bounded above by the integral of g, which is finite. We have proven that f is integrable.

Since $g + f_n$ is nonnegative, we can apply Fatou's lemma:

$$\int_E g(x)\,dx + \int_E f(x)\,dx = \int_E (g+f)(x)\,dx$$

$$\le \varliminf_{n\to\infty} \int_E (g+f_n)(x)\,dx$$

$$= \varliminf_{n\to\infty} \left(\int_E g(x)\,dx + \int_E f_n(x)\,dx \right)$$

$$= \int_E g(x)\,dx + \varliminf_{n\to\infty} \int_E f_n(x)\,dx.$$

Therefore,

$$\int_E f(x)\,dx \le \varliminf_{n\to\infty} \int_E f_n(x)\,dx. \qquad (6.31)$$

We also can apply Fatou's lemma to $g - f_n$, which also is bounded below by 0:

$$\int_E g(x)\,dx - \int_E f(x)\,dx = \int_E (g - f)(x)\,dx$$

$$\leq \varliminf_{n \to \infty} \int_E (g - f_n)(x)\,dx$$

$$= \int_E g(x)\,dx - \varlimsup_{n \to \infty} \int_E f_n(x)\,dx.$$

Therefore,

$$\varlimsup_{n \to \infty} \int_E f_n(x)\,dx \leq \int_E f(x)\,dx. \tag{6.32}$$

Combining inequalities (6.31) and (6.32) yields the desired equality. □

At this point, it is worth going back and comparing this to Osgood's proof of a much weaker result in Section 4.3. What should be most striking are the knots we had to tie ourselves into to deal with those sets on which the convergence was not nice. It is the ability to neatly excise troublesome sets of measure zero that makes all the difference.

Exercises

6.3.1. Show that if $m \neq n$ are positive integers, then

$$\left[\frac{1}{m}, \frac{1}{m} + \frac{1}{m^2}\right] \cap \left[\frac{1}{n}, \frac{1}{n} + \frac{1}{n^2}\right] = \emptyset.$$

6.3.2. Show that if f is integrable, then so is $|f|$ and

$$\left|\int_E f(x)\,dx\right| \leq \int_E |f(x)|\,dx.$$

Does the integrability of $|f|$ imply the integrability of f?

6.3.3. Prove that for all $n \geq 1$ and for all $x \in (0, 1]$,

$$\frac{2^{2/3}}{3} x^{-1/3} \geq \frac{n^2 x}{1 + n^3 x^2}.$$

6.3.4. Prove that if f is unbounded over $[a, b]$ and $f(x) \geq 0$ for $a \leq x \leq b$, but the improper Riemann integral $\int_a^b f(x)\,dx$ exists, then f is Lebesgue integrable over $[a, b]$.

6.3.5. Show that as an improper Riemann integral,

$$\int_0^1 \left(2x \sin(x^{-2}) - 2x^{-1} \cos(x^{-2})\right) dx$$

exists (and equals $\sin(1)$), but the Lebesgue integral does not exist in this case.

6.3.6. Show that the sequence of functions defined by $f_n(x) = (n+1)x^n$ over $E = [0, 1]$ is an example for which $\int_E \lim f_n(x)\, dx < \underline{\lim} \int_E f_n(x)\, dx$.

6.3.7. Define $g(x) = \sup_{n \geq 1}(n+1)x^n$ for $0 \leq x < 1$. Show that if this supremum is taken over all real values $n \geq 1$, then $g(x) = -e^{-1}/\ln x$, and this function is not Lebesgue integrable over $[0, 1]$.

6.3.8. Show that there is no sequence of functions on $[0, 2\pi]$ of the type

$$f_n(x) = a_n \sin(nx) + b_n \cos(nx),$$

which converges to the function 1 almost everywhere on $[-\pi, \pi]$, and where $|a_n| + |b_n| \leq 10$.

6.3.9. Let f be integrable. Show that

$$\lim_{h \to 0} \int_a^b |f(x+h) - f(x)|\, dx = 0.$$

6.3.10. Let (f_n) be a sequence of integrable functions such that

$$\sum_{n=1}^\infty |f_n(x)| \leq A < \infty \quad \text{for almost all } x \in [a, b].$$

Show that $\sum_{n=1}^\infty f_n$ converges almost everywhere to an integrable function and

$$\int_a^b \left(\sum_{n=1}^\infty f_n(x) \right) dx = \sum_{n=1}^\infty \left(\int_a^b f_n(x)\, dx \right).$$

6.3.11. Let (f_n) be a sequence of integrable functions and let f be an integrable function such that

$$\lim_{n \to \infty} \int_a^b |f_n(x) - f(x)|\, dx = 0.$$

Show that if $(f_n(x))_{n=1}^\infty$ converges almost everywhere, then $f_n(x)$ converges to $f(x)$ almost everywhere.

6.3.12. Let (f_n) be a sequence of nonnegative functions that converge to f on $\mathbb{R}$ and such that

$$\lim_{n \to \infty} \int_{-\infty}^\infty f_n(x)\, dx = \int_{-\infty}^\infty f(x)\, dx < \infty.$$

Show that for every measurable set E,

$$\lim_{n \to \infty} \int_E f_n(x)\, dx = \int_E f(x)\, dx.$$

6.3.13. Show that if (f_n) is a sequence of integrable functions that converges uniformly to f over $[a, b]$, then

$$\int_a^b f(x)\,dx = \lim_{n\to\infty} \int_a^b f_n(x)\,dx.$$

6.3.14. Let $f_n(x) = nx^{n-1} - (n+1)x^n$, $0 < x < 1$, Show that

$$\int_0^1 \left(\sum_{n=1}^\infty f_n(x)\right) dx \neq \sum_{n=1}^\infty \left(\int_0^1 f_n(x)\,dx\right)$$

and

$$\sum_{n=1}^\infty \left(\int_0^1 |f_n(x)|\,dx\right) = \infty.$$

6.3.15. Let (f_n) be a sequence of measurable functions on E such that

$$\sum_{n=1}^\infty \left(\int_E |f_n(x)|\,dx\right) < \infty.$$

Show that $\sum_{n=1}^\infty f_n$ is integrable and

$$\int_E \left(\sum_{n=1}^\infty f_n(x)\right) dx = \sum_{n=1}^\infty \left(\int_E f_n(x)\,dx\right).$$

6.3.16. Show that if f is integrable on $(-\infty, \infty)$, then

$$\lim_{n\to\infty} \int_{-\infty}^\infty f(x)\,\cos(nx)\,dx = 0.$$

6.3.17. Show that if f is integrable on $(-\infty, \infty)$ and g is bounded and measurable, then

$$\lim_{t\to 0} \int_{-\infty}^\infty \left| f(x)\left(g(x) - g(x+t)\right)\right| dx = 0.$$

6.4 Egorov's Theorem

In his book *Lectures on the Theory of Functions*, J. E. Littlewood explained that there are three principles that lie behind work in real analysis:

1. Every measurable set is almost a finite union of open intervals.
2. Every measurable function is almost a continuous function.
3. Every convergent sequence of measurable functions is almost uniformly convergent.

By "almost," we mean that it is true except for a set of measure less than ϵ where ϵ can be any positive number, no matter how small. This means that we can first try proving our theorem in the greatly simplified case where our sets are finite unions of open intervals, our functions are continuous, and our convergent sequences converge uniformly. We then use these "almost" statements to expand the range of situations in which our theorem holds.

The actual theorem summarized in the first principle is Theorem 5.11, that for any measurable set and any $\epsilon > 0$ we can find a finite union of open intervals so that the symmetric difference between the original set and this finite union has measure less than ϵ. The theorems that correspond to the second and third principles will be proven in this section. The second principle corresponds to Luzin's theorem, Theorem 6.26. The third principle is made explicit in Egorov's theorem, Theorem 6.21.

In Section 4.3 we saw how Osgood approached the justification of term-by-term integration by looking for a large subset of our interval on which convergence is uniform. Osgood sought to isolate the Γ-points, the points that are most problematic for uniform convergence.

As explained in the last section, we do not need to address uniform convergence directly in order to prove the dominated convergence theorem, but there is an implicit use of uniform convergence. In 1911 Dimitri Egorov would make explicit the connection between Lebesgue measure and uniform convergence. Egorov's student, Nikolaĭ N. Luzin, then used this result to prove that every measurable function is almost continuous. By this, we mean that given any $\epsilon > 0$ and any measurable function f, we can remove a set of measure $< \epsilon$ from the domain of f, and f will be continuous on what remains. Luzin was not the first to observe this. Lebesgue had stated this theorem – though without providing a proof – in 1903.[1] Vitali published a proof of this result in 1905.

Dimitri Fedorovich Egorov (1869–1931) began teaching at Moscow University in 1894 and earned his doctorate there in 1901. In addition to his work in real analysis, he is noted for his contributions to differential geometry. In 1923, he was appointed director of the Institute for Mechanics and Mathematics at Moscow State University. Egorov protested against the arrests and execution of clergy in the 1920s and also against the attempt to impose Marxist methodology in science. He was dismissed in 1929, arrested in 1930, and died in exile a year later.

Examples 4.6–4.8 from Section 4.3 are all nonuniformly convergent sequences. But if we remove any neighborhood of 0, no matter how small, each of these sequences converges uniformly on the interval that remains. These are all almost uniformly convergent.

---

[1] In a footnote, Lebesgue corrects a statement from a letter written to Borel in which he had claimed that one could remove a set of measure zero and have the function be continuous on the set that remained.

Definition: Almost uniform convergence

A sequence of functions (f_n) converges **almost uniformly** to f on the measurable set E if for each $\epsilon > 0$, there is a set $S \subseteq E$, $m(S) < \epsilon$, such that (f_n) converges uniformly on $E - S$.

Theorem 6.21 (Egorov's Theorem). *If (f_n) is a sequence of measurable functions that converges almost everywhere to f on $[a, b]$, then this sequence converges almost uniformly to f.*

Proof. By Theorem 6.5, f is a measurable function, and therefore so is $f - f_n$. The sequence (f_n) converges uniformly to f on a given set if and only if the sequence $(f - f_n)$ converges uniformly to 0 on that set. This implies that we lose no generality if we restrict our attention to sequences that converge to 0 almost everywhere.

We define g_n by

$$g_n(x) = \sup_{m \geq n} |f_m(x)|.$$

The sequence (g_n) converges uniformly to 0 on a given set if and only if (f_n) converges uniformly to 0 on that set (see Exercise 6.4.3). The advantage of working with (g_n) is that it is a sequence of monotonically decreasing functions. Let A be the subset of $[a, b]$ on which $g_n \nrightarrow 0$. By our assumption, $m(A) = 0$.

Define

$$S_{k,n} = \left\{ x \in [a, b] \,\middle|\, 0 \leq g_n(x) < 1/2^k \right\}.$$

Since our sequence (g_n) is monotonically decreasing and approaches 0 on $[a, b] - A$, we see that

$$S_{k,1} \subseteq S_{k,2} \subseteq \cdots, \quad \text{and} \quad \bigcup_{n=1}^{\infty} S_{k,n} \supseteq [a, b] - A.$$

By Corollary 5.12,

$$\lim_{n \to \infty} m(S_{k,n}) = m \left(\bigcup_{n=1}^{\infty} S_{k,n} \right) \geq m \left([a, b] - A \right) = b - a.$$

Given any $\epsilon > 0$, for each k, we choose an n that may depend on k, written $n(k)$, so that

$$m(S_{k,n(k)}) > b - a - \frac{\epsilon}{2^k}.$$

We set

$$S = [a, b] - \bigcap_{k=1}^{\infty} S_{k,n(k)} = \bigcup_{k=1}^{\infty} \left([a, b] - S_{k,n(k)} \right).$$

The measure of this set is bounded by

$$m(S) \leq \sum_{k=1}^{\infty} \frac{\epsilon}{2^k} = \epsilon. \qquad (6.33)$$

If $x \in [a, b]$ is not in S, then $x \in \bigcap_{k=1}^{\infty} S_{k,n(k)}$. Given any $k \geq 1$, we have that $x \in S_{k,n(k)}$, and therefore for any $m \geq n(k)$,

$$0 \leq g_m(x) \leq g_{n(k)}(x) < 1/2^k.$$

The convergence is uniform for all $x \in [a, b] - S$. $\qquad \square$

The converse of Egorov's theorem is easy to prove.

Theorem 6.22 (Egorov Converse). *If (f_n) is a sequence of measurable functions that converges almost uniformly to f on $[a, b]$, then this sequence converges almost everywhere to f.*

Proof. Let S_k be a set of measure $< 1/k$ for which f_n converges uniformly to f on $[a, b] - S_k$. In particular, it converges to f. The set on which it does not converge is contained in $\bigcap_{k=1}^{\infty} S_k$, which, by Corollary 5.12, has measure $\lim_{k \to \infty} m(S_k) = 0$. $\qquad \square$

Convergence in Measure

We have seen that convergence almost everywhere is equivalent to almost uniform convergence. Both of these are weaker than pointwise convergence which required convergence at every point. In the route we shall take to prove Luzin's theorem that measurable functions are continuous once we remove an arbitrarily small set, we make use of an even weaker type of convergence, convergence in measure.

Notice how similar this is to Kronecker's convergence (Theorem 4.6 on p. 110) in which the outer content (rather than the measure) of this set must converge to 0.

Definition: Convergence in measure

A sequence of functions (f_n) **converges in measure** to f on the measurable set E if for all $\sigma > 0$,

$$\lim_{n \to \infty} m \left(\{ x \in E \mid |f_n(x) - f(x)| \geq \sigma \} \right) = 0.$$

Since the outer content is always greater than or equal to the measure, convergence in measure is also considerably weaker than Kronecker's convergence.

Example 6.2. Consider the functions

$$f_{k,n}(x) = \begin{cases} 1, & (k-1)/n \le x \le k/n, \\ 0, & \text{otherwise.} \end{cases}$$

The function $f_{k,n}$ is 0 except on an interval of length $1/n$. The sequence

$$f_{1,1}, \; f_{1,2}, \; f_{2,2}, \; f_{1,3}, \; f_{2,3}, \; f_{3,3}, \; f_{1,4}, \cdots$$

converges to 0 in measure. However, for each $x \in [0, 1]$, there are infinitely many functions in this sequence at which $f_{k,n}(x) = 1$. This sequence does not converge at any x in $[0, 1]$.

Our first result shows what might be expected, that uniform convergence implies convergence in measure. The second result is perhaps more surprising. If a sequence converges in measure, it might not converge almost everywhere (as in our example), but there will *always* be a subsequence that converges almost everywhere.

Theorem 6.23 (Almost Uniform $\implies$ In Measure). *If the sequence $(f_n)_{n=1}^{\infty}$ converges almost uniformly on E, then it converges in measure on E.*

Proof. Choose any $\delta > 0$ and find a set S of measure less than δ so that (f_n) converges uniformly on $E - S$. Given $\sigma > 0$, there is a response N so that for any $n \ge N$ and any $x \in E - S$, $|f_n(x) - f(x)| < \sigma$. It follows that for $n \ge N$,

$$m\left(\{x \in E \mid |f_n(x) - f(x)| \ge \sigma\}\right) \le m(S) < \delta.$$

We see that

$$\lim_{n \to \infty} m\left(\{x \in E \mid |f_n(x) - f(x)| \ge \sigma\}\right) \le \delta.$$

Since this holds for every $\delta > 0$,

$$\lim_{n \to \infty} m\left(\{x \in E \mid |f_n(x) - f(x)| \ge \sigma\}\right) = 0.$$

$\square$

The next result, a kind of partial converse, is due to Frigyes Riesz (1880–1956) in 1909. He was a Hungarian mathematician who earned his doctorate in Budapest in 1902, working in projective geometry. He is known as one of the founders of functional analysis, the field that would provide the final answer to the problem of representability by Fourier series.

Theorem 6.24 (Riesz's Theorem). *If the sequence $(f_n)_{n=1}^{\infty}$ converges in measure to f on E, then it has a subsequence (f_{n_k}) that converges almost everywhere to f on E.*

Proof. Since the sequence converges in measure, we can find an n_1 so that

$$m\left(\left\{x \in E \mid |f_{n_1}(x) - f(x)| \geq 2^{-1}\right\}\right) < 2^{-1}.$$

For each $k > 1$, we find n_k so that $n_k > n_{k-1}$ and

$$m\left(\left\{x \in E \mid |f_{n_k}(x) - f(x)| \geq 2^{-k}\right\}\right) < 2^{-k}.$$

I claim that the subsequence $(f_{n_k})_{k=1}^{\infty}$ converges almost everywhere to f. $\square$

Let

$$S_k = \bigcup_{i=k}^{\infty} \left\{x \in E \mid |f_{n_i}(x) - f(x)| \geq 2^{-i}\right\}.$$

We see that

$$S_1 \supseteq S_2 \supseteq S_3 \supseteq \cdots.$$

Let

$$S = \bigcap_{k=1}^{\infty} S_k.$$

The measure of S_k is bounded above by

$$m(S_k) \leq 2^{-k} + 2^{-k-1} + 2^{-k-2} + \cdots = 2^{1-k},$$

and therefore the measure of S is

$$m(S) = \lim_{k \to \infty} m(S_k) = 0.$$

We need to show that if $x \in E - S$, then $f_{n_k}(x) \to f(x)$. Choose any $\epsilon > 0$ and find K so that $2^{-K} < \epsilon$ and $x \notin S_K$. Then for all $k \geq K$, $x \notin S_k$, so

$$\left|f_{n_k}(x) - f(x)\right| < 2^{-k} < \epsilon.$$

Limits of Step Functions

A **step function** is a simple function for which the measurable sets are intervals (open, closed, or half open). A single point is considered to be a closed interval, but recall that a simple function is a sum involving finitely many characteristic functions. A function that is constant on an interval is also continuous on the interior of that interval. A step function can be discontinuous only at the endpoints of the intervals, and therefore every step function is continuous almost everywhere. One of the consequences of Riesz's result is the following theorem.

Theorem 6.25 (Measurable $\Longleftrightarrow$ Limit of Step Functions). *A function f defined on $[a, b]$ is measurable if and only if it is the limit almost everywhere of step functions.*

Proof. One direction is easy. Step functions are measurable functions, so Theorem 6.5 implies that f must be measurable.

In the other direction, we know from Theorem 6.6 that f can be written as the limit almost everywhere of a sequence of simple functions. Let (ϕ_n) be such a sequence of simple functions, and let T be the set of measure zero on which this sequence does not converge to f. We write ϕ_n as

$$\phi_n = \sum_{k=1}^{m_n} l_{k,n} \chi_{S_{k,n}},$$

where the $S_{k,n}$ are measurable sets. By Theorem 5.11, we know that for each $S_{k,n}$, we can find a finite union of open intervals,

$$U_{k,n} = \bigcup_{i=1}^{N(k,n)} I_{k,n,i},$$

such that the measure of the symmetric difference between $S_{k,n}$ and $U_{k,n}$ is as small a positive value as we might wish. In particular, we can have

$$m\left(S_{k,n} \Delta U_{k,n}\right) < 2^{-k-n}.$$

We replace the simple function ϕ_n by

$$\psi_n = \sum_{k=1}^{m_n} l_{k,n} \chi_{U_{k,n}}.$$

Since each $U_{k,n}$ is a finite union of intervals, the function ψ_n is a step function. Note that the intervals in $\{U_{1,n}, U_{2,n}, \ldots, U_{m_n,n}\}$ might overlap, but each intersection is a finite union of intervals, and there are only finitely many such intersections, so we can always rewrite ψ_n as a simple function for which the measurable sets are nonoverlapping intervals.

Let $D_n = \bigcup_{k=1}^{m_n}(S_{k,n} \Delta U_{k,n})$. For every $x \in [a, b]$, x lies in exactly one of the $S_{k,n}$, $1 \leq k \leq m_n$. Therefore, if $x \notin D_n$, then $\phi_n(x) = \psi_n(x)$. We also know that

$$m(D_n) \leq \sum_{k=1}^{m_n} m\left(S_{k,n} \Delta U_{k,n}\right) < \sum_{k=1}^{m_n} 2^{-k-n} < 2^{-n}.$$

We have that

$$\left\{ x \in [a, b] \,\big|\, |\psi_n(x) - f(x)| \geq \epsilon \right\} \subseteq D_n \cup \left\{ x \in [a, b] \,\big|\, |\phi_n(x) - f(x)| \geq \epsilon \right\} \cup T.$$

Therefore, for $n \geq N$, we have

$$m \left(\left\{ x \in [a, b] \,\big|\, |\psi_n(x) - f(x)| \geq \epsilon \right\} \right)$$
$$\leq 2^{-n} + m \left(\left\{ x \in [a, b] \,\big|\, |\phi_n(x) - f(x)| \geq \epsilon \right\} \right) + 0.$$

Since (ϕ_n) converges almost everywhere to f, it also converges in measure to f, and therefore

$$\lim_{n \to \infty} m \left(\left\{ x \in [a, b] \,\big|\, |\phi_n(x) - f(x)| \geq \epsilon \right\} \right) = 0.$$

It follows that (ψ_n) converges in measure to f.

By Reisz's theorem (Theorem 6.24), we can find a subsequence of (ψ_n) that converges almost everywhere to f. A subsequence of step functions is still a sequence of step functions. $\square$

Luzin's Theorem

Nikolaĭ Nikolaevich Luzin (1883–1950) studied engineering at Moscow University from 1901 to 1905. Here Egorov spotted his talent and encouraged him to pursue mathematics. After graduation, Luzin began the study of medicine but returned to Moscow University in 1909 to study mathematics with Egorov. They began joint publications on function theory in 1910. Luzin's theorem comes from a paper published in 1912. In 1915, Luzin received his doctorate. He was appointed professor at Moscow University in 1917. In 1935 he became head of the Department of the Theory of Functions of Real Variables at the Steklov Institute. In 1936 he was denounced for publishing his mathematical results outside of the Soviet Union, activity that was viewed as anti-Soviet.[2] He came close to dismissal and possible imprisonment, but managed to survive. In addition to his work in function theory, Luzin is noted for his contributions to descriptive set theory (measure-theoretic and topological aspects of Borel sets and other σ-algebras) and to complex analysis.

We are going to use the fact that a convergent sequence of measurable functions is almost uniformly convergent to prove that for every measurable function, we can remove a set of arbitrarily small measure and the function will be continuous relative to what remains (see definition of continuity relative to a set on p. 114). We know that if f is measurable over E, then we can find a sequence of step functions that converge to f. We know that step functions are continuous almost everywhere.

---

[2] For more on the "Luzin affair," see //www-groups.dcs.st-and.ac.uk/~history/Extras/Luzin.html.

Theorem 6.26 (Luzin's Theorem). *If f is a measurable function on the set E, then given any $\epsilon > 0$, we can find a subset S of E so that $m(S) < \epsilon$ and f is continuous relative to the set $E - S$.*

Proof. Let $(\psi_n)_{n=1}^{\infty}$ be a sequence of step functions that converges almost everywhere to f on E. Let T be the set of all points of discontinuity of all of the ψ_n. Since each ψ_n has finitely many points of discontinuity, T is countable and its measure is 0. Every ψ_n is continuous on $E - T$. Given $\epsilon > 0$, let S be a set of measure $< \epsilon$ so that (ψ_n) converges uniformly on $E - S$. Since a uniformly convergent sequence of continuous functions converges to a continuous function, f is continuous relative to $E - T - S$. $\qquad\square$

Dirichlet's characteristic function of the rationals (Example 1.1), a function that is discontinuous at every point, is not so discontinuous after all. In this case, we can remove a set of measure 0, the rationals, and our function now is continuous on what remains. Not every measurable function is quite this close to being continuous. We cannot always remove a set of measure 0. But for any measurable function and any $\epsilon > 0$, we can remove a set of measure less than ϵ and our function will be continuous relative to what remains.

Exercises

6.4.1. Define the functions f_n by

$$f_n(x) = \begin{cases} 1, & \frac{1}{n} \le |x| \le \frac{2}{n}, \\ 0, & \text{otherwise.} \end{cases}$$

Show that $f_n \to 0$ on $[-2, 2]$ and there is no set S of measure 0 such that (f_n) converges uniformly on $[-2, 2] - S$.

6.4.2. Without using Egorov's theorem, show that for any $\epsilon > 0$, we can find a set S, $m(S) < \epsilon$, such that sequence of functions defined in exercise 6.4.1 converges uniformly on $[-2, 2] - S$.

6.4.3. Prove that if $g_n(x) = \sup_{m \ge n} \left| f_m(x) \right|$ over $[a, b]$, then (g_n) converges uniformly over $[a, b]$ if and only if (f_n) converges uniformly over $[a, b]$.

6.4.4. For the sequence given in Example 6.2, $f_{1,1}, f_{1,2}, f_{2,2}, f_{1,3}, \ldots$, find a subsequence that converges almost everywhere. Then apply the proof of Theorem 6.24

to this sequence and find the subsequence predicted by that proof. In other words, find the first pair (k_1, n_1) such that

$$m\left(\left\{x \in [0, 1] \mid |f_{k_1,n_1}(x)| \geq 2^{-1}\right\}\right) < 2^{-1}.$$

For each $j > 1$, find the first pair (k_j, n_j) after (k_{j-1}, n_{j-1}) such that

$$m\left(\left\{x \in [0, 1] \mid |f_{k_j,n_j}(x)| \geq 2^{-j}\right\}\right) < 2^{-j}.$$

6.4.5. Show that the sequence given in Example 6.2 (p. 195) also converges in Kronecker's sense (see Theorem 4.6 on p. 110). That is to say, for all $\sigma > 0$,

$$\lim_{n \to \infty} c_e\left(\left\{x \in [a, b] \mid |f_n(x) - f(x)| \geq \sigma\right\}\right) = 0.$$

Show that any sequence that converges in Kronecker's sense must converge in measure.

6.4.6. Find an example of a sequence of functions that converges in measure but does not converge in Kronecker's sense.

6.4.7. Give an example of a measurable function for which we cannot remove a set of measure 0 and have a function that is continuous relative to what remains. Justify your answer.

6.4.8. Consider $\chi_{SVC(4)}$, the characteristic function of SVC(4). Given any $\epsilon > 0$, describe how to construct a set S (the construction may depend on ϵ) with $m(S) < \epsilon$ and with the property that $\chi_{SVC(4)}$ is continuous relative to $[0, 1] - S$. Justify your answer.

6.4.9. Let $\mathcal{N}$ be the nonmeasurable set described in Theorem 5.13 on p. 151. Define the function f by $f(x) = q$, where q is the unique rational number chosen so that $x \in \mathcal{N} + q$. Prove that f is discontinuous at every value of x.

6.4.10. Let (f_n) be a sequence of measurable functions on E, $m(E) < \infty$. Show that

$$\lim_{n \to \infty} \int_E \frac{|f_n(x)|}{1 + |f_n(x)|} \, dx = 0$$

if and only if (f_n) converges to 0 in measure. Show that this result is false if we omit the assumption $m(E) < \infty$.

6.4.11. In equation (6.33) in the proof of Egorov's theorem, Theorem 6.21, we showed that $m(S) < \epsilon$. We chose ϵ to be an arbitrary positive integer. Where is the flaw in our reasoning if we conclude from this statement that $m(S) = 0$?

6.4.12. Egorov's theorem does not claim that there exists a subset $E \subset [a, b]$ with $m(E) = 0$ such that (f_n) converges uniformly to f on $[a, b] - E$. However, show that it does imply that there exists a sequence of measurable sets, (E_n), in $[a, b]$ such that

$$m\left([a, b] - \bigcup_{n=1}^{\infty} E_n\right) = 0$$

and (f_n) converges uniformly to f on each E_n.

6.4.13. Show that if (f_n) converges in measure to f, then (f_n) converges in measure to g whenever $f = g$ almost everywhere.

6.4.14. Prove that if for each $\epsilon > 0$ there is a measurable set E, $m(E) < \epsilon$, such that f is continuous on $[a, b] - E$, then f is measurable on $[a, b]$.

6.4.15. Let (f_n) be a sequence of measurable functions. Show that the set E of points at which this sequence converges must be a measurable set.

6.4.16. Prove that if (f_n) is a sequence of nonnegative, measurable functions on $[a, b]$ such that $\lim_{n\to\infty} \int_a^b f_n(x) \, dx = 0$, then (f_n) converges to 0 in measure. Show by example that we cannot replace the conclusion with the assertion that (f_n) converges to 0 almost everywhere.

6.4.17. We say that the measurable functions f_n, $n \geq 1$, are **equi-integrable** over the set E if for every $\epsilon > 0$ there is a response $\delta > 0$ such that if $S \subseteq E$ and $m(S) < \delta$, then

$$\int_S |f_n(x)| \, dx < \epsilon \quad \text{for all } n \geq 1.$$

Show that if (f_n) is a convergent sequence of equi-integrable functions over a set E of finite measure, then

$$\lim_{n\to\infty} \int_E f_n(x) \, dx = \int_E \lim_{n\to\infty} f_n(x) \, dx.$$

6.4.18. Prove the following version of Lebesgue's dominated convergence theorem: Let (f_n) be a sequence of measurable functions that converges in measure to f on $[a, b]$. If there exists an integrable function on $[a, b]$, call it g, such that $|f_n(x)| \leq g(x)$ for all $n \geq 1$ and all $x \in [a, b]$, then

$$\lim_{n\to\infty} \int_a^b f_n(x) \, dx = \int_a^b f(x) \, dx.$$

6.4.19. Show that if (f_n) is a sequence of equi-integrable functions (see Exercise 6.4.17) that converges in measure over a set E of finite measure, then

$$\lim_{n\to\infty} \int_E f_n(x) \, dx = \int_E \lim_{n\to\infty} f_n(x) \, dx.$$

6.4.20. Let (f_n) be a sequence that converges in measure to f on $[a, b]$ and for which $|f_n(x)| < C$ for all $n \geq 1$ and all $x \in [a, b]$. Show that if g is continuous on $[-C, C]$, then

$$\lim_{n \to \infty} \int_a^b g(f_n(x)) \, dx = \int_a^b g(f(x)) \, dx.$$

6.4.21. Let (f_n) be a sequence that converges in measure to f on $[a, b]$. Show that

$$\lim_{n \to \infty} \int_a^b \sin(f_n(x)) \, dx = \int_a^b \sin(f(x)) \, dx.$$

7

The Fundamental Theorem of Calculus

The derivative of Volterra's function oscillates between $+1$ and -1 in every neighborhood of a point in SVC(4), the Smith–Volterra–Cantor set with measure 1/2. Since the set of discontinuities has positive measure, the derivative of Volterra's function cannot be Riemann integrable. As we have seen, this violates the assumption, encoded in the evaluation part of the fundamental theorem of calculus, that if we differentiate a function, we can then integrate that derivative to get back to our original function.

The derivative of Volterra's function may not be Riemann integrable, but it is not too hard to show that it *is* Lebesgue integrable. As we shall see in this chapter, Lebesgue integration saves the evaluation part of the functional theorem of calculus.

The antiderivative part of the fundamental theorem faces a more fundamental obstacle. The idea here is that if we integrate a function, we can then differentiate that integral to get back to the original function. We have a problem if the function with which we start does not satisfy the intermediate value property (see p. 16). By Darboux's theorem (Theorem 1.7), such a function might be integrable, but it cannot be possible to get back to the original function by differentiating the integral.

Even here, Lebesgue enables us to say something very close to what we might wish. We shall see that if a function f is Lebesgue integrable, then the derivative of $\int_a^x f(t)\,dt$ exists and is equal to $f(x)$ almost everywhere. Even without assuming that f is continuous, the antidifferentiation part of the fundamental theorem of calculus holds almost everywhere.

In this chapter, we need to take a closer look at differentiation. One of the surprising outcomes will be Lebesgue's result that continuity plus monotonicity implies differentiability almost everywhere. Weierstrass's example of a continuous, monotonic function that is not differentiable at any point of an arbitrary countable set (such as the set of all algebraic numbers) is essentially as nondifferentiable as any continuous, monotonic function can be.

7.1 The Dini Derivatives

Ulisse Dini (1845–1918) grew up in Pisa and studied at the university in that city, working under the direction of Enrico Betti. Riemann was in Pisa for extended periods during 1863–1865, there for his health but working with Betti. Dini, who was Betti's student at that time, must have learned much directly from Riemann. In 1871, Dini took over Betti's chair in Analysis and Higher Geometry as Betti switched his interests to physics. That same year, Dini entered politics and was elected to the Pisa City Council. He was elected to parliament in 1880, became rector of the University of Pisa in 1888, and was elected to the Italian senate in 1892. He was appointed director of the Scuola Normale Superiore in 1908.

Dini had begun work on the existence of and representability by Fourier series in the early 1870s, inspired by the work of Heine, Cantor, and Hankel. In 1878, he published an influential book, *Fondamenti per la teorica delle funzioni di variabili reali* (Foundations for the theory of functions of real variables). Here was the first statement and proof that any first species set has outer content zero. Here was the first rigorous proof of a result that gave conditions under which continuity implies differentiability: If f is continuous on $[a, b]$ and if $f(x) + Ax$ is piecewise monotonic for all but at most finitely many values of A, then f is differentiable on a dense subset of $[a, b]$ (for Dini's proof with the slightly stricter condition that $f(x) + Ax$ is piecewise monotonic for all A, see Exercises 7.1.14 through 7.1.20). But the most important contribution in this book was Dini's work on nondifferentiable functions.

A function f is differentiable at c if and only if

$$\lim_{x \to c} \frac{f(x) - f(c)}{x - c}$$

exists. There is not much more to say about this limit for a function that is differentiable at c. For a nondifferentiable function, there are many ways in which this limit might fail to exist. It was Dini's insight to focus separately on the lim sup and the lim inf of this ratio, on what happens as we approach c from the right, and what happens as we approach c from the left. He defined four derivatives.

Definition: Dini derivatives

The four **Dini derivatives** of f at c are

$$D^+ f(c) = \overline{\lim_{x \to c^+}} \frac{f(x) - f(c)}{x - c}, \quad D_+ f(c) = \underline{\lim_{x \to c^+}} \frac{f(x) - f(c)}{x - c},$$

$$D^- f(c) = \overline{\lim_{x \to c^-}} \frac{f(x) - f(c)}{x - c}, \quad D_- f(c) = \underline{\lim_{x \to c^-}} \frac{f(x) - f(c)}{x - c}.$$

As examples, for $f(x) = |x|$, the Dini derivatives at 0 are

$$D^+ f(0) = D_+ f(0) = 1, \quad D_- f(0) = D_- f(0) = -1.$$

For g defined by $g(x) = x \sin(1/x)$, $x \neq 0$, $g(0) = 0$, the Dini derivatives at 0 are

$$D^+ g(0) = D^- g(0) = 1, \quad D_+ g(0) = D_- g(0) = -1.$$

All of the Dini derivatives exist, provided the function is defined in a neighborhood of c. A function is differentiable at c if and only if the four Dini derivatives at c are finite and equal. It is possible to have a continuous and strictly increasing function for which all four Dini derivatives at c are different (Exercise 7.1.1). There is no necessary relationship among these derivatives at a single point except the fact that the lim sup is always greater than or equal to the lim inf,

$$D^+ f(c) \geq D_+ f(c), \qquad D^- f(c) \geq D_- f(c). \tag{7.1}$$

One of Dini's important realizations is that if any one of these derivatives is integrable, then they all must be integrable, and the values of the definite integrals will be identical. In partricular, he proved the following result, which helps to patch up the fundamental theorem of calculus in the case where the integral is not differentiable.

Theorem 7.1 (Dini's Theorem). *Let f be Riemann integrable on $[a, b]$. For $x \in [a, b]$ define*

$$F(x) = \int_a^x f(t)\, dt.$$

Let DF be any of the Dini derivatives of F. Then DF is bounded and integrable on $[a, b]$, and

$$\int_a^b DF(t)\, dt = F(b) - F(a).$$

This is a nice result with which to introduce Dini derivatives because its proof illustrates some of their characteristics that will be useful later.

Proof. We begin by noting that if $g(x) = -f(x)$, $h(x) = f(-x)$, and $k(x) = -f(-x)$, then (Exercise 7.1.2)

$$D_+ f(c) = -D^+ g(c), \quad D^- f(c) = D^+ k(-c), \quad D_- f(c) = -D^+ h(-c). \tag{7.2}$$

It follows that if this theorem holds for D^+, then it holds for each of the Dini derivatives.

Since f is Riemann integrable on $[a, b]$, it is bounded on this interval. For each $x \in [a, b]$, define

$$l(x) = \varliminf_{t \to x} f(t), \quad L(x) = \varlimsup_{t \to x} f(t).$$

The functions l and L are bounded. For $x > c$, we have that

$$\left(\inf_{t \in [c,x]} f(t) \right)(x - c) \le \int_c^x f(t)\,dt \le \left(\sup_{t \in [c,x]} f(t) \right)(x - c),$$

and, therefore,

$$D^+ F(c) = \varlimsup_{x \to c^+} \frac{1}{x - c} \int_c^x f(t)\,dt \le \varlimsup_{x \to c^+} \left(\sup_{t \in [c,x]} f(t) \right) \le L(c),$$

$$D^+ F(c) = \varliminf_{x \to c^+} \frac{1}{x - c} \int_c^x f(t)\,dt \ge \varliminf_{x \to c^+} \left(\inf_{t \in [c,x]} f(t) \right) \ge l(c).$$

Therefore, $D^+ F$ is also bounded. We also see that

$$\varlimsup_{x \to c} D^+ F(x) \le \varlimsup_{x \to c} \varlimsup_{y \to x^+} \left(\sup_{t \in [x,y]} f(t) \right) = L(c), \quad \text{and}$$

$$\varliminf_{x \to c} D^+ F(x) \ge \varliminf_{x \to c} \varliminf_{y \to x^+} \left(\inf_{t \in [x,y]} f(t) \right) = l(c).$$

The oscillation of $D^+ F$ at each $x \in [a, b]$ is less than or equal to the oscillation of f, so $D^+ F$ is Riemann integrable. Furthermore, if the oscillation of f at c is zero, then $D^+ F(c) = f(c)$.

Since f is integrable, it is continuous almost everywhere, so the set of points of continuity of f is dense in $[a, b]$. No matter how fine the partition, every interval contains points at which f and $D^+ F$ are equal. This implies that for any partition we can find Riemann sums for f and $D^+ F$ that are equal. Their integrals must be equal.   □

Bounded Variation

One of Dini's observations in his 1878 book was that if a function f has Dini derivatives that are either bounded above or bounded below, then f can be written as a difference of two monotonically increasing functions (see Exercise 7.1.3). This is significant because Dirichlet's theorem that prescribed sufficient conditions under which a function can be represented by its Fourier series included piecewise monotonicity as one of the conditions. Three years later, in his 1881 paper *Sur la série de Fourier*, Camille Jordan found a simple characterization that is equivalent

---

Definition: Total variation, bounded variation

Given a function f defined on $[a, b]$ and a partition $P = (a = x_0 < x_1 < \cdots < x_m = b)$, the **variation** of f with respect to P is

$$V(P, f) = \sum_{i=1}^{n} |f(x_i) - f(x_{i-1})|.$$

The **total variation** of f over $[a, b]$ is the supremum of the variation over all partitions,

$$V_a^b(f) = \sup_P V(P, f).$$

We say that f has **bounded variation** on this interval if the total variation is finite.

---

to being representable as the difference of two monotonically increasing functions, a property he called **bounded variation**.

If a function is unbounded, it cannot have bounded variation. There are also bounded functions that fail to have bounded variation. The classic example is the function defined to be $\sin(1/x)$ for $x \neq 0$ and 0 at $x = 0$. We can make the variation as large as we want by taking sufficiently small intervals near 0.

Example 7.1. A more subtle example is the continuous function g defined by

$$g(x) = x \cos(1/x), \ x \neq 0, \quad g(0) = 0, \quad \text{over } [0, 1/\pi].$$

If we take the partition $(0, 1/(N\pi), 1/((N-1)\pi), \ldots, 1/2\pi, 1/\pi)$, then the variation is

$$\frac{1}{N\pi} + \sum_{i=1}^{N-1} \left(\frac{1}{(N-i+1)\pi} + \frac{1}{(N-i)\pi} \right) = \frac{2}{\pi} \left(1 + \frac{1}{2} + \cdots + \frac{1}{N} \right) - \frac{1}{\pi}.$$

We can make this as large as we want by taking N sufficiently large, so this function does not have bounded variation.

Example 7.2. Still more subtle is the differentiable function h defined by

$$h(x) = x^2 \cos(1/x), \ x \neq 0, \quad h(0) = 0, \quad \text{over } [0, 1/\pi].$$

Given any partition of $[0, 1/\pi]$, we find the smallest N so that $1/N\pi$ is less than x_1, the first point of the partition that lies to the right of 0. The variation that corresponds to this partition is bounded above by

$$\frac{1}{(N\pi)^2} + \sum_{i=1}^{N-1} \left(\frac{1}{(N-i+1)^2\pi^2} + \frac{1}{(N-i)^2\pi^2} \right)$$

$$= \frac{2}{\pi^2} \left(1 + \frac{1}{2^2} + \cdots + \frac{1}{N^2} \right) - \frac{1}{\pi^2}.$$

This quantity is bounded for all N by

$$\frac{2}{\pi^2} \cdot \frac{\pi^2}{6} - \frac{1}{\pi^2} = \frac{1}{3} - \frac{1}{\pi^2}.$$

This function has bounded variation, even though it oscillates infinitely often in any neighborhood of 0.

We are now ready to state Jordan's characterization of functions that are a difference of monotonically increasing functions.

Theorem 7.2 (Jordan Decomposition Theorem). *A function defined on $[a, b]$ is the difference of two monotonically increasing functions if and only if it has bounded variation.*

Proof. We begin with the assumption that $f = g - h$, where g and h are monotonically increasing. It follows that

$$\sum_{i=1}^{n} |f(x_i) - f(x_{i-1})| = \sum_{i=1}^{n} |g(x_i) - h(x_i) - g(x_{i-1}) + h(x_{i-1})|$$

$$\leq \sum_{i=1}^{n} |g(x_i) - g(x_{i-1})| + \sum_{i=1}^{n} |h(x_i) - h(x_{i-1})|$$

$$\leq \big(g(b) - g(a)\big) + \big(h(b) - h(a)\big).$$

In the other direction, we assume that f has bounded variation. We define the function T by

$$T(x) = V_a^x(f),$$

the total variation of x over the interval from a to x. The function T is monotonically increasing. Since $f = f - (T - f)$, we only need to show that $T - f$ is also monotonically increasing. We leave it as Exercise 7.1.8 to verify that for $a < b < c$, we have

$$V_a^c(f) = V_a^b(f) + V_b^c(f).$$

Take any pair $x, y, a \leq x < y \leq b$. We have that

$$(T(y) - f(y)) - (T(x) - f(x)) = V_a^y(f) - V_a^x(f) - f(y) + f(x)$$
$$= V_x^y - (f(y) - f(x)).$$

This is greater than or equal to 0, because

$$V_x^y \geq |f(y) - f(x)|. \qquad \square$$

Corollary 7.3 (Continuity of Variation). *The function f is continuous and of bounded variation on $[a, b]$ if and only if it is equal to the difference of two continuous, monotonically increasing functions.*

Proof. We assume that f is continuous and of bounded variation. To show that f is the difference of two continuous, monotonically increasing functions, we only need to establish that $T(x) = V_a^x(f)$ is continuous on $[a, b]$. We assume that T is not continuous at some $c \in [a, b]$ and show that this leads to a contradiction. Since T is monotonically increasing and bounded, $\lim_{x \to c^-} T(x)$ and $\lim_{x \to c^+} T(x)$ exist. The only way T could be discontinuous at c is if at least one of these limits does not equal $T(c)$. By symmetry (we can replace $f(x)$ by $g(x) = f(-x)$ so that a limit from the right becomes a limit from the left) and the fact that T is monotonically increasing, we can assume that

$$T(c) = \lim_{x \to c^-} T(x) + \epsilon, \quad \text{for some } \epsilon > 0.$$

Since f is continuous at c, we can find a response $\delta > 0$ so that $|x - c| < \delta$ implies that $|f(x) - f(c)| < \epsilon/2$. Take any partition of $[a, c]$ for which the last subinterval has length less than δ, say $a = x_0 < x_1 < \cdots < x_n = c, c - x_{n-1} < \delta$. The variation is bounded by

$$\sum_{i=1}^{n} |f(x_i) - f(x_{i-1})| < \sum_{i=1}^{n-1} |f(x_i) - f(x_{i-1})| + \frac{\epsilon}{2}$$
$$\leq T(x_{n-1}) + \frac{\epsilon}{2}$$
$$\leq \lim_{x \to c^-} T(x) + \frac{\epsilon}{2}$$
$$= T(c) - \frac{\epsilon}{2}.$$

This means that every sufficiently fine partition of $[a, c]$ has a variation strictly less than $T(c) - \epsilon/2$, but this contradicts the definition of $T(c)$ as the supremum of the variations over all partitions.

The other direction is left as Exercise 7.1.10. $\qquad \square$

Exercises

7.1.1. Find a strictly increasing, continuous function for which $D^+ f(0)$, $D_+ f(0)$, $D^- f(0)$, and $D_- f(0)$ are all different.

7.1.2. Show that if $g(x) = -f(x)$, $h(x) = f(-x)$ and $k(x) = -f(-x)$, then

$$D_+ f(c) = -D^+ g(c), \quad D^- f(c) = D^+ k(-c), \quad D_- f(c) = -D^+ h(-c).$$

7.1.3. Show that if all four Dini derivatives are bounded below by A and if c is any constant larger than $|A|$, then $f(x) + cx$ is a monotonically increasing function of x. It follows that f is the difference of two monotonically increasing functions.

7.1.4. Show that Dirichlet's function, the characteristic function of the rationals, does not have bounded variation.

7.1.5. Prove that if a function has bounded variation on $[a, b]$, then it is bounded on $[a, b]$.

7.1.6. Give two examples of continuous functions on $[0, 1]$ that do not have bounded variation and whose difference does not have bounded variation.

7.1.7. Show that the set of points of discontinuity of a monotonic function is countable. Using this result, prove that any function of bounded variation has at most countably many points of discontinuity.

7.1.8. Show that for $a < b < c$, we have

$$V_a^c(f) = V_a^b(f) + V_b^c(f).$$

7.1.9. Show that

$$V_a^b(f + g) \leq V_a^b(f) + V_a^b(g) \quad \text{and} \quad V_a^b(cf) = |c| \, V_a^b(f).$$

7.1.10. Show that if $T(x) = V_a^x(f)$ is continuous, then so is f.

7.1.11. Show that if (f_n) is a sequence of functions that converges pointwise to f on $[a, b]$, then

$$V_a^b(f) \leq \lim_{n \to \infty} V_a^b(f_n).$$

7.1.12. Let f be defined by $f(0) = 0$ and $f(x) = x^2 \sin(1/x^2)$ for $x \neq 0$. Does f have bounded variation on $[0, 1]$? Justify your answer.

7.1.13. For positive constants α and β, define $f_{\alpha,\beta}$ by $f_{\alpha,\beta}(0) = 0$ and $f_{\alpha,\beta}(x) = x^\alpha \sin(\pi/x^\beta)$ for $x \neq 0$. Prove that $f_{\alpha,\beta}$ has bounded variation on $[0, 1]$ if and only if $\alpha > \beta$.

Exercises 7.1.14–7.1.20 will lead you through Dini's proof that if f is continuous on $[a, b]$ and $f(x) + Ax$ is piecewise monotonic for all $A \in \mathbb{R}$, then f is differentiable on a dense set of points in $[a, b]$. In these exercises, by "interval" we mean a closed interval with distinct endpoints.

7.1.14. For each pair of real numbers $p < q$ in $[a, b]$, consider the function defined by

$$g(x; p, q) = f(x) - f(p) - \frac{f(p) - f(q)}{p - q}(x - p).$$

Show that $g(p; p, q) = g(q; p, q) = 0$. Explain why g also is continuous and piecewise monotonic.

7.1.15. Given any interval $I \subseteq [a, b]$, define

$$l_I = \inf \left\{ \frac{f(x) - f(y)}{x - y} \,\middle|\, x \neq y \in I \right\}.$$

Show that for each $\epsilon > 0$, there exists a pair $r, s \in I, r < s$ such that

$$l_I \leq \frac{f(r) - f(s)}{r - s} < l_I + \epsilon.$$

7.1.16. Show that there is a subinterval $J \subseteq I$ on which $g(x; r, s)$ (defined in Exercise 7.1.14) is decreasing.

7.1.17. Show that if g is decreasing on the interval $J \subseteq I$, then for any $x \neq y \in J$, we have that

$$l_I \leq \frac{f(x) - f(y)}{x - y} \leq \frac{f(r) - f(s)}{r - s} < l_I + \epsilon.$$

7.1.18. Explain how to construct a nested sequence of intervals, $I \supseteq I_1 \supseteq I_2 \supseteq \cdots$, such that $x \neq y \in I_n$ implies that

$$l_n \leq \frac{f(x) - f(y)}{x - y} < l_n + \frac{1}{2^n}$$

for some sequence $(l_1, l_2, \ldots)$

7.1.19. Show that the sequence (l_n) is increasing and for all $n \geq 1$,

$$l_{n+1} \leq l_1 + \frac{1}{2} + \frac{1}{2^2} + \frac{1}{2^3} + \cdots + \frac{1}{2^n} < l_1 + 1.$$

It follows that this sequence converges. Denote this limit by l_0.

7.1.20. Let x_0 be any element of $\bigcap_{n=1}^{\infty} I_n$. Show that f, the function with which we began in Exercise 7.1.14, is differentiable at x_0 and that $f'(x_0) = l_0$.

7.1.21. Show that the devil's staircase, $DS(x)$ (Example 4.1 on p. 86), is not differentiable at $1/4$.

7.2 Monotonicity Implies Differentiability Almost Everywhere

We are ready to tackle the proof that a monotonic, continuous function is differentiable almost everywhere. Since every function of bounded variation is a difference of monotonic functions, this implies that every continuous function of bounded variation is differentiable almost everywhere.

The connection between continuity and differentiability had long been debated. As soon as we had the modern definition of continuity, the Bolzano–Cauchy definition, it was realized that differentiability necessarily implied continuity. It was always clear that the implication did not work in the other direction, $|x|$ being the classic example of a function that is continuous at $x = 0$ but not differentiable here. But for many years it was believed that continuity on a closed and bounded interval implied differentiability except possibly on a finite set. This was stated as a theorem in J. L. Raabe's calculus text of 1839, *Die Differential- und Integralrechnung*. In Section 2.2, we discussed Weierstrass's example of a function that is everywhere continuous and nowhere differentiable, as well as his example of a monotonic, continuous function that is not differentiable at any of a countable set of points. Weierstrass believed, a belief shared by others, that it was only a matter of time before someone found a monotonic, continuous function that was nowhere differentiable.

Dini's 1878 result – if $f(x) + Ax$, defined on a closed and bounded interval, is piecewise monotonic for all but finitely many A, then f is differentiable on a dense set – had cast doubt on the existence of a nowhere differentiable, continuous, monotonic function, but it did not rule it out. Lebesgue laid the matter to rest when he proved that any monotonic, continuous function must be differentiable almost everywhere. In 1910, Georg Faber showed that continuity was not necessary. Every monotonic function – and thus every function of bounded variation – is differentiable almost everywhere. A year later, the husband and wife team of William H. Young and Grace Chisholm Young published an independent proof that continuity is not needed. In retrospect, the fact that continuity is not needed should not be surprising. A function that is monotonic on a closed, bounded interval is quite limited in how discontinuous it can be.

Georg Faber (1877–1966) received his doctorate in Munich in 1902, his *Habilitation* in Würzburg in 1905. He taught at the Technische Hochschule (Institute of Technology) in Munich and became rector of this university at the end of World War II, overseeing the resumption of its activities. William Henry Young (1863–1942) and Grace Chisolm Young (1868–1944) were a husband and wife team of British mathematicians who made many important contributions to analysis. Although William held a succession of positions at various British universities, the Youngs made their home in Göttingen, Germany, until 1908, in Geneva, Switzerland until 1915, and finally in Lausanne.

The easiest way to prove the Faber–Chisholm–Young result is to first prove Lebesgue's theorem and then to deal with discontinuous monotonic functions. Our proof will follow that expounded by Chae, which, in turn, is based on the elementary proof of Lebesgue's theorem by Riesz in 1932 and the proof that continuity is not required that was given by Lee Rubel in 1963.

Theorem 7.4 (Continuity + Bounded Variation $\Longrightarrow$ Differentiable AE). *If f is continuous and has bounded variation on $[a, b]$, then f is differentiable almost everywhere on $[a, b]$*

Outlining the Proof

Riesz's proof may be elementary in the sense that it does not require the introduction of sophisticated mathematical tools, but it is not simple. I shall discuss the approach we are to take and the main obstacles and shall then break the proof into a sequence of lemmas. In what follows, we assume that f is monotonically increasing. As we saw in the previous section, any continuous function of bounded variation is a difference of two continuous, monotonically increasing functions.

We use the Dini derivatives. Since f is monotonically increasing, the Dini derivatives are never negative. We allow the value $+\infty$ to signify that the limit diverges to ∞. With this understanding, all Dini derivatives have values at every $x \in [a, b]$. We need to show that the Dini derivatives are finite and equal almost everywhere.

We already know, by the definition of lim sup and lim inf, that

$$D^+ f(x) \geq D_+ f(x) \quad \text{and} \quad D^- f(x) \geq D_- f(x).$$

If we can establish that

$$D_- f(x) \geq D^+ f(x) \quad \text{and} \quad D_+ f(x) \geq D^- f(x),$$

then all four Dini derivatives are equal. If we can establish that the first of these inequalities, $D_- f(x) \geq D^+ f(x)$, holds almost everywhere provided that f is continuous and monotonically increasing, then this inequality also holds almost everywhere on $[-b, -a]$ for $k(x) = -f(-x)$: $D_- k(x) \geq D^+ k(x)$. But as we saw in the previous section

$$D_- k(-x) = D_+ f(x) \quad \text{and} \quad D^+ k(-x) = D^- f(x).$$

It follows that

$$D_+ f(x) \geq D^- f(x).$$

Definition: Shadow point

Given a continuous function g on $[a, b]$, we say that $x \in [a, b]$ is a **shadow point** of g if there exists $z \in [a, b]$, $z > x$, such that $g(z) > g(x)$.

We have shown that it is enough to prove that

$$D^+ f(x) \leq D_- f(x), \quad \text{almost everywhere, and} \tag{7.3}$$

$$D^+ f(x) < \infty, \quad \text{almost everywhere.} \tag{7.4}$$

We start with inequality (7.3). Riesz observed that if $D_- f(x) < D^+ f(x)$ for a given value of x, then we can find two rational numbers, r and R, such that

$$D_- f(x) < r < R < D^+ f(x).$$

For each pair of rational numbers $0 < r < R < \infty$, we define the set

$$E_r^R = \left\{ x \in (a, b) \,\big|\, D_- f(x) < r < R < D^+ f(x) \right\}.$$

There are a countable number of such pairs. If we can show that each set E_r^R has measure zero, then inequality (7.3) holds almost everywhere.

The set E_r^R is the intersection of $E^R = \{x \in (a, b) \mid D^+ f(x) > R\}$ and $E_r = \{x \in (a, b) \mid D_- f(x) < r\}$. We need to limit the size of these sets. Notice that the set of x for which $D^+ f(x) = \infty$ is equal to $\bigcap_R E^R$. Using the flipping operation, $h(x) = f(-x)$, we see that $D^+ h(x) = -D_- f(x)$, so whatever we can say about E^R can be translated into a comparable result for E_r. The key is to be able to limit the size of E^R.

If $x \in E^R$, then we can find a $z > x$ such that

$$\frac{f(z) - f(x)}{z - x} > R, \quad \text{or, equivalently,}$$

$$f(z) - Rz > f(x) - Rx.$$

If we define $g(x) = f(x) - Rx$, then the set E^R is contained in the set of shadow points of g. The shadow points correspond to points in the valleys (see Figure 7.1). A point x is a shadow point if the point $(x, g(x))$ on the graph of g lies in the shadow of the rising sun. The next lemma provides the key to bounding the size of E^R.

Lemma 7.5 (Rising Sun Lemma). *Let g be continuous on $[a, b]$. The set of shadow points of g that lie in (a, b) is a countable union of pairwise disjoint open intervals (a_k, b_k) for which*

$$g(a_k) \leq g(b_k) \quad \text{for all } k. \tag{7.5}$$

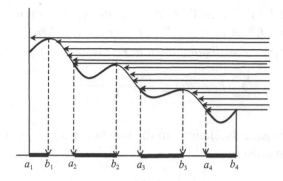

Figure 7.1. Shadow points.

Proof. For each shadow point x, we can use the continuity of g to find a small neighborhood of x that is left of z and over which the value of the function stays less than $g(z)$. This tells us that the set of shadow points is an open set. By Theorem 3.5, the set of shadow points is a countable union of pairwise disjoint open intervals (a_k, b_k). What may seem obvious from looking at Figure 7.1 but actually takes some work is that $g(a_k) \leq g(b_k)$ (inequality (7.5)). This is a critical part of the rising sum lemma.

We shall prove that $g(x) \leq g(b_k)$ for every $x \in (a_k, b_k)$. The inequality for $x = a_k$ then follows from the continuity of g (see Exercise 7.2.4). For $a_k < x < b_k$, let

$$A_x = \left\{ y \in [x, b_k] \,\middle|\, g(x) \leq g(y) \right\}.$$

Since $x \in A_x$, A_x is nonempty. It is bounded by b_k, so A_x has a least upper bound, $t = \sup A_x$, $g(t) \geq g(x)$ (see Exercise 7.2.5). If $t < b_k$, then $g(b_k) < g(x)$. Also, if $t < b_k$, then $t \in (a_k, b_k)$, so t is a shadow point. We can find a $z > t$ so that $g(z) > g(t)$. We have the inequalities

$$g(z) > g(t) \geq g(x) > g(b_k). \tag{7.6}$$

Since t is the least upper bound of $y \in [x, b_k]$ for which $g(y) \geq g(x)$, z must be larger than b_k, and, as we have seen, $g(z) > g(b_k)$. This means that b_k is a shadow point, a contradiction. Therefore $b_k = \sup A_x$ and $g(b_k) \geq g(x)$. □

Notice that the shadow points of $g(x) = f(x) - Rx$ include much more than just the points in E^R. The reason for using shadow points is so that we can work with a countable union of intervals, as we shall see in the next result that tells us about the size of E^R.

The Proof of Theorem 7.4

The next three lemmas complete the proof of Theorem 7.4.

Lemma 7.6 (Size of E^R). *If $R > 0$, f is a monotonically increasing continuous function on (α, β), and $E^R = \{x \in (\alpha, \beta) \mid D^+ f(x) > R\}$, then E^R is contained in a countable union of pairwise disjoint open intervals, (α_k, β_k), for which*

$$\sum_k (\beta_k - \alpha_k) \le \frac{f(\beta) - f(\alpha)}{R}.$$

Proof. We apply the rising sun lemma to the function $g(x) = f(x) - Rx$. As we have seen, E^R is contained in

$$\bigcup_k (\alpha_k, \beta_k),$$

where the open intervals are pairwise disjoint and $g(\alpha_k) \le g(\beta_k)$. This tells us that

$$f(\alpha_k) - R\alpha_k \le f(\beta_k) - R\beta_k, \quad \text{or, equivalently,} \tag{7.7}$$

$$\beta_k - \alpha_k \le \frac{1}{R} \left(f(\beta_k) - f(\alpha_k) \right). \tag{7.8}$$

Since f is monotonically increasing,

$$\sum_k (\beta_k - \alpha_k) \le \sum_k \frac{1}{R} \left(f(\beta_k) - f(\alpha_k) \right) \le \frac{f(\beta) - f(\alpha)}{R}. \qquad \square$$

The set of x for which $D^+ f(x) = \infty$ is the intersection of E^R taken over all $R \in \mathbb{N}$. We see that $E^1 \supseteq E^2 \supseteq \cdots$. By Corollary 5.12, we can conclude that

$$m \left(\{x \in [a, b] \mid D^+ f(x) = \infty \} \right) = \lim_{R \to \infty} m(E^R) \le \lim_{R \to \infty} \frac{f(b) - f(a)}{R} = 0. \tag{7.9}$$

To find the size of E_r^R and finish our proof, we need a slightly different result for E_r.

Lemma 7.7 (Size of E_r). *If f is a monotonically increasing continuous function on (α, β), and $E_r = \{x \in (\alpha, \beta) \mid D_- f(x) < r\}$, then E_r is contained in a countable union of pairwise disjoint open intervals, (α_k, β_k), for which*

$$f(\beta_k) - f(\alpha_k) \le r(\beta_k - \alpha_k).$$

Proof. We follow the proof of Lemma 7.6 with f replaced by $h(x) = f(-x)$ and R replaced by $-r$. Notice that we did not need the fact that R is positive until we divided by R to get inequality (7.8). We observe that

$$D^+ h(x) > -r \quad \Longleftrightarrow \quad D_- f(x) < r,$$

so

$$E_r = \{x \in (\alpha, \beta) \mid D_- f(x) < r\} = \{x \in (-\beta, -\alpha) \mid D^+ h(x) > -r\}.$$

Following the proof of Lemma 7.6 up to inequality (7.7), we see that

$$h(-\beta_k) - (-r)(-\beta_k) \leq h(-\alpha_k) - (-r)(-\alpha_k). \tag{7.10}$$

This is equivalent to

$$f(\beta_k) - f(\alpha_k) \leq r(\beta_k - \alpha_k). \tag{7.11}$$

$\square$

The next lemma completes the proof of Theorem 7.4. It uses a very ingenious trick.

Lemma 7.8 (Size of E_r^R). *If f is a monotonically increasing continuous function on (a, b), then*

$$m\left(\{x \in (a, b) \mid D_- f(x) < r < R < D^+ f(x)\}\right) = 0.$$

Proof. Let (α, β) be any open interval in (a, b), and consider

$$E_r^R \cap (\alpha, \beta) = \left(E_r \cap (\alpha, \beta)\right) \cap E^R.$$

By Lemma 7.7, $E_r \cap (\alpha, \beta)$ is contained in a countable union of disjoint open intervals (α_k, β_k) for which

$$f(\beta_k) - f(\alpha_k) \leq r(\beta_k - \alpha_k).$$

For each interval (α_k, β_k), we know from Lemma 7.6 that $E^R \cap (\alpha_k, \beta_k)$ is contained in a countable union of open intervals, $(a_{k,n}, b_{k,n})$, for which

$$\sum_n (b_{k,n} - a_{k,n}) \leq \frac{f(\beta_k) - f(\alpha_k)}{R}.$$

We have shown that $E_r^R \cap (\alpha, \beta) \subseteq \bigcup_{k,n}(a_{k,n}, b_{k,n})$, a union of pairwise disjoint open intervals for which

$$\sum_k \sum_n (b_{k,n} - a_{k,n}) \leq \sum_k \frac{f(\beta_k) - f(\alpha_k)}{R}$$

$$\leq \sum_k \frac{r(\beta_k - \alpha_k)}{R}$$

$$\leq \frac{r}{R}(\beta - \alpha). \tag{7.12}$$

For each pair k, n, we apply this result to $E_r^R \cap (a_{k,n}, b_{k,n})$. This set is contained in a countable union of disjoint open intervals, $(a_{k,n,s}, b_{k,n,s})$, for which

$$\sum_s (b_{k,n,s} - a_{k,n,s}) \leq \frac{r}{R}(b_{k,n} - a_{k,n}). \tag{7.13}$$

Combining equations (7.12) and (7.13), we see that $E_r^R \cap (\alpha, \beta)$ is contained in the countable union of pairwise disjoint open intervals $\bigcup_k \bigcup_n \bigcup_s (a_{k,n,s}, b_{k,n,s})$ for which

$$\sum_k \sum_n \sum_s (b_{k,n,s} - a_{k,n,s}) \leq \frac{r}{R} \sum_{k,n} (b_{k,n} - a_{k,n}) \leq \frac{r^2}{R^2}(\beta - \alpha). \tag{7.14}$$

For each triple k, n, s, we apply equation (7.12) to $E_r^R \cap (a_{k,n,s}, b_{k,n,s})$. We are able to put $E_r^R \cap (\alpha, \beta)$ inside a countable collection of pairwise disjoint open intervals $(a_{n,k,s,t}, b_{n,k,s,t})$ for which

$$\sum_k \sum_n \sum_s \sum_t (b_{k,n,s,t} - a_{k,n,s,t}) \leq \frac{r^3}{R^3}(\beta - \alpha). \tag{7.15}$$

Proceeding by induction, for any positive integer N, we can put E_r^R inside a countable collection of pairwise disjoint open intervals for which the sum of the lengths is less than or equal to

$$\frac{r^N}{R^N}(\beta - \alpha).$$

Since $r/R < 1$, this implies that the outer measure of E_r^R is 0. $\qquad\square$

The Faber–Chisholm–Young Theorem

We now show that we can do without the assumption of continuity.

Theorem 7.9 (Bounded Variation $\implies$ Differentiable AE). *If f has bounded variation on $[a, b]$, then f is differentiable almost everywhere on $[a, b]$.*

Again, we can assume that f is monotonically increasing. A function of bounded variation is simply a difference of two such functions. If two functions are differentiable almost everywhere, then so is their difference.

The discontinuities of an increasing function are jumps, places where

$$\lim_{x \to c^-} f(x) < \lim_{x \to c^+} f(x).$$

Consider the subintervals of $(f(a), f(b))$ that are *not* in the range of f. Each contains a distinct rational number. Therefore, the number of such intervals is

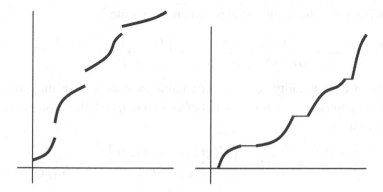

Figure 7.2. A strictly increasing function with discontinuities. Its inverse.

countable, and the set of points of $[a, b]$ at which f is discontinuous is either empty or a countable set.

Let $g(x) = f(x) + x$. The function f is differentiable if and only if g is differentiable, and g is slightly easier to work with because it is strictly increasing. The inverse of a strictly increasing function is continuous and monotonically increasing on its domain and can be uniquely extended to a continuous, monotonically increasing function on $(f(a), f(b))$ (see Figure 7.2). This should be obvious from the picture. We can make it rigorous by defining

$$G(x) = \sup\{t \in [a, b] \mid g(t) \leq x\}, \quad g(a) \leq x \leq g(b).$$

This function is well defined, monotonically increasing, and – because g is strictly increasing – $G(g(x)) = x$ (see Exercise 7.2.6). Recall from Theorem 3.1 that a function is continuous if and only if its inverse maps open sets to open sets. Since g is strictly increasing, it maps the open interval (α, β) to the open interval $(g(\alpha), g(\beta))$, and therefore G is continuous.

We can apply Theorem 7.4 to the function G. This will be the key to our proof, which will come in two parts. We use the existence of the derivative of G to prove that the Dini derivatives of g are equal (possibly $+\infty$) almost everywhere. We then prove that D^+g is finite almost everywhere.

Lemma 7.10 (Equality of Dini Derivatives). *Let g be a strictly increasing function on $[a, b]$. The Dini derivatives of g are equal (possibly $+\infty$) almost everywhere.*

Proof. Let G be the continuous, monotonically increasing function obtained by extending the inverse of g to the entire interval $[g(a), g(b)]$. We know by Theorem 7.4

that $G'(x)$ exists and is finite almost everywhere. We write

$$\frac{g(y) - g(x)}{y - x} = \frac{g(y) - g(x)}{G(g(y)) - G(g(x))} = \left[\frac{G(g(y)) - G(g(x))}{g(y) - g(x)} \right]^{-1}.$$

The points of discontinuity of g are countable, and thus have measure zero. Where g is continuous, $y \to x$ implies that $g(y) \to g(x)$, and, therefore, except on this set of measure zero,

$$\lim_{y \to x} \frac{g(y) - g(x)}{y - x} = \lim_{g(y) \to g(x)} \left[\frac{G(g(y)) - G(g(x))}{g(y) - g(x)} \right]^{-1} = \frac{1}{G'(g(x))}. \qquad \square$$

Lemma 7.11 (D^+g Finite AE). *Let g be a strictly increasing function on $[a, b]$. The Dini derivative D^+g is finite almost everywhere.*

Proof. We can restrict our attention to those x for which $D^+g(x) = D^-g(x)$ because the set of x on which they differ has measure zero. Let $E^\infty = \{x \in (a, b) \mid D^+g(x) = D^-g(x) = +\infty\}$. If $x \in E^\infty$, then for any positive N we can find s and t, $s < x < t$, such that

$$g(t) - g(x) > N(t - x),$$
$$g(x) - g(s) > N(x - s).$$

Therefore, $g(t) - g(s) > N(t - s)$. We define S_N to be the set of all $x \in (a, b)$ for which we can find s and t, $a < s < x < t < b$, such that $g(t) - g(s) > N(t - s)$. By what we have just shown about E^∞, we know that it is a subset of S_N. For each $x \in S_N$, we select s_x and t_x so that $s_x < x < t_x$,

$$f(t_x) - f(s_x) > N(t_x - s_x). \tag{7.16}$$

The intervals (s_x, t_x), taken over all $x \in S_N$, provide an open cover of S_N.

The set S_N is open (Exercise 7.2.7). By Theorem 3.5, S_N is the union of a countable collection of pairwise disjoint open intervals, $S_N = \bigcup_k (a_k, b_k)$. We choose a closed interval inside each (a_k, b_k) whose length is exactly half $b_k - a_k$,

$$[\alpha_k, \beta_k] \subseteq (a_k, b_k), \quad \beta_k - \alpha_k = \frac{1}{2}(b_k - a_k). \tag{7.17}$$

Each closed interval $[\alpha_k, \beta_k]$ is contained in the open cover $\bigcup_{x \in S_N} (s_x, t_x)$. By the Heine–Borel theorem, Theorem 3.6, we can find a finite collection of these open intervals that covers $[\alpha_k, \beta_k]$,

$$[\alpha_k, \beta_k] \subseteq \bigcup_{j=1}^{n_k} \left(s_{x(k,j)}, t_{x(k,j)} \right).$$

This finite open cover can be ordered so that $s_{x(k,1)} < s_{x(k,2)} < \cdots < s_{x(k,n_k)}$. We can assume that $t_{x(k,1)} < t_{x(k,2)} < \cdots < t_{x(k,n_k)}$, because otherwise there is an open interval that can be eliminated from the cover. Furthermore, if the right endpoint of one interval is strictly greater than the left endpoint of the second interval to its right, $s_{x(k,j+2)} < t_{x(k,j)}$, then the interval in the middle is contained in their union,

$$\left(s_{x(k,j+1)}, t_{x(k,j+1)}\right) \subseteq \left(s_{x(k,j)}, t_{x(k,j)}\right) \cup \left(s_{x(k,j+2)}, t_{x(k,j+2)}\right)$$

(see Exercise 7.2.8), so we can eliminate $\left(s_{x(k,j+1)}, t_{x(k,j+1)}\right)$ from the cover. Thus, after removing these superfluous intervals, the intervals in odd position are pairwise disjoint, as are the intervals in even position.

We can use the fact that g is strictly increasing, together with equation 7.16, to put a bound on the length of $[\alpha_k, \beta_k]$,

$$
\begin{aligned}
\beta_k - \alpha_k &\leq \sum_{1 \leq j \leq n_k/2} \left(t_{x(k,2j-1)} - s_{x(k,2j-1)}\right) + \sum_{1 \leq j \leq n_k/2} \left(t_{x(k,2j)} - s_{x(k,2j)}\right) \\
&\leq \frac{1}{N} \sum_{1 \leq j \leq n_k/2} \left(g(t_{x(k,2j-1)}) - g(s_{x(k,2j-1)})\right) \\
&\quad + \frac{1}{N} \sum_{1 \leq j \leq n_k/2} \left(g(t_{x(k,2j)}) - g(s_{x(k,2j)})\right) \\
&\leq \frac{1}{N} \left(g(\beta_k) - g(\alpha_k)\right) + \frac{1}{N} \left(g(\beta_k) - g(\alpha_k)\right) = \frac{2}{N} \left(g(\beta_k) - g(\alpha_k)\right).
\end{aligned}
$$

It follows that $E^\infty \subseteq S_N \subseteq \bigcup_{k=1}^\infty (a_k, b_k)$ where these intervals are pairwise disjoint. The measure of E^∞ is less than or equal to

$$
\begin{aligned}
\sum_{k=1}^\infty (b_k - a_k) &= 2 \sum_{k=1}^\infty (\beta_k - \alpha_k) \\
&\leq \frac{4}{N} \sum_{k=1}^\infty \left(g(\beta_k) - g(\alpha_k)\right) \\
&\leq \frac{4}{N} \left(g(b) - g(a)\right).
\end{aligned}
$$

Since this is true for all $N > 0$, no matter how large, we can conclude that $m(E^\infty) = 0$. $\square$

Exercises

7.2.1. Let f be the function defined by $f(0) = 0$ and $f(x) = x \sin(1/x)$ for $x \neq 0$. Find $D^+ f(0)$, $D^- f(0)$, $D_+ f(0)$, and $D_- f(0)$.

7.2.2. Show that if a function f assumes its maximum at c, then $D^+f(c) \leq 0$ and $D_-f(c) \geq 0$.

7.2.3. Show that if f is continuous on $[a, b]$ and any one of its Dini derivatives (say D^+) is everywhere nonnegative on $[a, b]$, then $f(b) \geq f(a)$.

7.2.4. Prove that if f is continuous on $[a_k, b_k]$ and if $f(x) \leq f(b_k)$ for all $x \in (a_k, b_k)$, then $f(a_k) \leq f(b_k)$.

7.2.5. Prove that if f is continuous on $[a_k, b_k]$ and if

$$t = \sup\left\{y \in [a_k, b_k] \,\middle|\, f(x) \leq f(y)\right\},$$

then $f(t) \geq f(x)$.

7.2.6. Show that if g is strictly increasing and

$$G(x) = \sup\{t \in [a, b] \mid g(t) \leq x\}, \quad g(a) \leq x \leq g(b),$$

then $G(g(x)) = x$.

7.2.7. Let f be a strictly increasing function and let E_N be the set of all $x \in (a, b)$ for which we can find s and t, $a < s < x < t < b$, such that $f(t) - f(s) > N(t - s)$. Prove that E_N is open.

7.2.8. Given three intervals (a_1, b_1), (a_2, b_2), (a_3, b_3) that cover (a_1, b_3) with $a_1 < a_2 < a_3$ and $b_1 < b_2 < b_3$, show that if $a_3 < b_1$, then

$$(a_2, b_2) \subseteq (a_1, b_1) \cup (a_3, b_3).$$

7.2.9. Let f be monotonically increasing on $[a, b]$, and c an arbitrary value in (a, b). Show that

$$\sup_{a < t < c} f(t) = \lim_{x \to c^-} f(x) \leq f(c) \leq \lim_{x \to c^+} f(x) = \inf_{c < t < b} f(t).$$

Explain how this implies that $\lim_{x \to c^-} f(x)$ and $\lim_{x \to c^+} f(x)$ exist.

7.2.10. Given an arbitrary sequence $(x_n) \subseteq [a, b]$ and a sequence of positive numbers (c_n) such that $\sum_{n=1}^{\infty} c_n < \infty$, define the function f by

$$f(x) = \sum_{x_n < x} c_n.$$

Show that

1. f is monotonically increasing on $[a, b]$,
2. f is discontinuous at each x_n, and
3. f is continuous at each $x \in [a, b] - \{x_n\}$.

7.2.11. Verify that in the rising sun lemma (Lemma 7.5), we have $f(a_k) = f(b_k)$ except possibly when $a_k = a$.

7.3 Absolute Continuity

Once mathematicians ceased to define integration as the inverse process of differentiation, they were faced with the two sets of questions that constitute the fundamental theorem of calculus,

1. **Antiderivative part:**
   (a) When is a function integrable?
   (b) If the integral exists, when can that integral be differentiated?
   (c) When does differentiating the integral take us back to the original function?
2. **Evaluation part:**
   (a) When is a function differentiable?
   (b) If the derivative exists, when can that derivative be integrated?
   (c) When does integrating the derivative take us back to the original function?

For question 1(a), we have a simple characterization of Lebesgue integrable functions: measurable functions for which the integrals of f^+ and f^- are finite. Question 1(b) asks when the resulting function $F(x) = \int_a^x f(t)\,dt$ is differentiable. As the next theorem implies, the answer is "always, almost everywhere." We shall answer question 1(c) in the next section.

Theorem 7.12 (Properties of Integral). *If f is integrable on $[a, b]$, then $F(x) = \int_a^x f(t)\,dt$ is uniformly continuous and of bounded variation on $[a, b]$.*

By Theorem 7.9, it follows that F is differentiable almost everywhere.

Proof. We are looking at F on a closed and bounded interval, so continuity will imply uniform continuity. Given an $\epsilon > 0$, we seek a response δ so that $|x - y| < \delta$ implies that

$$\left| F(x) - F(y) \right| = \left| \int_y^x f(t)\,dt \right| < \epsilon.$$

Since f is integrable, by Corollary 6.17 we can find a simple function ϕ such that

$$\int_a^b \left| f(x) - \phi(x) \right| dx < \frac{\epsilon}{2}.$$

Since every simple function takes on only finitely many values, it is bounded, say $|\phi(x)| < B$ for all $x \in [a, b]$. Choose $\delta = \epsilon/2B$, then $|x - y| < \delta$ implies that

$$\left| \int_x^y f(t)\,dt \right| \leq \int_x^y |f(t)|\,dt$$

$$\leq \int_x^y |f(t) - \phi(t)|\,dt + \int_x^y |\phi(t)|\,dt$$

$$\leq \int_a^b |f(t) - \phi(t)|\,dt + \int_x^y |\phi(t)|\,dt$$

$$< \frac{\epsilon}{2} + \frac{\epsilon}{2B} \cdot B = \epsilon.$$

To see that F has bounded variation, we observe that

$$\int_a^x f^+(t)\,dt \quad \text{and} \quad \int_a^x f^-(t)\,dt$$

are monotonically increasing functions, and therefore

$$F(x) = \int_a^x f(t)\,dt = \int_a^x f^+(t)\,dt - \int_a^x f^-(t)\,dt$$

is a difference of monotonically increasing functions. By the Jordan decomposition theorem (Theorem 7.2), it has bounded variation. $\qquad \square$

The Evaluation Part

The results of Section 7.2 give us a good answer to 2(a): If f has bounded variation, then it is differentiable almost everywhere. Bounded variation is not necessary, but it is a strong sufficient condition. We also have another approach. We can use one of the Dini derivatives, which always exist, and then ask when a function that is a Dini derivative can be integrated.

We are going to skip over 2(b) for the moment and go straight to 2(c). If we have a function that can be differentiated almost everywhere, and that derivative can be integrated, do we get back to the original function? If we do, then that says that our original function could be represented as a definite integral. As we saw in Theorem 7.12, that means that the original function must be continuous and of bounded variation on $[a, b]$. We see that bounded variation is not only sufficient to answer the first question affirmatively, it is necessary if we are to answer the last question in the affirmative. What about continuity? If f is a continuous function with bounded variation, then does it always follow that

$$f(x) = \int_a^x f'(t)\,dt?$$

The answer is no.

> **Definition: Absolute continuity**
>
> A function f is **absolutely continuous** on $[a, b]$ if given any $\epsilon > 0$, there is a response δ so that given any finite collection of pairwise disjoint open intervals in $[a, b]$, $\{(a_k, b_k)\}_{k=1}^{n}$, for which
> $$\sum_{k=1}^{n} (b_k - a_k) < \delta,$$
> we have that
> $$\sum_{k=1}^{n} |f(b_k) - f(a_k)| < \epsilon.$$

To see why not, we consider the devil's staircase, $\mathrm{DS}(x)$, (Example 4.1 on p. 86). The total variation of this function is 1, so it has bounded variation. It is a continuous function that is constant on the open intervals that form the complement of SVC(3). Since SVC(3), the Cantor ternary set, has measure 0, we have that

$$\frac{d}{dx} \mathrm{DS}(x) = 0, \quad \text{almost everywhere.}$$

The integral of any function that is 0 almost everywhere is a constant function, and $\mathrm{DS}(x)$ is not constant. If we start with DS, differentiate, and then define

$$F(x) = \int_0^x \mathrm{DS}'(t) \, dt,$$

then $F(x) = 0$ for all $x \in [0, 1]$.

We need something stronger than continuity to characterize functions that are integrals. We need **absolute continuity**.

To see that the devil's staircase is not absolutely continuous, let us take $\epsilon = 1/2$. Our function increases by 1/2 from $x = 0$ to $x = 1/3$, so our response δ must be less than 1/3. But the increase of 1/2 also occurs in an increase of 1/4 over $[0, 1/9]$ and an increase of 1/4 over $[2/9, 1/3]$. Our δ must be less than 2/9. But these increases actually occur over four intervals, each of length 1/27. The response δ is less than 4/27. Continuing in this way, we see that for each positive integer n,

$$\delta < \frac{2^n}{3^{n+1}} \xrightarrow{n \to \infty} 0.$$

There is no δ response.

A Little History

This property of definite integrals, absolute continuity, was first observed by Axel Harnack in 1884. The name was coined by Vitali in 1905, but several mathematicians were aware of it and using it by the 1890s, including Charles de la Vallée Poussin, Camille Jordan, Otto Stolz, and E. H. Moore. As we shall prove later in this section, if a function F can be defined as a definite integral, $F(x) = \int_a^x f(t)\, dt$ (using either the Lebesgue or Riemann definition of the integral), then F is absolutely continuous.

What about the other direction? If a function is absolutely continuous, does that imply that it is an integral? The Riemann integral is intractable, but we can do this for the Lebesgue integral. Because we are not limited to bounded functions, it will take more work to verify that any function defined as a definite integral must be absolutely continuous. But it will be possible to show that the implication also runs in the opposite direction; every absolutely continuous function is a definite integral. This result was observed by Lebesgue in 1904, but he gave no proof. The first proof was published by Vitali in 1905, the same paper in which this property received its name. The next two propositions will move us toward the theorem that F can be written as

$$F(x) = \int_a^x f(t)\, dt$$

for some function f if and only if F is absolutely continuous.

Lebesgue Integral and Absolute Continuity

Proposition 7.13 (Lebesgue Integral Is Absolutely Continuous). *If f is Lebesgue integrable on $[a, b]$, then $F(x) = \int_a^x f(t)\, dt$ is absolutely continuous on $[a, b]$.*

Since every Riemann integrable function is also Lebesgue integrable, it follows that if f is Riemann integrable on $[a, b]$, then $F(x) = \int_a^x f(t)\, dt$ is absolutely continuous on $[a, b]$. There are, however, other ways of defining the integral – see Appendix A.2 – for which $\int_a^x f(t)\, dt$ might not be absolutely continuous.

Proof. The proof is almost identical to the first part of the proof of Theorem 7.12. Given $\epsilon > 0$, we find a simple function ϕ such that

$$\int_a^b \left| f(x) - \phi(x) \right| dx < \frac{\epsilon}{2}.$$

Since ϕ is simple, it is bounded, say $|\phi(x)| < B$ for all $x \in [a, b]$. Choose the response $\delta = \epsilon/2B$. Let

$$S = \bigcup_{k=1}^{n} (a_k, b_k),$$

where $a \leq a_1 < b_1 \leq a_2 < b_2 \leq a_3 < \cdots < b_n \leq b$ and $\sum_{k=1}^{n}(b_k - a_k) < \delta = \epsilon/2B$. The set S is a union of finitely many pairwise disjoint intervals in $[a, b]$ for which the sum of the length is less than $\epsilon/2B$. This means that $m(S) < \epsilon/2B$. It follows that

$$\sum_{k=1}^{n}\left|\int_{a_k}^{b_k} f(t)\,dt\right| \leq \sum_{k=1}^{n}\int_{a_k}^{b_k}\left|f(t)\right|dt = \int_{S}\left|f(t)\right|dt$$

$$\leq \int_{S}\left|f(t) - \phi(t)\right|dt + \int_{S}\left|\phi(t)\right|dt$$

$$\leq \int_{a}^{b}\left|f(t) - \phi(t)\right|dt + \int_{S}\left|\phi(t)\right|dt$$

$$< \frac{\epsilon}{2} + \frac{\epsilon}{2B} \cdot B = \epsilon. \qquad \square$$

What about the other direction? If F is absolutely continuous, can we find an integrable function f for which

$$F(x) = \int_{a}^{x} f(t)\,dt.$$

The natural candidate for f is F', but does F' exist? The next proposition guarantees that it does, almost everywhere.

Proposition 7.14 (Absolute Continuity $\Longrightarrow$ Bounded Variation). *If F is absolutely continuous on $[a, b]$, then it has bounded variation on $[a, b]$.*

Proof. We let δ be the response to $\epsilon = 1$. Given any finite collection of pairwise disjoint intervals, $\left((a_k, b_k)\right)_{k=1}^{n}$, for which the sum of the lengths is less than δ, we have that

$$\sum_{k=1}^{n}\left|f(b_k) - f(a_k)\right| < 1.$$

Let $N = \lceil (b - a)/\delta \rceil$, the smallest integer greater than or equal to $(b - a)/\delta$. Let $P = (a = x_0, x_1, \ldots, x_m = b)$ be any partition of $[a, b]$ into intervals of length $< \delta$, $m \geq N$. For $1 \leq j < N$, choose $l(j)$ to be the largest integer such that

---

Definition: Lipschitz condition of order α

A function f defined on $[a, b]$ is said to satisfy a **Lipschitz condition of order** α if there is a constant $M > 0$ such that

$$\left| f(x) - f(y) \right| \leq M \left| x - y \right|^{\alpha}$$

for all $x, y \in [a, b]$.

$x_{l(j)} < a + j\delta$. The interval from $x_{l(j)}$ to $x_{l(j)+1}$ is one of the intervals in P, so it has length less than δ. Since

$$a + j\delta \leq x_{l(j)+1} \leq x_{l(j+1)} < a + (j+1)\delta,$$

the interval from $x_{l(j)+1}$ to $x_{l(j+1)}$ also has length less than δ. On each of these intervals, the variation of F with respect to P is less than 1. Counting the initial and final intervals, $[a, x_{l(1)}]$ and $[x_{l(N-1)+1}, b]$, there are at most $2N$ such intervals, so the variation of F with respect to P is strictly less than $2N$,

$$V(P, f) < 2N = 2 \left\lceil \frac{b - a}{\delta} \right\rceil.$$

Since every partition has a refinement with intervals of length less than δ and since refining a partition can only increase the variation, we see that the total variation is bounded by $2\lceil (b - a)/\delta \rceil$.   □

For the evaluation part of the fundamental theorem of calculus, if we start with F, differentiate it, and then integrate, we wind up with an absolutely continuous function. If we want any hope that we end with the same function with which we started, then we need to have started with an absolutely continuous function. This condition is necessary. As we shall see in the final section of this chapter, it is also sufficient.

A Hierarchy of Functions

Absolute continuity implies bounded variation, but as the devil's staircase illustrates, bounded variation does not imply absolute continuity. With one more definition in place, we can describe a nice hierarchy of functions defined on a closed bounded interval. In Section 8.1 we shall see how Lipschitz's condition arose and how he used it.

A function is said to be C^1 or **continuously differentiable** on $[a, b]$ if it is differentiable and its derivative is continuous on this interval. All of the following statements hold for functions on a closed and bounded interval:

1. If a function is C^1, then it has a bounded derivative.

2. If a function is differentiable with a bounded derivative, then it satisfies a Lipschitz condition of order 1.
3. If a function satisfies a Lipschitz condition of order 1, then it is absolutely continuous.
4. If a function is absolutely continuous, then it has bounded variation.
5. If a function has bounded variation, then it is differentiable almost everywhere.

The proofs of the first three statements are left as exercises. All of these implications go only one way, a fact that is also left for the exercises.

Absolute Continuity and Monotonicity

To conclude this section, we observe that any absolutely continuous function is the difference of two absolutely continuous and monotonically increasing functions, a key observation that we shall use in the next section.

Proposition 7.15 (Absolute Continuity of Variation). *The function f is absolutely continuous on [a, b] if and only if it is equal to the difference of two absolutely continuous, monotonically increasing functions.*

Proof. One direction is easy. The difference of absolutely continuous functions is absolutely continuous (see Exercise 7.3.13).

By Proposition 7.14, if f is absolutely continuous on $[a, b]$, then it has bounded variation. We need to prove that $T(x) = V_a^x(f)$ is absolutely continuous on this interval. Given $\epsilon > 0$, let $\delta > 0$ be the response to $\epsilon/2$: If the pairwise disjoint intervals (a_k, b_k), $1 \leq k \leq n$, have combined length less than δ, then $\sum_{k=1}^n |f(b_k) - f(a_k)|$ is strictly less than $\epsilon/2$. We shall see that $\sum_{k=1}^n |T(b_k) - T(a_k)| < \epsilon$.

We observe that

$$|T(b_k) - T(a_k)| = V_{a_k}^{b_k} = \sup \left\{ \sum_{j=1}^{m_k} |f(x_{k,j}) - f(x_{k,j-1})| \right\},$$

where the supremum is taken over all partitions, $a_k = x_{k,0} < x_{k,1} < \cdots < x_{k,m_k} = b_k$, of $[a_k, b_k]$. It follows that

$$\sum_{k=1}^n |T(b_k) - T(a_k)| = \sup \left\{ \sum_{k=1}^n \sum_{j=1}^{m_k} |f(x_{k,j}) - f(x_{k,j-1})| \right\}. \tag{7.18}$$

Since the set of intervals $(x_{k,j-1}, x_{k,j})$, $1 \leq j \leq m_k$, $1 \leq k \leq n$, is a finite collection of pairwise disjoint intervals of total length less than δ, each double sum on the right side of equation (7.18) is less than $\epsilon/2$, and so the supremum of these sums is strictly less than ϵ. □

Exercises

7.3.1. Find an example of a simple function ϕ such that

$$\int_0^\pi \left| \sin x - \phi(x) \right| dx < 0.1.$$

7.3.2. Let f be defined by $f(0) = 0$, $f(x) = x^2 \sin(1/x^2)$ for $x \neq 0$. Show that f does not have bounded variation in any neighborhood of 0, but it is differentiable at 0.

7.3.3. If a function is continuous and of bounded variation, does it necessarily follow that it is absolutely continuous?

7.3.4. Let f be absolutely continuous on the interval $[\epsilon, 1]$ for every $\epsilon > 0$ and continuous on $[0, 1]$. Does it necessarily follow that f is absolutely continuous on $[0, 1]$? If we add the restriction that f has bounded variation on $[0, 1]$, does it now follow that f is absolutely continuous on $[0, 1]$?

7.3.5. In the definition of absolute continuity, we restricted ourselves to a finite collection of pairwise disjoint open intervals. Show that this is equivalent to the following definition: A function f is absolutely continuous on $[a, b]$ if given any $\epsilon > 0$, there is a response δ so that given any countable collection (finite or infinite) of pairwise disjoint open intervals in $[a, b]$, $\{(a_k, b_k)\}$, for which

$$\sum (b_k - a_k) < \delta,$$

we have that

$$\sum \left| f(b_k) - f(a_k) \right| < \epsilon.$$

7.3.6. Show that if a function is C^1 on a closed and bounded interval, then it has a bounded derivative on this interval. Give an example of a function that is differentiable with a bounded derivative on a closed and bounded interval but which is not C^1 on this interval.

7.3.7. Show that if f is differentiable with bounded derivative on $[a, b]$, then f satisfies the Lipschitz condition of order 1. Show that if a differentiable function f satisfies a Lipschitz condition of order 1 on $[a, b]$, then the derivative of f is bounded.

7.3.8. Give an example of a function that satisfies a Lipschitz condition of order 1 on some closed and bounded interval but is not differentiable at every interior point of that interval.

7.3.9. Consider the function f defined on $[0, 1]$ as follows: If $x \in SVC(3)$, then $f(x) = 0$. If x is in a removed interval of length 3^{-n} with center at $a/(2 \cdot 3^{n-1})$, then $f(x) = \left| x - a/(2 \cdot 3^{n-1}) \right| - 1/(2 \cdot 3^n)$. This is a continuous function with a graph that looks like a lot of v's. Show that this function is Lipschitz of order 1 but that it is not differentiable at any point in $SVC(3)$.

7.3.10. Using the idea from Exercise 7.3.9, show how to construct a bounded, continuous function that is differentiable almost everywhere but does not have bounded variation.

7.3.11. Give an example of a function that is differentiable at every point of $[0, 1]$ but does not satisfy the Lipschitz condition on this interval.

7.3.12. Show that if f satisfies a Lipschitz condition of order 1 on $[a, b]$, then f is absolutely continuous on $[a, b]$.

7.3.13. Let f and g be absolutely continuous functions on $[a, b]$, $c \in \mathbb{R}$. Show that cf, $f + g$, and fg are absolutely continuous on $[a, b]$.

7.3.14. For positive constants α and β, define $f_{\alpha,\beta}$ by $f_{\alpha,\beta}(0) = 0$ and $f_{\alpha,\beta}(x) = x^\alpha \sin(\pi/x^\beta)$ for $x \neq 0$. Prove that $f_{\alpha,\beta}$ is absolutely continuous on $[0, 1]$ if and only if $\alpha > \beta$.

7.3.15. Let f be absolutely continuous on $[a, b]$, let $[c, d]$ be the range of f, and let g satisfy a Lipschitz condition on $[c, d]$. Show that $g \circ f$, the composition of g with f, is absolutely continuous on $[a, b]$.

7.3.16. Define the function f by $f(0) = 0$ and $f(x) = x^2 \left| \sin(1/x) \right|$ for $x > 0$. Let $g(x) = \sqrt{x}$, $x > 0$. Show that f, g, and $f \circ g$ are absolutely continuous on $[0, 1]$, but $g \circ f$ is not absolutely continuous on $[0, 1]$.

7.3.17. Show that if f and g are absolutely continuous and g is monotonically increasing, then $f \circ g$ is absolutely continuous.

7.3.18. Let f be an absolutely continuous function on $[0, 1]$ and $S \subseteq [0, 1]$ a subset of measure zero. Show that $f(S)$ has measure zero.

7.3.19. Show that any absolutely continuous function maps measurable sets to measurable sets.

7.4 Lebesgue's FTC

Finally, we are prepared to prove the fundamental theorem of calculus. As we have seen, for the antiderivative part, we must start with an integrable function, a measurable function f for which the Lebesgue integrals of both f^+ and f^- are finite. For the evaluation part, we must start with an absolutely continuous function. It turns out that these restrictions are not only necessary, they are sufficient.

Theorem 7.16 (FTC, Antiderivative). *If f is integrable on $[a, b]$, then*

$$F(x) = \int_a^x f(t)\, dt$$

is differentiable almost everywhere, and $F'(x) = f(x)$ almost everywhere.

Proof. We saw in Propositions 7.13 and 7.14 that F has bounded variation and thus is differentiable almost everywhere. We shall show that

$$F'(x) \le f(x), \quad \text{almost everywhere.}$$

This will complete the proof because if f is integrable, then so is $-f$. The definite integral of $-f$ is $-F$, and the derivative of $-F$ is $-F'$. Thus once we have proven that $F' \le f$ for every integrable f, it also follows that

$$-F'(x) \le -f(x), \quad \text{almost everywhere.}$$

We get the second inequality for free. But we pay dearly for the first inequality.

The proof is very reminiscent of the proof of Theorem 7.4. Let S be the subset of $[a, b]$ on which F is differentiable and begin by considering the set

$$E_p^q = \left\{ x \in [a, b] \cap S \,\middle|\, f(x) < p < q < F'(x) \right\},$$

where $p, q \in \mathbb{Q}$. The set of x for which $f(x) < F'(x)$ is the union over all pairs $p < q$ of E_p^q. This is a countable union. If we can show that $m(E_p^q) = 0$, then it follows that $F'(x) \le f(x)$ almost everywhere.

For $x \in E_p^q$, $f(x) < p$, and therefore

$$\int_{E_p^q} f(t)\, dt \le p\, m(E_p^q). \tag{7.19}$$

On the other hand, if F' exists at $x \in E_p^q$, then

$$\lim_{y \to x} \frac{\int_x^y f(t)\, dt}{y - x} = \lim_{y \to x} \frac{F(y) - F(x)}{y - x} = F'(x) > q.$$

If E_p^q were a closed interval, then we could turn this inequality (see Exercise 7.4.2) into the statement

$$\int_{E_p^q} f(t)\, dt \ge q\, m(E_p^q).$$

Together with equation (7.19), this would imply that $m(E_p^q) = 0$. Of course, E_p^q may not be a closed interval, but we can use the rising sun lemma to approximate E_p^q by intervals. This will be our approach.

Given any $\epsilon > 0$, Corollary 6.18 guarantees a response δ so that for any measurable set $A \subseteq [a, b]$ for which $m(A) < \delta$, we have that

$$\left| \int_A f(x)\,dx \right| < \epsilon. \tag{7.20}$$

We first find an open set U that contains E_p^q and such that

$$m\left(U - E_p^q\right) < \delta.$$

We can find such a set U because E_p^q is measurable. As an open set, U is a countable union of pairwise disjoint open intervals,

$$U = \bigcup_{k=1}^{\infty} (a_k, b_k).$$

Let

$$U_k = E_p^q \cap (a_k, b_k) \subseteq \left\{ x \in (a_k, b_k) \,\middle|\, q < F'(x) \right\}.$$

If $x \in U_k$, then we can find a $z > x$ such that

$$\frac{F(z) - F(x)}{z - x} > q,$$
$$F(z) - qz > F(x) - qx.$$

This tells us that x is a shadow point for the function defined by $F(x) - qx$. By the rising sum lemma (Lemma 7.5), U_k is contained within a countable union of pairwise disjoint open intervals $\left((\alpha_{k,j}, \beta_{k,j}) \right)_{j=1}^{\infty}$, each contained in (a_k, b_k), with

$$F(\beta_{k,j}) - q\beta_{k,j} \geq F(\alpha_{k,j}) - q\alpha_{k,j}.$$

It follows that

$$q(\beta_{k,j} - \alpha_{k,j}) \leq F(\beta_{k,j}) - F(\alpha_{k,j}) = \int_{\alpha_{k,j}}^{\beta_{k,j}} f(t)\,dt. \tag{7.21}$$

Let

$$T = \bigcup_{k=1}^{\infty} \bigcup_{j=1}^{\infty} (\alpha_{k,j}, \beta_{k,j}).$$

If we fix k, then the intervals $(\alpha_{k,j}, \beta_{k,j})$ are pairwise disjoint. Since $(\alpha_{k,j}, \beta_{k,j}) \subseteq (a_k, b_k)$ and the intervals (a_k, b_k) are pairwise disjoint, all of the intervals over all pairs k, j are pairwise disjoint. We also have that

$$E_p^q \subseteq T \subseteq U, \text{ and therefore } m(T - E_p^q) \leq m(U - E_p^q) < \delta.$$

We are now ready to put the pieces back together, using equations (7.19)–(7.21):

$$q\, m(E_p^q) \le q \sum_{k=1}^{\infty} \sum_{j=1}^{\infty} (\beta_{k,j} - \alpha_{k,j})$$

$$\le \sum_{k=1}^{\infty} \sum_{j=1}^{\infty} \int_{\alpha_{k,j}}^{\beta_{k,j}} f(t)\, dt$$

$$= \int_T f(t)\, dt$$

$$= \int_{E_p^q} f(t)\, dt + \int_{T - E_p^q} f(t)\, dt$$

$$< p\, m(E_p^q) + \epsilon.$$

Since this holds for every $\epsilon > 0$, we have that

$$q\, m(E_p^q) \le p\, m(E_p^q).$$

Since p is strictly less than q and measure is nonnegative, this can happen only if $m(E_p^q) = 0$. $\qquad\square$

Before proving the evaluation part of the fundamental theorem of calculus, we need to give a precise statement and proof of the result that if $f' = 0$, then f is constant.

Lemma 7.17 (Zero Derivative $\Longrightarrow$ Constant). *If f is absolutely continuous and monotonically increasing and if $f'(x) = 0$ almost everywhere on $[a, b]$, then f is constant.*

Note that without absolute continuity, f could be the devil's staircase, in which case the conclusion to this lemma would be false.

Proof. Let $E = \{x \in [a, b] \mid f'(x) = 0\}$. Since $f' = 0$ almost everywhere, we know that $m(E) = b - a$. Since f is monotonically increasing, we know that $f(E)$, the image of E, is contained in $[f(a), f(b)]$.

Let $Z = (a, b) - E$, $m(Z) = 0$. We shall use absolute continuity to show that $f(Z)$ has measure 0. We begin by choosing any $\epsilon > 0$. By absolute continuity, we can find a response $\delta > 0$ such that given any finite collection of pairwise disjoint open intervals, the sum of whose lengths is less than δ, the sum of the images of those intervals will be less than ϵ.

Since $Z \subseteq (a, b)$ and it has measure zero, we can find a countable collection of pairwise disjoint open intervals $\big((a_k, b_k)\big)_{k=1}^{\infty}$, whose union contains Z and for which

$$m(Z) \leq \sum_{k=1}^{\infty} (b_k - a_k) < \delta.$$

If we take any finite collection of these, $\bigcup_{k=1}^{n}(a_k, b_k)$, the sum of these lengths is less than δ, so f maps $\bigcup_{k=1}^{n}(a_k, b_k)$ to

$$\bigcup_{k=1}^{n} \big(f(a_k), f(b_k)\big),$$

a set of measure less than ϵ. Since the intervals $\big(f(a_k), f(b_k)\big)$ are pairwise disjoint,

$$\sum_{k=1}^{n} \big(f(b_k) - f(a_k)\big) < \epsilon.$$

Since this is true for all n, we see that

$$m(f(Z)) \leq \sum_{k=1}^{\infty} \big(f(b_k) - f(a_k)\big) \leq \epsilon.$$

Since ϵ is an arbitrary positive constant, we can conclude that $m\big(f(Z)\big) = 0$.

So far, so good. We want to prove that $f(a) = f(b)$. We know that

$$0 \leq f(b) - f(a) = m\big(f(E) \cup f(Z)\big) \leq m\big(f(E)\big) + m\big(f(Z)\big) = m\big(f(E)\big).$$

This lemma comes down to proving that $m\big(f(E)\big) = 0$.

We again choose any $\epsilon > 0$. For each $x \in E$, $x < b$, select a $z > x$ such that

$$\frac{f(z) - f(x)}{z - x} < \epsilon.$$

In what by now should look like a very familiar move, we rewrite this inequality as

$$\epsilon z - f(z) > \epsilon x - f(x).$$

The set E is contained in the set of shadow points of the function defined by $\epsilon x - f(x)$. Let $\big((\alpha_k, \beta_k)\big)_{k=1}^{\infty}$ be a collection of pairwise disjoint open intervals whose union contains E and for which

$$\epsilon \alpha_k - f(\alpha_k) \leq \epsilon \beta_k - f(\beta_k),$$
$$f(\beta_k) - f(\alpha_k) \leq \epsilon(\beta_k - \alpha_k).$$

Since f is monotonically increasing, $f(E)$ is contained in $\bigcup_{k=1}^{\infty}(f(\alpha_k), f(\beta_k))$, a union of pairwise disjoint open intervals. We have shown that

$$m\left(f(E)\right) \leq \sum_{k=1}^{\infty}\left(f(\beta_k) - f(\alpha_k)\right) \leq \epsilon \sum_{k=1}^{\infty}(\beta_k - \alpha_k) \leq \epsilon(b - a).$$

This is true for all $\epsilon > 0$; therefore, $m\left(f(E)\right) = 0$ and $f(a) = f(b)$. Since f is monotonically increasing, it is constant on $[a, b]$. $\square$

We are now prepared to state and prove the second half of the fundamental theorem of calculus. As we have seen, absolute continuity is not just sufficient; it is also necessary.

Theorem 7.18 (FTC, Evaluation). *If f is absolutely continuous on $[a, b]$, then it is differentiable almost everywhere, f' is integrable on $[a, b]$, and*

$$\int_a^b f'(t)\,dt = f(b) - f(a). \tag{7.22}$$

Proof. We have already shown that f has bounded variation on $[a, b]$ and therefore is differentiable almost everywhere. We can extend f' however we wish so that it is defined on all of $[a, b]$. In particular, we can use any of the Dini derivatives in place of f'. Changing the value of the integrand on a set of measure zero does not affect the value of the Lebesgue integral. We need to prove that f' is integrable and to establish equation (7.22).

As shown in Proposition 7.15, if f is absolutely continuous, then it is the difference of two absolutely continuous and monotonically increasing functions. It is enough to prove our theorem with the added assumption that f is monotonically increasing.

To prove that f' is integrable, we define a sequence of functions f_n by

$$f_n(x) = \frac{f(x + 1/n) - f(x)}{1/n} = n\left(f(x + 1/n) - f(x)\right),$$

where we extend f to the right of $x = b$ by defining $f(x) = f(b)$ for $x > b$. Each f_n is nonnegative, and $(f_n)_{n=1}^{\infty}$ converges to f' almost everywhere. By Fatou's lemma

(Theorem 6.20), if the integral of f_n over $[a, b]$ has a bound independent of n, then f' is integrable. The bound on the integral of f_n follows from the monotonicity of f,

$$\int_a^b f_n(x)\,dx = n\int_a^b f(x + 1/n)\,dx - n\int_a^b f(x)\,dx$$

$$= n\int_{a+1/n}^{b+1/n} f(x)\,dx - n\int_a^b f(x)\,dx$$

$$= n\int_b^{b+1/n} f(x)\,dx - n\int_a^{a+1/n} f(x)\,dx$$

$$\leq n\cdot\frac{1}{n}\cdot f(b + 1/n) - n\cdot\frac{1}{n}\cdot f(a) = f(b) - f(a).$$

We can replace a by any lower limit $x \in [a, b]$ and b by any upper limit $y \in [a, b]$. We see that for $a \leq x \leq y \leq b$,

$$\int_x^y f'(t)\,dt \leq f(y) - f(x). \tag{7.23}$$

We now define g by

$$g(x) = f(x) - \int_a^x f'(t)\,dt.$$

We want to apply Lemma 7.17 to the function g. From its definition, $g(a) = f(a) - 0 = f(a)$. If we can show that g is constant on $[a, b]$, then that constant is $f(a)$ and equation (7.22) is proven. We only need to show that g is absolutely continuous and monotonically increasing, and that its derivative is 0 almost everywhere.

Since g is a difference of two absolutely continuous functions, it is absolutely continuous. By equation (7.23), $x < y$ implies that

$$g(y) - g(x) = f(y) - f(x) - \int_x^y f'(t)\,dt \geq 0,$$

so g is monotonically increasing. By the antiderivative part of the fundamental theorem of calculus, Theorem 7.16,

$$\frac{d}{dx}\int_a^x f'(t)\,dt = f'(x), \quad \text{almost everywhere,}$$

and therefore $g' = 0$ almost everywhere. By Lemma 7.17, g is the constant function equal to $f(a)$. For all $x \in [a, b]$,

$$f(x) - \int_a^x f'(t)\,dt = f(a), \qquad \int_a^x f'(t)\,dt = f(x) - (a). \qquad \square$$

We have now answered four of our five original questions. We have found the right way to define integration. In Lebesgue's dominated convergence theorem we have found a condition which, though not necessary, is a strong and useful sufficient condition that allows term-by-term integration of a series. We have learned that the connection between continuity and differentiability is stronger than we might have expected. And in this section, we have explained the exact relationship between integration and differentiation.

That still leaves one question, our very first question, the question that started us asking all of these other questions,

When does a function have a Fourier series expansion that converges to that function?

We now have the tools to make serious progress. One of the most surprising insights of the early twentieth century was that this is not quite the right way to pose the problem. As we shall see in the next chapter, there is a better, more useful question that will have a very elegant answer.

Exercises

7.4.1. Give an example of a function, f, integrable on $[0, 1]$, such that for $F(x) = \int_0^x f(t)\, dt$, there is a $c \in (0, 1)$ such that F is differentiable at c but $F'(c) \neq f(c)$.

7.4.2. Show that if

$$\lim_{y \to x} \frac{\int_x^y f(t)\, dt}{y - x} > q$$

for all $x \in [a, b]$, then we can find a $\delta > 0$ so that $x, y \in [a, b]$ and $0 < y - x < \delta$ implies that

$$\int_x^y f(t)\, dt > q(y - x).$$

It follows that

$$\int_a^b f(t)\, dt \geq q(b - a).$$

7.4.3. Let f be differentiable with a bounded derivative on $[a, b]$. Show that for all $x \in [a, b]$,

$$\int_a^x f'(t)\, dt = f(x) - f(a).$$

7.4.4. Let f be integrable on $[a, b]$ with

$$\int_a^x f(t)\, dt = 0$$

for all $x \in [a, b]$. Using Proposition 6.13 but *not* using Theorem 7.16, show that $f = 0$ almost everywhere.

7.4.5. Use the result of Exercise 7.4.4 and the evaluation part of the fundamental theorem of calculus, Theorem 7.18, to prove the antiderivative part of the fundamental theorem of calculus, Theorem 7.16.

7.4.6. Let f and g be absolutely continuous on $[a, b]$ with $f' = g'$ almost everywhere. Show that $f = g + c$ for some constant c.

7.4.7. Show that if f is absolutely continuous on $[a, b]$, then

$$V_a^b(f) = \int_a^b |f'(x)|\, dx,$$

where $V_a^b(f)$ is the total variation of f on $[a, b]$. Show that this is not necessarily true if f is not absolutely continuous.

7.4.8. A monotonic function f defined on $[a, b]$ is said to be **singular** if $f' = 0$ almost everywhere. Show that any monotonically increasing function is the sum of an absolutely continuous function and a singular function.

7.4.9. Let g be a strictly increasing, absolutely continuous function on $[a, b]$ with $g(a) = c$, $g(b) = d$.

1. Show that for any measurable set $S \subseteq [a, b]$,

$$m\left(g(S)\right) = \int_S g'(x)\, dx.$$

2. Show that if $A = \{x \in [a, b] \mid g'(x) \neq 0\}$, and B is any subset of $[c, d]$ of measure zero, then

$$m\left(A \cap g^{-1}(B)\right) = 0.$$

3. Show that if A is the set defined in part 2 and C is any measurable subset of $[c, d]$, then

$$m(C) = \int_{A \cap g^{-1}(C)} g'(x)\, dx = \int_a^b \chi_C\left(g(x)\right) g'(x)\, dx.$$

7.4.10. [Change of Variable] Prove the change of variable formula for Lebesgue integrals: If g is strictly increasing and absolutely continuous on $[a, b]$ with $g(a) = c$ and $g(b) = d$ and if f is integrable on $[c, d]$, then

$$\int_c^d f(t)\, dt = \int_a^b f\left(g(x)\right) g'(x)\, dx. \tag{7.24}$$

7.4.11. Let f and g be integrable on $[a, b]$ and define

$$F(x) = \alpha + \int_a^x f(t)\, dt, \quad G(x) = \beta + \int_a^x g(t)\, dt.$$

Prove that

$$\int_a^b G(t)f(t)\,dt + \int_a^b g(t)F(t)\,dt = F(b)G(b) - F(a)G(a). \qquad (7.25)$$

7.4.12. [Integration by Parts] Prove the formula for integration by parts for Lebesgue integrals and absolutely continuous functions: If f and g are absolutely continuous on $[a, b]$, then

$$\int_a^b f(t)g'(t)\,dt + \int_a^b f'(t)g(t)\,dt = f(b)g(b) - f(a)g(a). \qquad (7.26)$$

7.4.13. Let f be integrable on $[a, b]$. We say that $c \in (a, b)$ is a **Lebesgue point** if $f(c) \neq \pm\infty$ and

$$\lim_{h \to 0} \frac{1}{h} \int_c^{c+h} |f(t) - f(c)|\,dt = 0.$$

Show that if c is a Lebesgue point for f, then $F(x) = \int_a^x f(t)\,dt$ is differentiable at c and $F'(c) = f(c)$.

7.4.14. Show that if f is integrable on $[a, b]$, then each point of continuity of f is a Lebesgue point for f.

7.4.15. Show that if f is integrable on $[a, b]$, then almost every point of $[a, b]$ (all but a set of measure zero) is a Lebesgue point for f.

7.4.16. Let f, not necessarily a measurable function, be defined on $[a, b]$. For each $x_0 \in [a, b]$ and $h, \epsilon > 0$, let $S(x_0, h, \epsilon)$ be the set of points $x \in [x_0 - h, x_0 + h] \cap [a, b]$ for which $|f(x) - f(x_0)| \geq \epsilon$. We say that $x_0 \in [a, b]$ is a **point of approximate continuity** of f if for each $\epsilon > 0$,

$$\lim_{h \to 0} \frac{m_e\left(S(x_0, h, \epsilon)\right)}{2h} = 0.$$

Show that any point of continuity is also a point of approximate continuity. Give an example of a function for which there is a point of approximate continuity that is not a point of continuity. Justify your example.

7.4.17. Prove that if f is measurable on $[a, b]$, then almost all points of $[a, b]$ (all but a set of measure zero) are points of approximate continuity.

8

Fourier Series

The development of measure theory and Lebesgue integration did not come about because mathematicians decided they needed a new definition of the integral. It happened because they were trying to develop and use tools of analysis to solve real and practical problems. These included solutions to partial differential equations, extensions of calculus to higher dimensions and to complex-valued functions, and generalizations of the concepts of area and volume. Fourier series were not unique in motivating work in analysis, but they constitute a very useful lens through which to view the development of analysis because these series often were the principal source of the questions that would prove most troublesome and insightful. As progress was made in our understanding of analysis, these insights often translated directly into answers about Fourier series.

This is true especially of Lebesgue's work on the integral. In 1905, armed with the power of his new integral, he gave a definitive answer to the question of when the Fourier series of a function converges pointwise to that function. We shall see his answer in this first section.

The story does not stop there. Once we are using the Lebesgue integral, we can change the values of the function on any set of measure zero without changing the value of the integral. Therefore, two functions that are equal almost everywhere will have the same Fourier coefficients, and so the same Fourier series. The best we can hope for from a theorem with the weak assumption that f is integrable is that the Fourier series of f converges to f almost everywhere. In fact, this is not quite true, though we can come close to it by either strengthening the assumption just a little (Theorem 8.9) or slightly weakening the conclusion (Theorem 8.2). If we want the Fourier series to converge to f at every point, we shall need to be quite restrictive about the kind of function with which we start.

If we are content with convergence almost everywhere, then we really need to think of equivalence classes of functions where $f \sim g$ if $f = g$ almost everywhere, or, equivalently, $\int_a^b |f(x) - g(x)| \, dx = 0$. This integral defines a

241

natural distance between equivalence classes of integrable functions on $[a, b]$, $D(f, g) = \int_a^b |f(x) - g(x)| \, dx$. Suddenly, we are looking at a geometric space in which each point is an equivalence class of functions. In fact, this is a vector space equipped with a definition of distance. We can think of the partial sums of the Fourier series of f as points in this space and ask if these points converge to the point represented by f. Is every point in this space the limit of the partial sums of its Fourier series? Is there a unique trigonometric series that converges to our equivalence class? As we shall see, with some simple assumptions the answers to these questions are "yes" and "yes."

It is very important to realize that these clean "yes's" do not answer the original question about pointwise convergence. Real progress in mathematics often consists of realizing when you are asking the wrong question. Asking the right question in this case opened an entire world of new and powerful mathematics, functional analysis.

8.1 Pointwise Convergence

Because the Fourier coefficients are given in terms of integrals of the original function, we cannot even speak of a Fourier series unless the function is integrable. When Dirichlet introduced the characteristic function of the rationals, he intended it as an illustration that not all functions are integrable. While this characteristic function *is* integrable in the Lebesgue sense, not all functions are. The question is "how much more than integrability is needed in order to guarantee that the resulting Fourier series converges to the original function, either at every point or almost everewhere?"

In 1829, Peter Gustav Lejeune Dirichlet (1805–1859) gave us the first proof of sufficient conditions. For simplicity, we assume that we are only considering functions on the interval $[-\pi, \pi]$.

Dirichlet's Conditions. The following conditions collectively imply that the Fourier series of f converges pointwise to f on $(-\pi, \pi)$:

1. f is integrable in the sense of Cauchy and Riemann (and thus bounded),
2. $f(x) = \frac{1}{2} \left(\lim_{y \to x^-} f(y) + \lim_{y \to x^+} f(y) \right)$,
3. f is piecewise monotonic on $[-\pi, \pi]$, and
4. f is piecewise continuous on $[-\pi, \pi]$.

Actually, Dirichlet's proof allows for an infinite number of points of discontinuity, provided the set of points of discontinuity is nowhere dense. Dirichlet believed that monotonicity was not necessary. While the requirement of piecewise monotonicity can be weakened, Paul du Bois-Reymond would show in 1876 that there

---

Definition: Absolutely integrable

A function f is **absolutely integrable** on $[a, b]$ if it is Riemann integrable on this interval or if the improper Riemann integrals of f and $|f|$ exist and are finite.

---

are functions that satisfy conditions 1, 2, and 4 but for which the Fourier series does not even converge at all values of $x \in (-\pi, \pi)$. See Exercises 8.1.10–8.1.15 for Fejér's example of a continuous function whose Fourier series fails to converge at any rational multiple of π.

Riemann would show that the function f does not need to be bounded. It is enough that f is absolutely integrable.

Assumption 2 is essential. To say that the Fourier series for f converges to $f(x)$ at a given value of x is to say that

$$\lim_{n \to \infty} \left(f(x) - \frac{a_0}{2} - \sum_{k=1}^{n} \left[a_k \cos(kx) + b_k \sin(kx) \right] \right) = 0,$$

where

$$a_k = \frac{1}{\pi} \int_{-\pi}^{\pi} f(x) \cos(kx) \, dx,$$

$$b_k = \frac{1}{\pi} \int_{-\pi}^{\pi} f(x) \sin(kx) \, dx.$$

We substitute these integrals, interchange the finite summation and the integration, use the sum of angles formula, and employ the trigonometric identity[1]

$$\frac{1}{2} + \cos u + \cos 2u + \cdots + \cos nu = \frac{\sin[(2n + 1)u/2]}{2 \sin[u/2]} \tag{8.1}$$

to rewrite the limit as

$$\lim_{n \to \infty} \left(f(x) - \frac{1}{\pi} \int_{-\pi}^{\pi} \frac{\sin[(2n + 1)(t - x)/2]}{2 \sin[(t - x)/2]} f(t) \, dt \right) = 0.$$

With a change of variable and the continuation of f outside $(-\pi, \pi)$ by assuming $f(x + 2\pi) = f(x)$, we can rewrite this as

$$\lim_{n \to \infty} \left(f(x) - \frac{1}{\pi} \int_{0}^{\pi/2} \frac{\sin[(2n + 1)u]}{\sin u} \left[f(x - 2u) + f(x + 2u) \right] \, du \right) = 0.$$

---

[1] This process leading to the derivation of equation (8.3) is done at a more leisurely pace in Section 6.1 of *A Radical Approach to Real Analysis*. It includes a proof of the assertion that equation (8.3) implies equation (8.4).

We use the integral identity

$$\int_0^{\pi/2} \frac{\sin[(2n+1)u]}{\sin u} \, du = \frac{\pi}{2} \tag{8.2}$$

to rewrite $f(x)$ as

$$\frac{1}{\pi} \int_0^{\pi/2} \frac{\sin[(2n+1)u]}{\sin u} \, 2f(x) \, du.$$

We have reduced the problem of proving convergence of the Fourier series to f to proving that

$$\lim_{n\to\infty} \int_0^{\pi/2} \frac{\sin[(2n+1)u]}{\sin u} \left[f(x-2u) + f(x+2u) - 2f(x) \right] du = 0. \tag{8.3}$$

A necessary condition for this to be true is that

$$\lim_{u\to 0} \left[f(x-2u) + f(x+2u) - 2f(x) \right] = 0. \tag{8.4}$$

This is equivalent to condition 2:

$$f(x) = \frac{1}{2} \left(\lim_{y\to x^-} f(y) + \lim_{y\to x^+} f(y) \right).$$

If we define

$$\phi_x(u) = f(x-2u) + f(x+2u) - 2f(x),$$

then we can replace condition 2 with the requirement that ϕ_x is continuous at $u = 0$. How much more do we need?

Rudolph Otto Sigismund Lipschitz (1832–1903) was the first mathematician after Dirichlet to make significant progress. In 1864, he published his doctoral thesis on representation by Fourier series. He showed that beyond the two necessary conditions of absolute integrability and the continuity of ϕ_x at $u = 0$ for all x, one more condition on ϕ_x would be enough to guarantee pointwise convergence. In fact, that additional condition implies the continuity of ϕ_x.

Lipschitz's Conditions. The following conditions imply that the Fourier series of f converges pointwise to f on $(-\pi, \pi)$:

1. f is absolutely integrable.
2. For each x there exist positive constants A, ϵ, and α such that for any $u, t \in N_\epsilon(x)$,

$$\left| \phi_x(t) - \phi_x(u) \right| < A|t - u|^\alpha. \tag{8.5}$$

Any function that satisfies inequality (8.5) is said to satisfy a **Lipschitz condition of order** α. It implies continuity at $u = 0$. Notice that it is not strong enough to imply differentiability at $u = 0$ unless $\alpha > 1$.

The next advances were made by Ulisse Dini. In 1872, he showed that the bound $A|t - u|^\alpha$ on the right side of inequality (8.5) could be replaced by $A/\big|\log|t - u|\big|$. In 1880, he found a single condition that implies pointwise convergence of the Fourier series of f to f.

Dini's Condition. The following condition implies that the Fourier series of f converges pointwise to f on $(-\pi, \pi)$:

1. $|\phi_x(u)/u|$ is absolutely integrable over $[0, \pi]$.

Dini's condition is simple to state, but not always simple to check. The following year, Camille Jordan published his criteria. Note that bounded variation implies Riemann integrability (see Exercise 8.1.2).

Jordan's Conditions. The following conditions imply that the Fourier series of f converges pointwise to f on $(-\pi, \pi)$:

1. f has bounded variation on $(-\pi, \pi)$.
2. ϕ_x is continuous at $u = 0$.

Cesàro Convergence

In 1890, the Italian Ernesto Cesàro (1859–1906) broadened the definition of convergence.

For example, the series $1 - 1 + 1 - 1 + 1 - \cdots$ corresponds to the sequence of partial sums $(1, 0, 1, 0, 1, 0, \ldots)$. This sequence does not converge. But if we take the sum of the first n terms of this sequence and divide it by n, we get

$$\text{for } n \text{ odd: } \frac{(n+1)/2}{n} = \frac{1}{2} + \frac{1}{2n}, \quad \text{for } n \text{ even: } \frac{n/2}{n} = \frac{1}{2}.$$

The limit of this average value does exist. It equals $1/2$. We say that the Cesàro limit of $1 - 1 + 1 - 1 + 1 - \cdots$ is $1/2$.

Cesàro limits are particularly useful for Fourier series. Consider the Fourier cosine series expansion of the constant function $\pi/4$, valid for $-\pi/2 < x < \pi/2$:

$$f(x) = \cos(x) - \frac{1}{3}\cos(3x) + \frac{1}{5}\cos(5x) - \frac{1}{7}\cos(7x) + \cdots.$$

Definition: Cesàro limit

The sequence $(a_k)_{k=1}^{\infty}$ is said to have **Cesàro limit** A if

$$\lim_{n\to\infty} \frac{a_1 + a_2 + \cdots + a_n}{n} = A.$$

The derivative of f is 0 for all $x \in (-\pi/2, \pi/2)$, but if we try to differentiate term by term, we get a series that does not converge except at $x = 0$:

$$-\sin(x) + \sin(3x) - \sin(5x) + \sin(7x) - \cdots.$$

Now consider the Cesàro sum of this series. We first need to find the kth partial sum (see Exercise 8.1.5)

$$-\sin(x) + \sin(3x) - \sin(5x) + \cdots + (-1)^k \sin\big((2k-1)x\big) = \frac{(-1)^k \sin(2kx)}{2\cos(x)}. \tag{8.6}$$

We now compute the average of the first n partial sums (see Exercise 8.1.6)

$$\frac{1}{n}\left(\frac{-\sin(2x)}{2\cos x} + \frac{\sin(4x)}{2\cos x} - \frac{\sin(6x)}{2\cos x} + \cdots + (-1)^n \frac{\sin(2nx)}{2\cos x}\right)$$

$$= \frac{(\tan x)(-1 + (-1)^n \cos(2nx)) + (-1)^n \sin(2nx)}{4n \cos x}. \tag{8.7}$$

We fix $x \in (-\pi/2, \pi/2)$ and take the limit $n \to \infty$. Since the numerator stays bounded as the denominator approaches ∞, the Cesàro limit of the series obtained by term-by-term differentiation is 0, regardless of the value of x.

What if the limit of a sequence exists? Is the Cesàro limit the same? Fortunately, the answer is "yes."

Proposition 8.1 (Limit $\Longrightarrow$ Cesàro Limit). *If $a_n \to A$, then the Cesàro limit of (a_n) exists and also equals A.*

Proof. Given any $\epsilon > 0$, we can find an N such that $n \geq N$ implies that $|a_n - A| < \epsilon$. We take the average of the first n terms, $n \geq N$, and compare it to A,

$$\left|\frac{a_1 + a_2 + \cdots + a_n}{n} - A\right|$$

$$= \left|\frac{(a_1 + \cdots + a_N) - NA}{n} + \frac{a_{N+1} - A}{n} + \cdots + \frac{a_n - A}{n}\right|$$

$$\leq \left|\frac{(a_1 + \cdots + a_N) - NA}{n}\right| + \left|\frac{a_{N+1} - A}{n}\right| + \cdots + \left|\frac{a_n - A}{n}\right|$$

$$< \left|\frac{(a_1 + \cdots + a_N) - NA}{n}\right| + (n - N)\frac{\epsilon}{n}$$

$$< \left|\frac{(a_1 + \cdots + a_N) - NA}{n}\right| + \epsilon.$$

For n sufficiently large,

$$\left|\frac{(a_1 + \cdots + a_N) - NA}{n}\right| < \epsilon,$$

and therefore

$$\left| \frac{a_1 + a_2 + \cdots + a_n}{n} - A \right| < 2\epsilon$$

for all n sufficiently large. Since this is true for every $\epsilon > 0$, the Cesàro limit is A. $\qquad\qquad\square$

In 1900, the Hungarian mathematician Lipót Fejér (1880–1959) proved that if f is continuous, then the Fourier series of f converges at least in the Cesàro sense pointwise to f. For most of his career, Fejér taught at the University of Budapest. Among his doctoral students are Paul Erdős, George Pólya, Gabor Szegő, and John von Neumann.

In 1905, Lebesgue used his insights into integration to prove the following theorem.

Theorem 8.2 (Lebesgue on Fourier). *If f is integrable (in the Lebesgue sense) on the interval $[-\pi, \pi]$, then the Fourier series of f converges to f almost everywhere, at least in the Cesàro sense of convergence.*

In some sense, we could not possibly ask for a better result. The only assumption we need to make about f is that it is integrable, an assumption needed before we can even define the coefficients of the Fourier series. On the other hand, the conclusion is weaker than we might have wished: almost everywhere instead of for all $x \in (-\pi, \pi)$, convergence in the Cesàro sense rather than strict convergence.

Yet, as mathematicians were beginning to realize, asking for certain properties to hold for all x introduces unnecessary complications. For many purposes, it makes sense to consider two functions to be equivalent if they agree almost everywhere. If we work with equivalence classes and f is integrable, then the Cesàro limit of its Fourier series exists and is equivalent to f. If the Fourier series of f converges in the usual sense, then it converges to a function that is equivalent to f.

Allowing for Cesàro convergence does introduce its own complications. The series

$$\sin x - \sin(3x) + \sin(5x) - \cdots$$

Cesàro converges to the constant function 0. We have lost the uniqueness of the representation by a trigonometric series. That is a high price to pay. When it is worth paying depends on how we want to use the trigonometric representation. Sometimes existence is more important than uniqueness; sometimes it is not.

Exercises

8.1.1. Show that if

$$\lim_{n\to\infty} \int_0^{\pi/2} \frac{\sin[(2n+1)u]}{\sin u} \left[f(x-2u) + f(x+2u) - 2f(x) \right] du = 0,$$

where the integral is the Riemann integral, then

$$\lim_{u\to 0} \left[f(x-2u) + f(x+2u) - 2f(x) \right] = 0.$$

8.1.2. Show that if f has bounded variation on $[a, b]$, then it is Riemann integrable on this interval.

8.1.3. Consider the function f defined by $f(0) = 0$, $f(x) = 1/\ln(|x|/2\pi)$ for $x \in [-\pi, \pi] - \{0\}$. Show that this function satisfies Jordan's conditions but does not satisfy Lipschitz's conditions.

8.1.4. Consider the function g defined by $g(0) = 0$, $g(x) = x \cos(\pi/(2x))$ for $x \in [-\pi, \pi] - \{0\}$. Show that this function satisfies Lipschitz's conditions at $x = 0$ but does not satisfy Jordan's conditions in any neighborhood of $x = 0$.

8.1.5. Use geometric series to find the sum of

$$-e^y + e^{3y} - e^{5y} + \cdots + (-1)^k e^{(2k-1)y}.$$

Set $y = ix$ and use the fact that the imaginary part of e^{ix} is $i \sin x$ to prove equation (8.6),

$$-\sin(x) + \sin(3x) - \sin(5x) + \cdots + (-1)^k \sin\left((2k-1)x\right) = \frac{(-1)^k \sin(2kx)}{2\cos(x)}.$$

8.1.6. Using the approach of Exercise 8.1.5, prove equation (8.7).

8.1.7. Let $(a_n)_{n=1}^\infty$ be a sequence of real numbers and define

$$a_n^{(1)} = \frac{1}{n} \left(a_1 + a_2 + \cdots + a_n \right).$$

Show that

$$\varliminf_{n\to\infty} a_n \leq \varliminf_{n\to\infty} a_n^{(1)} \leq \varlimsup_{n\to\infty} a_n^{(1)} \leq \varlimsup_{n\to\infty} a_n.$$

8.1.8. Cesàro convergence is also known as $(C, 1)$-**convergence**. If the sequence $(a_n^{(1)})_{n=1}^\infty$ (see Exercise 8.1.7) does not converge, but it does converge in the Cesàro sense, then we say that the original sequence $(a_n)_{n=1}^\infty$ has $(C, 2)$-convergence. In general, for $k \geq 1$ define

$$a_n^{(k+1)} = \frac{1}{n} \left(a_1^{(k)} + a_2^{(k)} + \cdots + a_n^{(k)} \right).$$

If the sequence $\left(a_n^{(k-1)}\right)_{n=1}^{\infty}$ does not converge but $\left(a_n^{(k)}\right)_{n=1}^{\infty}$ does converge, then we say that the original sequence has (C, k)-convergence. Find examples of sequences with (C, k) convergence for each k, $2 \leq k \leq 4$.

8.1.9. We can use the symbol

$$x_n \xrightarrow{C} x_0$$

to mean that x_0 is the Cesàro limit of (x_n). We say that a function f is **Cesàro continuous** at $x = x_0$ if

$$x_n \xrightarrow{C} x_0 \qquad \text{implies} \qquad f(x_n) \xrightarrow{C} f(x_0).$$

Note that we have weakened the conclusion, but we have also weakened the hypothesis. Is every continuous function also Cesàro continuous? Is every Cesàro continuous function also continuous? Is $f(x) = x^2$ Cesàro continuous? Is it Cesàro continuous for any values of x?

Exercises 8.1.10–8.1.15 develop Fejér's example of a continuous function whose Fourier series does not converge at any point of a dense subset of $[-\pi, \pi]$.

8.1.10. Show that the Fourier sine series for the constant function $\pi/4$ on $(0, 2\pi)$ is given by

$$\frac{\pi}{4} = \sum_{k=1}^{\infty} \frac{\sin\left((k - \frac{1}{2})x\right)}{2k - 1}, \qquad 0 < x < 2\pi. \tag{8.8}$$

8.1.11. Define the function ϕ by

$$\phi(n, r, x) = \frac{\cos\left((r + 1)x\right)}{2n - 1} + \frac{\cos\left((r + 2)x\right)}{2n - 3} + \cdots + \frac{\cos\left((r + n)x\right)}{1}$$
$$- \frac{\cos\left((r + n + 1)x\right)}{1} - \frac{\cos\left((r + n + 2)x\right)}{3}$$
$$- \cdots - \frac{\cos\left((r + 2n)x\right)}{2n - 1}.$$

Show that

$$\phi(n, r, x) = 2\sin\left((r + n + 1/2)x\right) \sum_{k=1}^{n} \frac{\sin\left((k - \frac{1}{2})x\right)}{2k - 1}, \tag{8.9}$$

and therefore there is a bound B, independent of n, r, or x, such that $\left|\phi(n, r, x)\right| < B$.

8.1.12. Let $(\lambda_j)_{j=1}^\infty$ be an increasing sequence of positive integers. Define the function f by

$$f(x) = \phi(\lambda_1, 0, x) + \sum_{n=1}^\infty \frac{\phi(\lambda_n, 2\lambda_1 + 2\lambda_2 + \cdots + 2\lambda_{n-1}, x)}{n^2}. \qquad (8.10)$$

Show that this series converges absolutely and uniformly, regardless of the choice of sequence (λ_j). Therefore, f is continuous on $\mathbb{R}$.

8.1.13. Using the uniqueness of the Fourier series expansion of a continuous function, show that the Fourier series for f on $[-\pi, \pi]$ is given by

$$f(x) = \sum_{n=1}^\infty \alpha_n \cos(nx),$$

where

$$\alpha_n = \frac{1}{m^2(2\lambda_m + 1 - 2k)},$$

for m and k, the unique positive integers that satisfy

$$n = 2\lambda_1 + 2\lambda_2 + \cdots + 2\lambda_{m-1} + k, \quad 1 \le k \le 2\lambda_m.$$

8.1.14. Show that the sum of the first $2\lambda_1 + \cdots + 2\lambda_{m-1} + \lambda_m$ terms of the Fourier series for f at $x = 0$ equals

$$\sum_{n=1}^{2\lambda_1 + \cdots + 2\lambda_{m-1} + \lambda_m} \alpha_n \cos(n \cdot 0) = \frac{1}{m^2}\left(1 + \frac{1}{3} + \cdots + \frac{1}{2\lambda_m - 3} + \frac{1}{2\lambda_m - 1}\right),$$

and this is asymptotically equal to

$$\frac{\ln(\lambda_m)}{2m^2},$$

as m approaches ∞. Show that if $\lambda_m = m^{m^2}$, then the Fourier series does not converge at $x = 0$. Explain the difference between the series in equation (8.10) that is used to define f and the Fourier series for f.

8.1.15. Show that if we define g by

$$g(x) = \phi(\lambda_1, 0, x) + \sum_{n=1}^\infty \frac{\phi(\lambda_n, 2\lambda_1 + 2\lambda_2 + \cdots + 2\lambda_{n-1}, n!\, x)}{n^2},$$

then g is continuous and the Fourier series for g does not converge at any point of the form $k\pi/n$, $k \in \mathbb{Z}$, $n \in \mathbb{N}$.

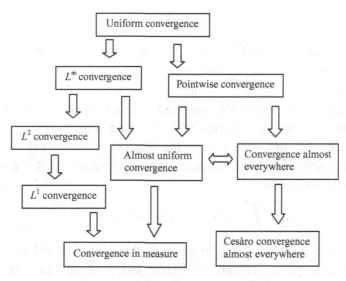

Figure 8.1. Types of convergence on $[a, b]$.

8.2 Metric Spaces

We have seen several types of convergence for a sequence of functions on a closed and bounded interval, $[a, b]$. We shall see several more in this section (see Figure 8.1). Uniform convergence is the strictest. It implies pointwise convergence. This, in turn, implies convergence almost everywhere, which we have seen is equivalent to almost uniform convergence. Convergence almost everywhere implies convergence in measure.

In the early twentieth century, mathematicians began to think of functions as objects with various means of measuring the distance between them. This in turn would lead to new ways of defining convergence, and an ultimately satisfying answer to the question of when a function has a representation as a trigonometric series that converges to that function.

In a remarkable feat of anticipation that came too early to be successful, Axel Harnack in 1882 came up with an original approach to the question of the convergence of Fourier series. He observed that if $(a_0, a_1, a_2, \ldots, b_1, b_2, \ldots)$ are the Fourier coefficients of a function f whose square is integrable on $[-\pi, \pi]$, then

$$\frac{a_0^2}{2} + \sum_{n=1}^{N} (a_n^2 + b_n^2) \leq \int_{-\pi}^{\pi} f^2(x) dx$$

for all $N \geq 1$. It follows that if we let S_n be the partial sum of the Fourier series,

$$S_n(x) = \frac{a_0}{2} + \sum_{k=1}^{n} (a_k \cos(kx) + b_k \sin(kx)),$$

then

$$\int_{-\pi}^{\pi} \left(S_n(x) - S_m(x)\right)^2 dx = \sum_{k=m+1}^{n} \left(a_k^2 + b_k^2\right), \quad m < n,$$

which converges to 0 as m approaches infinity.

Although Harnack never explicitly expressed it as such, integrating the square of the difference of these two partial sums gives us a kind of distance between these functions. From the fact that we can force the distance to be as small as we wish by going out sufficiently far along the sequence, it can be shown that this sequence has a limit g in the sense that

$$\lim_{n \to \infty} \int_{-\pi}^{\pi} \left(g(x) - S_n(x)\right)^2 dx = 0.$$

Initially, Harnack claimed that $f = g$ "in general," that is to say at all but an isolated set of points (a set of points with empty derived set). Later that year, he realized that he was wrong.

That same year, George Halphen found a function f for which the integral of $(S_n(x) - S_m(x))^2$ converges to 0 – where S_n is the partial sum of the trigonometric series formed from the Fourier coefficients of f – but the sequence S_n fails to converge at any but a single point. In other words, if we define a distance between two function as

$$D(f, g) = \int_{-\pi}^{\pi} \left(f(x) - g(x)\right)^2 dx,$$

and if f^2 is integrable, then the partial sums of the Fourier series form a Cauchy sequence under this notion of distance, and they converge, relative to this notion of distance, to a well-defined function. But, as Halphen showed, we can find a function (necessarily one for which the square is not integrable) for which the sequence of partial sums of its Fourier series is still a Cauchy sequence, relative to distance D, but the sequence of partial sums converges pointwise at only one point.

The truth, as we shall see, is that if we start with a function whose square is integrable, then the limit function g exists and equals f almost everywhere. The very statement of the result requires Lebesgue measure. The Lebesgue integral is critical to its proof.

Maurice Fréchet (1878–1973), the next player in our story, had the great fortune to be taught his high school mathematics by Jacques Hadamard at the Lycée Buffon in Paris. Even after Hadamard went on to a university position in Bordeaux, they continued to correspond, and after Hadamard returned to Paris, he became Fréchet's doctoral adviser. Fréchet served during the First World War as an interpreter attached to the British army. After teaching at the University of Strasbourg for many years, he eventually become professor of analysis and mechanics at the

École Normale Supérieure. In addition to his work in analysis, he is noted for his contributions to probability and statistics.

Real progress toward our modern understanding of convergence came in 1906, with the publication of Fréchet's doctoral dissertation, *Sur quelques points du calcul fonctionnel* (On several aspects of functional calculus). It made use of and demonstrated the power of thinking of the set of continuous functions on a closed and bounded interval as points. The distance between two continuous functions was defined as the maximum of the absolute value of their difference.

Frigyes Riesz read Fréchet's thesis with great interest, and in that same year of 1906, showed how Fréchet's use of distance between functions could be used to prove a result of Erhard Schmidt on orthogonal systems of functions, a concept that David Hilbert had devised for solving integral equations, about which there will be more to say in Section 8.4. For now, suffice it to say that Riesz, who was familiar with Harnack's attempts in 1882, recognized that with the Lebesgue integral he could attain Harnack's goal and develop a very powerful tool for analysis in the process. For functions f and g whose squares are Lebesgue integrable over the interval $[a, b]$, he defined the distance between these functions to be $(\int_a^b (f(x) - g(x))^2 dx)^{1/2}$. Riesz's proof of the convergence result for Fourier series was published in 1907. Ernst Fischer (1874–1954) discovered the same result in the same year.

In 1910, Riesz published his groundbreaking generalization, *Untersuchungen über Systeme integrierbarer Funktionen* (Analysis of a system of integrable functions), extending his analysis to the general space of functions f for which $|f|^p$ is integrable over $[a, b]$, the L^p spaces, $p \geq 1$. The term **metric space** would not come until 1914 when Felix Hausdorff laid the foundations for topology in the seminal work *Grundzüge der Mengenlehre* (A basic course in set theory).

L^p Spaces

The set of vectors in $\mathbb{R}^n$ has a lot of structure. It is closed under addition and under multiplication by any scalar:

$$(a_1, a_2, \ldots, a_n) + (b_1, b_2, \ldots, b_n) = (a_1 + b_1, a_2 + b_2, \ldots, a_n + b_n),$$
$$c(a_1, a_2, \ldots, a_n) = (ca_1, ca_2, \ldots, ca_n).$$

The set of all functions defined on $[a, b]$ also is closed under addition and scalar multiplication. Both sets have zero elements and additive inverses and satisfy the basic properties of addition and scalar multiplication.

Given any vector in $\mathbb{R}^n$, we can define its length or **norm** by

$$\|(a_1, a_2, \ldots, a_n)\| = \sqrt{a_1^2 + a_2^2 + \cdots + a_n^2}.$$

This is simply the square root of the dot product of the vector with itself. If θ is the angle between vectors $\vec{a}$ and $\vec{b}$, then

$$\vec{a} \cdot \vec{b} = \|\vec{a}\| \, \|\vec{b}\| \cos\theta, \quad \vec{a} \cdot \vec{a} = \|\vec{a}\|^2.$$

The distance between two vectors is the norm of their difference, $\|\vec{a} - \vec{b}\|$, which satisfies the following equation, also known as the **law of cosines**,

$$\begin{aligned}
\|\vec{a} - \vec{b}\|^2 &= (\vec{a} - \vec{b}) \cdot (\vec{a} - \vec{b}) \\
&= \vec{a} \cdot \vec{a} + \vec{b} \cdot \vec{b} - 2\vec{a} \cdot \vec{b} \\
&= \|\vec{a}\|^2 + \|\vec{b}\|^2 - 2\|\vec{a}\| \, \|\vec{b}\| \cos\theta.
\end{aligned} \tag{8.11}$$

The dot product is basic to working with vectors in $\mathbb{R}^n$. What makes it so useful is that it maps a pair of vectors to a real number so that two vectors that are orthogonal map to 0 and two identical unit vectors map to 1. If we define the natural basis unit vectors by

$$\vec{e}_1 = (1, 0, \ldots, 0), \; \vec{e}_2 = (0, 1, 0, \ldots, 0), \; \ldots, \; \vec{e}_n = (0, \ldots, 0, 1),$$

then the dot product satisfies

$$\vec{e}_j \cdot \vec{e}_k = \begin{cases} 0, & \text{if } j \neq k, \\ 1, & \text{if } j = k. \end{cases} \tag{8.12}$$

Combining this with a distributive law and the ability to factor out scalars, equation (8.12) uniquely defines the dot product of any two vectors,

$$\begin{aligned}
(a_1, a_2, &\ldots, a_n) \cdot (b_1, b_2, \ldots, b_n) \\
&= (a_1\vec{e}_1 + a_2\vec{e}_2 + \cdots + a_n\vec{e}_n) \cdot (b_1\vec{e}_1 + b_2\vec{e}_2 + \cdots + b_n\vec{e}_n) \\
&= a_1b_1\vec{e}_1 \cdot \vec{e}_1 + a_1b_2\vec{e}_1 \cdot \vec{e}_2 + \cdots + a_1b_n\vec{e}_1 \cdot \vec{e}_n \\
&\quad + a_2b_1\vec{e}_2 \cdot \vec{e}_1 + \cdots + a_nb_n\vec{e}_n \cdot \vec{e}_n \\
&= a_1b_1 + a_2b_2 + \cdots + a_nb_n.
\end{aligned}$$

Does this have an analog among the set of functions defined on $[a, b]$? Characteristic functions play the role of unit vectors, so we need a mapping that takes a pair of characteristic functions to a real number so that the image is 0 if and only if the characteristic functions are "orthogonal." A natural characterization of orthogonality would be if the sets are disjoint. Specifically, we define an **inner product** of the characteristic functions for sets S and T as the measure of $S \cap T$,

$$\langle \chi_S, \chi_T \rangle = m(S \cap T).$$

Definition: Inner product

If f and g are integrable functions over $[a, b]$, then we define the **inner product** of f and g, denoted $\langle f, g \rangle$, by

$$\langle f, g \rangle = \int_a^b f(x) \, g(x) dx.$$

We can use linearity to extend this to inner products of simple functions. If $\phi = \sum a_i \chi_{S_i}$, $\psi = \sum b_j \chi_{T_j}$, then

$$\langle \phi, \psi \rangle = \sum_{i,j} a_i b_j m \left(S_i \cap T_j \right) = \int \phi(x) \psi(x) \, dx.$$

If f and g are measurable, then Theorem 6.6 guarantees sequences of simple functions that converge to f and g, $\phi_n \to f$, $\psi_n \to g$. We define

$$\langle f, g \rangle = \lim_{n \to \infty} \langle \phi_n, \psi_n \rangle = \lim_{n \to \infty} \int \phi_n(x) \psi_n(x) \, dx.$$

If fg is integrable, then this limit exists and equals

$$\int \left(\lim_{n \to \infty} \phi_n(x) \psi_n(x) \right) dx = \int f(x) g(x) \, dx.$$

If f and g are each integrable, then so is their product.

Now that we have an inner product, we can define the norm of a function and the distance between two functions. If f^2 is integrable on $[a, b]$, we define the **norm** of f over $[a, b]$ by

$$\|f\| = \langle f, f \rangle^{1/2} = \left(\int_a^b f(x)^2 \, dx \right)^{1/2}.$$

We define the **distance** between two functions f and g, both integrable over $[a, b]$, by

$$d(f, g) = \|f - g\|.$$

Notice that if $f = g$ almost everywhere, then the distance between f and g is zero. If the distance between f and g is zero, then $(f - g)^2 = 0$ almost everywhere, so $f = g$ almost everywhere. Because it is common to insist that two objects are the same if the distance between them is zero, we shall assume that f and g are the same if they are equal almost everywhere. To be specific, we work with equivalence classes of integrable functions over $[a, b]$ where two functions are equivalent if and only if they are equal almost everywhere.

Definition: L^p

The space L^p, $p \geq 1$, consists of all functions f for which f^p is integrable over $[a, b]$, together with a norm defined by

$$\|f\|_p = \left(\int_a^b |f(x)|^p \, dx \right)^{1/p},$$

and a distance defined by

$$d_p(f, g) = \|f - g\|_p.$$

We consider f and g to be identical if they are equal almost everywhere.

Definition: L^∞

The space L^∞ consists of all functions f that are bounded almost everywhere over $[a, b]$, together with a norm defined by

$$\|f\|_\infty = \inf \left\{ \alpha \mid |f(x)| < \alpha \text{ almost everywhere} \right\},$$

and a distance defined by

$$d_\infty(f, g) = \|f - g\|_\infty.$$

Again, f and g are considered identical if they are equal almost everywhere.

The set of functions f for which f^2 is integrable over $[a, b]$, equipped with the distance function we have just defined, is denoted by L^2, or $L^2[a, b]$ if we need to specify the interval. More generally, Riesz defined L^p as shown above.

We need to check that these really are vector spaces. The properties of vector spaces are given on p. 18. Most of these properties are easily seen to be satisfied, and we leave these for Exercise 8.2.18. We shall prove closure under addition.

Proposition 8.3 (L^p Closed under Addition). *If $f, g \in L^p$, then so is $f + g$.*

Proof. Since f and g are measurable, so is $f + g$. We only need to verify that the integral of $|f + g|^p$ is finite. This follows because

$$|f + g|^p \leq (|f| + |g|)^p \leq (2 \max\{|f|, |g|\})^p$$
$$\leq 2^p \max\{|f|^p, |g|^p\} \leq 2^p(|f|^p + |g|^p),$$

which is integrable. $\square$

In addition to $p = 2$, the most important cases of L^p spaces are $p = 1$ and the limit as $p \to \infty$.

Definition: Norm

Given a vector space V and a mapping N from V to $\mathbb{R}$, we say that N is a **norm** if it satisfies the following properties:

1. $N(x) \geq 0$ for all $x \in V$.
2. $N(x) = 0$ if and only if $x = 0$.
3. $N(\alpha x) = |\alpha| N(x)$ for all $x \in V$ and $\alpha \in \mathbb{R}$.
4. $N(x + y) \leq N(x) + N(y)$ for all $x, y \in V$.

Definition: Metric space

A **metric space** is a set M together with a distance function d that assigns a real number to each pair of elements of M such that

1. $d(x, y) \geq 0$ for all $x, y \in M$ (positivity),
2. $d(x, y) = 0$, if and only if $x = y$ (nondegeneracy),
3. $d(x, y) = d(y, x)$ for all $x, y \in M$ (symmetry), and
4. $d(x, y) + d(y, z) \geq d(x, z)$ for all $x, y \in M$ (triangle inequality).

In this case, it is easy to see that this space is closed under addition and scalar multiplication (see Exercise 8.2.19).

We have constructed examples of norms on sets of functions and then used these norms to define distance. Our definition of an L^p space can be used with $0 < p < 1$ to define a vector space. The problem with these values of p is that the resulting norm does not satisfy the fourth of the required properties of a norm, the triangle inequality (see Exercises 8.2.26–8.2.28).

The last of these conditions, the **triangle inequality**, is the only property of the L^p norms that does not follow immediately from the definition. Later in this section we shall see how to prove it.

Convergence

Convergence of (f_k) to f in L^p means that given any $\epsilon > 0$, we can find a response K so that for any $k \geq K$, we have that

$$d_p(f_k, f) = \| f_k - f \|_p < \epsilon.$$

In order to be able to work with such a definition, we need some basic assumptions about how distance works, given in the following definition of a **metric space**.

Whenever distance is defined in terms of a norm, $d(x, y) = N(x - y)$, it will satisfy these conditions. Therefore, L^p is a metric space if we can show that its norm satisfies the triangle inequality.

For $p = 1$, the triangle inequality is just the real number inequality $|a + b| \le |a| + |b|$. I leave the case $p = \infty$ for Exercise 8.2.3. For finite $p \ge 1$, the triangle inequality is equivalent to the following result of Hermann Minkowski (1864–1909), who proved it for series in 1896, and Riesz, who proved it for integrals in 1910.

Theorem 8.4 (Minkowski–Riesz Inequality). *If $f, g \in L^p$, $1 \le p < \infty$, then*

$$\left(\int_a^b |f(x) + g(x)|^p \, dx \right)^{1/p} \le \left(\int_a^b |f(x)|^p \, dx \right)^{1/p} + \left(\int_a^b |g(x)|^p \, dx \right)^{1/p}.$$

$$(8.13)$$

As we shall see, this will follow from the next inequality.

Lemma 8.5 (Hölder–Riesz Inequality). *For $p, q > 1$ such that $1/p + 1/q = 1$, we take any $f \in L^p$, $g \in L^q$. We shall always have that $fg \in L^1$ and*

$$\int_a^b |f(x)g(x)| \, dx \le \left(\int_a^b |f(x)|^p \, dx \right)^{1/p} \left(\int_a^b |g(x)|^q \, dx \right)^{1/q}. \qquad (8.14)$$

Proof. Since f and g are measurable, so is fg. Inequality (8.14) implies that fg is integrable. We only need to prove inequality (8.14).

In Exercise 8.2.21, you are asked to verify that the equation $1/p + 1/q = 1$ is equivalent to $p - 1 = 1/(q - 1)$, to $(q - 1)p = q$, and to $(p - 1)q = p$. Using this equivalence, we see that for positive x, the function $f(x) = xy - x^p/p$ has its maximum at $x = y^{1/(p-1)} = y^{q-1}$ (see Exercise 8.2.22). Therefore,

$$xy - \frac{x^p}{p} \le y^q - \frac{y^{p(q-1)}}{p} = y^q \left(1 - \frac{1}{p} \right) = \frac{y^q}{q}.$$

Substituting α for x and β for y with $\alpha, \beta > 0$, we have that

$$\alpha\beta \le \frac{\alpha^p}{p} + \frac{\beta^q}{q}. \qquad (8.15)$$

We now set

$$A = \left(\int_a^b |f(x)|^p \, dx \right)^{1/p}, \quad B = \left(\int_a^b |g(x)|^q \, dx \right)^{1/q},$$

$$A^p = \int_a^b |f(x)|^p \, dx, \quad B^q = \int_a^b |g(x)|^q \, dx.$$

We use inequality (8.15) with $\alpha = |f(x)|/A$, $\beta = |g(x)|/B$:

$$\frac{1}{AB} \int_a^b |f(x)g(x)| \, dx = \int_a^b \frac{|f(x)|}{A} \frac{|g(x)|}{B} \, dx$$

$$\leq \int_a^b \frac{|f(x)|^p}{p A^p} \, dx + \int_a^b \frac{|g(x)|^q}{q B^q} \, dx$$

$$= \frac{1}{p} + \frac{1}{q} = 1.$$

This is precisely the inequality that we set out to prove. $\qquad\square$

Now we can prove Theorem 8.4.

Proof. **Minkowski–Riesz Inequality.** We recall that $(p-1)q = p$. Let M be the right-hand side of our inequality,

$$M = \left(\int_a^b |f(x)|^p \, dx \right)^{1/p} + \left(\int_a^b |g(x)|^p \, dx \right)^{1/p}.$$

We begin with the observation that

$$\int_a^b |f(x) + g(x)|^p \, dx$$

$$\leq \int_a^b |f(x) + g(x)|^{p-1} |f(x)| \, dx + \int_a^b |f(x) + g(x)|^{p-1} |g(x)| \, dx.$$

We apply the Hölder–Riesz inequality to each integral on the right side of this inequality,

$$\int_a^b |f(x) + g(x)|^p \, dx$$

$$\leq \left(\int_a^b |f(x) + g(x)|^{(p-1)q} \, dx \right)^{1/q} \left(\int_a^b |f(x)|^p \, dx \right)^{1/p}$$

$$+ \left(\int_a^b |f(x) + g(x)|^{(p-1)q} \, dx \right)^{1/q} \left(\int_a^b |g(x)|^p \, dx \right)^{1/p}$$

$$= M \left(\int_a^b |f(x) + g(x)|^{(p-1)q} \, dx \right)^{1/q}$$

$$= M \left(\int_a^b |f(x) + g(x)|^p \, dx \right)^{1/q}.$$

To get the Minkowski–Riesz inequality, we divide each side by

$$\left(\int_a^b |f(x) + g(x)|^p \, dx \right)^{1/q}$$

and remember that $1 - 1/q = 1/p$. □

Ordering L^p Spaces

The space L^1 consists of all Lebesgue integrable functions. This includes many unbounded functions such as $f(x) = x^{-1/2}$ on $[0, 1]$. The space L^∞ admits only those measurable functions that are bounded almost everywhere, a much more restricted class of functions. In general, if $p \le q$ then $L^p \supseteq L^q$. Any function in L^q is also in L^p.

Proposition 8.6 (Containment of L^p Spaces). *If $1 \le p \le q \le \infty$ and L^p, L^q are defined over the finite interval $[a, b]$, then $L^p \supseteq L^q$. Furthermore, if $q < \infty$ then*

$$\|f\|_p \le (b - a)^{(q-p)/pq} \|f\|_q. \tag{8.16}$$

Proof. Assume that $f \in L^q$. This implies that f is measurable. It is in L^p if and only if $\int_a^b |f(x)|^p \, dx$ is finite. Let $r = q/p > 1$ and define s so that $1/r + 1/s = 1$; that is, $s = r/(r - 1) = q/(q - p)$. We use the Hölder–Riesz inequality, Lemma 8.5, with $|f|^p$ and the constant function 1,

$$\int_a^b |f(x)|^p \, dx = \int_a^b |f(x)|^p \cdot 1 \, dx$$

$$\le \left(\int_a^b |f(x)|^{pr} \, dx \right)^{1/r} \left(\int_a^b 1^s \, dx \right)^{1/s}$$

$$= \left(\int_a^b |f(x)|^q \, dx \right)^{p/q} (b - a)^{(q-p)/q},$$

$$\left(\int_a^b |f(x)|^p \, dx \right)^{1/p} \le (b - a)^{(q-p)/pq} \left(\int_a^b |f(x)|^q \, dx \right)^{1/q}. \quad □$$

Exercises

8.2.1. For $1 \le p < q < \infty$, find an example of a function in L^p that is not in L^q.

8.2.2. For $1 \le p < \infty$, find an example of a function in L^p that is not in L^∞.

8.2.3. Show that if f and g are bounded almost everywhere, then

$$\inf\left\{\alpha \mid |f(x) + g(x)| < \alpha \text{ almost everywhere}\right\}$$
$$\leq \inf\left\{\alpha \mid |f(x)| < \alpha \text{ almost everywhere}\right\}$$
$$+ \inf\left\{\alpha \mid |g(x)| < \alpha \text{ almost everywhere}\right\}.$$

8.2.4. Show that for $f \in L^\infty[a, b]$,

$$\|f\|_\infty = \sup\left\{B \mid m\left(\{x \in [a, b] \mid |f(x)| \geq B\}\right) \neq 0\right\}.$$

8.2.5. Show that for $f \in L^\infty[a, b]$,

$$\|f\|_\infty = \inf_{g \sim f}\left\{\sup_{x \in [a,b]} |g(x)|\right\},$$

where $g \sim f$ means that $g = f$ almost everywhere.

8.2.6. Let f be a continuous function on (a, b), and for any c, $a < c < b$, let $g = \chi_{(a,c)}$, the characteristic function of (a, c). Show that

$$\|f - g\|_\infty \geq \frac{1}{2}.$$

8.2.7. Show that convergence in the L^∞ norm is uniform convergence almost everywhere. That is to say, show that if $f_n \to f$ in $L^\infty[a, b]$, then there exists a set S of measure zero in $[a, b]$ such that $f_n \to f$ uniformly on $[a, b] - S$, and if $f_n \to f$ uniformly almost everywhere in $[a, b]$, then $f_n \to f$ in $L^\infty[a, b]$.

8.2.8. Prove that if $f_n \to f$ in $L^\infty[a, b]$, then $f_n \to f$ in $L^p[a, b]$ for every finite $p \geq 1$.

8.2.9. Prove that if $1 \leq p < q < \infty$ and $f_n \to f$ in $L^q[a, b]$, then $f_n \to f$ in $L^p[a, b]$.

8.2.10. Prove that if $f_n \to f$ in L^1, then (f_n) converges to f in measure (see definition on p. 194).

8.2.11. Let (f_n) be a sequence in L^p that converges to f in the L^p norm. Show that if (f_n) converges pointwise almost everywhere to g, then $f = g$ almost everywhere.

8.2.12. Consider the sequence of functions

$$(f_{1,1}, f_{1,2}, f_{2,2}, f_{1,3}, f_{2,3}, f_{3,3}, f_{1,4}, \ldots),$$

where $f_{k,n}$ is the characteristic function of the open interval $\big((k-1)/n, k/n\big)$,

$$f_{k,n} = \chi_{\left(\frac{k-1}{n}, \frac{k}{n}\right)}.$$

Show that if $1 \leq p < \infty$, then this sequence converges in the L^p norm. Show that it does not converge in the L^∞ norm.

8.2.13. Show that the sequence given in Exercise 8.2.12 converges in the Cesàro sense almost everywhere.

8.2.14. Define $f_n(x) = nxe^{-nx^2}$. Show that f_n converges pointwise to the constant function 0, but this sequence does not converge in any L^p norm, $1 \le p \le \infty$. Compare to Example 4.6 on p. 100. Explain why this sequence does *not* converge to the constant function $1/2$ in the L^1 norm.

8.2.15. Find a sequence of functions that converges in the L^∞ norm but does not converge pointwise.

8.2.16. Find a sequence of functions that Cesàro converges almost everywhere but does not converge in measure.

8.2.17. Find a sequence of functions that converges in measure but does not Cesàro converge almost everywhere.

8.2.18. Verify that $L^p[a, b]$ satisfies the definition of a vector space as given on p. 18.

8.2.19. Show that the set of function in $L^\infty[a, b]$ is closed under addition and scalar multiplication.

8.2.20. Show that if x and y are nonnegative, then

$$\max\{x, y\} = \lim_{p \to \infty} \left(x^p + y^p\right)^{1/p},$$

and, in general, for nonnegative $x_1, x_2, \ldots, x_n$,

$$\max\{x_1, x_2, \ldots, x_n\} = \lim_{p \to \infty} \left(x_1^p + x_2^p + \cdots + x_n^p\right)^{1/p}.$$

8.2.21. Show that if $p, q > 1$, then $1/p + 1/q = 1$ is equivalent to

$$p - 1 = \frac{1}{q - 1} \quad \text{and} \quad (p - 1)q = p.$$

8.2.22. Show that the function $f(x) = xy - x^p/p$, $x > 0$ has its maximum at $x = y^{1/(p-1)} = y^{q-1}$.

8.2.23. Show that equality in the Hölder–Riesz inequality holds if and only if there exist nonnegative numbers r and s such that

$$r \left|f(x)\right|^p = s \left|g(x)\right|^q.$$

8.2.24. The $p = 1$ case of the Hölder–Riesz inequality is

$$\int_a^b \left|f(x)g(x)\right| dx \le \int_a^b |f(x)| \, dx \, \|g\|_\infty. \tag{8.17}$$

Prove this inequality.

8.2.25. If we take the limit $q \to \infty$ of each side of inequality (8.16), we get an inequality between $\|f\|_p$ and $\|f\|_\infty$, $1 \le p < \infty$. State and prove this inequality.

Exercises 8.2.26–8.2.28 establish the fact that the triangle inequality does not hold for L^p spaces when $0 < p < 1$.

8.2.26. Let $f \in L^p[a, b]$, where $0 < p < 1$, and $g \in L^q[a, b]$, $q = p/(p-1) < 0$ and, where $f \ge 0$ and $g > 0$ on $[a, b]$. Show that

$$\int_a^b |f(x)g(x)| \, dx \ge \left(\int_a^b |f(x)|^p \, dx \right)^{1/p} \left(\int_a^b |g(x)|^q \, dx \right)^{1/q}.$$

8.2.27. Let $f, g \in L^p$, where $0 < p < 1$ and $f, g \ge 0$ on $[a, b]$. Show that

$$\|f + g\|_p \ge \|f\|_p + \|g\|_p.$$

8.2.28. For $0 < p < 1$, show that there exist functions $f, g \in L_p[a, b]$ such that

$$\|f + g\|_p > \|f\|_p + \|g\|_p.$$

8.2.29. For $n \in \mathbb{N}$, define

$$f_n(x) = \left(n(n+1) \right)^{1/p} \chi_{\left(\frac{1}{n+1}, \frac{1}{n} \right)}.$$

Show that for each pair of distinct integers, m, n, the L^p distance between f_m and f_n is 2, $\|f_m - f_n\|_p = 2$, and therefore this is a bounded sequence in L^p that does not have a limit point in L^p.

8.3 Banach Spaces

The L^p spaces are equipped with a norm and therefore a definition of distance. We can begin to explore questions of convergence. As we have seen for sequences of real numbers, we often find ourselves in the situation where we want to prove convergence but we do not know the limit. Therefore, we cannot prove convergence by showing that the terms of the sequence will eventually stay as close as we wish to this limit. We need to resort to the Cauchy criterion, that if we can force the terms of the sequence to be as close together as we wish by going far enough out, there must be something to which the sequence converges.

We never really proved that the Cauchy criterion is valid for sequences of real numbers. It is a property we expect of the real number line, and we take it as an axiom that every Cauchy sequence converges. We would like this Cauchy criterion to be true for sequences in our L^p spaces, but we can no longer just assume that it is true. Now we need to prove it. We shall show that every L^p space is a **Banach space**.

Definition: Banach space

A vector space equipped with a norm is called a **Banach space** if it is complete, that is to say, if every Cauchy sequence converges.

The name used to designate complete metric spaces was chosen to honor one of the founders of functional analysis, Stefan Banach (1892–1945). He was born in Kraków in what was then Austria-Hungary, today Poland. In 1920, he began teaching at Lvov Technical University in what was then Poland and is now Ukraine. Until the Nazi occupation of Lvov in 1941, he was a prolific and important mathematician. Imprisoned briefly by the Nazis, he spent the remainder of the war feeding lice in Rudolf Stefan Weigl's Typhus Institute.[2] Banach died of lung cancer shortly after the war ended.

The proof that L^p is complete for every p, $1 \leq p \leq \infty$, is known as the Riesz–Fischer theorem. Riesz and Fischer each proved it independently for $p = 2$ in 1907. The remaining cases were established by Riesz in 1910.

Theorem 8.7 (L^p **is Banach**). *For $1 \leq p \leq \infty$, L^p is a Banach space.*

Proof. We shall first prove the case $p = \infty$. To say that a sequence (f_n) is Cauchy in the L^∞ norm means that given any $\epsilon > 0$, we can find a response N so that $m, n \geq N$ implies that $\| f_m - f_n \|_\infty < \epsilon$. This means that $|f_m(x) - f_n(x)| < \epsilon$ almost everywhere. We eliminate the values of x in the following sets:

$$A_k = \left\{ x \in [a, b] \,\middle|\, |f_k(x)| > \| f_k \|_\infty \right\},$$
$$B_{m,n} = \left\{ x \in [a, b] \,\middle|\, |f_m(x) - f_n(x)| > \| f_m - f_n \|_\infty \right\}.$$

From the definition of the L^∞ norm, each of these sets has measure zero (see Exercise 8.3.2). There are countably many of these sets, so their union also has measure zero,

$$F = \left(\bigcup_{k=1}^\infty A_k \right) \cup \left(\bigcup_{1 \leq m < n < \infty} B_{m,n} \right), \quad m(F) = 0.$$

For each x in what remains, $[a, b] - F$, we have that $m, n \geq N$ implies that

$$|f_m(x) - f_n(x)| \leq \| f_m - f_n \|_\infty < \epsilon.$$

---

[2] For more information on this unpleasant occupation and how it was used to save the lives of a number of Polish and Ukrainian intellectuals during World War II, see www.lwow.home.pl/Weigl.html.

The sequence $(f_k(x))_{k=1}^\infty$ is a Cauchy sequence, and so it converges to a finite value we can call $f(x)$. We can define f however we wish for $x \in F$, say $f(x) = 0$. It follows that $f_k \to f$ almost everywhere. It only remains to show that $f \in L^\infty$. Since f is a limit almost everywhere of measurable functions, it is measurable. We have to show that there is a bound on f that holds almost everywhere.

Let N be the Cauchy criterion response to $\epsilon = 1$ and let

$$\beta = \max \{\|f_1\|_\infty, \|f_2\|_\infty, \ldots, \|f_N\|_\infty\}.$$

Then for $x \in [a, b] - F$ and for all $k \geq 1$

$$|f_k(x)| \leq \|f_k\|_\infty \leq \beta + 1.$$

It follows that $|f(x)| \leq \beta + 1$ almost everywhere, and therefore $f \in L^\infty$.

We now consider the case of finite p. To say that $(f_n)_{n=1}^\infty$ is Cauchy in the L^p norm means that for any $\epsilon > 0$, we can find a response N so that $m, n \geq N$ implies that $\|f_m - f_n\|_p < \epsilon$. We construct a subsequence of the f_k as follows. Choose n_1 so that $n \geq n_1$ implies that

$$\|f_n - f_{n_1}\|_p < \frac{1}{2}.$$

We choose $n_2 > n_1$ so that $n \geq n_2$ implies that

$$\|f_n - f_{n_2}\|_p < \frac{1}{2^2}.$$

In general, once we have found n_{k-1}, we choose $n_k > n_{k-1}$ so that $n \geq n_k$ implies that

$$\|f_n - f_{n_k}\|_p < \frac{1}{2^k}.$$

It follows that

$$\|f_{n_1}\|_p + \sum_{j=1}^\infty \|f_{n_{j+1}} - f_{n_j}\|_p \leq \|f_{n_1}\|_p + 1 < \infty.$$

We shall use the fact that this sum is bounded.

We create a new sequence of functions by

$$g_k = |f_{n_1}| + \sum_{j=1}^k |f_{n_{j+1}} - f_{n_j}|.$$

By the Minkowski–Riesz inequality, Theorem 8.4 on p. 258, we have a finite bound on $\int_a^b g_k^p(x)\,dx$ that does not depend on k,

$$
\begin{aligned}
\int_a^b \left(g_k(x)\right)^p dx &= \int_a^b \left(|f_{n_1}(x)| + \sum_{j=1}^k |f_{n_{j+1}}(x) - f_{n_j}(x)|\right)^p dx \\
&\le \left(\|f_{n_1}\|_p + \sum_{j=1}^k \|f_{n_{j+1}} - f_{n_j}\|_p\right)^p \\
&\le \left(\|f_{n_1}\|_p + 1\right)^p < \infty.
\end{aligned}
\tag{8.18}
$$

Using Fatou's lemma, Theorem 6.20 on p. 187,

$$
\int_a^b \left(|f_{n_1}(x)| + \sum_{j=1}^\infty |f_{n_{j+1}}(x) - f_{n_j}(x)|\right)^p dx \le \lim_{k\to\infty} \int_a^b \left(g_k(x)\right)^p dx < \infty.
\tag{8.19}
$$

By the monotone convergence theorem, Theorem 6.14 on p. 174, the sequence of functions

$$
g_k^p = \left(|f_{n_1}| + \sum_{j=1}^k |f_{n_{j+1}} - f_{n_j}|\right)^p
$$

converges almost everywhere, and therefore so does the sequence of functions

$$
g_k = |f_{n_1}| + \sum_{j=1}^k |f_{n_{j+1}} - f_{n_j}|.
$$

We know that if a sequence of real values converges absolutely, then it converges (see Theorem 1.9). Therefore the sequence of functions

$$
f_{n_1} + \sum_{j=1}^k \left(f_{n_{j+1}} - f_{n_j}\right) = f_{n_{k+1}}
$$

converges almost everywhere.

We have found a subsequence of (f_n) that converges almost everywhere. As before, we define f by $f(x) = \lim_{k\to\infty} f_{n_k}(x)$ where our sequence converges, and define f any way we want, say $f(x) = 0$, on the set of measure zero where the subsequence does not converge. From equation (8.19),

$$\int_a^b |f(x)|^p \, dx = \int_a^b \left| f_{n_1}(x) + \sum_{j=1}^\infty \left(f_{n_{j+1}}(x) - f_{n_j}(x) \right) \right|^p \, dx$$

$$\leq \int_a^b \left(|f_{n_1}(x)| + \sum_{j=1}^\infty |f_{n_{j+1}}(x) - f_{n_j}(x)| \right)^p \, dx < \infty,$$

and therefore $f \in L^p$. It only remains to prove that f_n converges to f in the sense of the L^p norm.

We observe that

$$f(x) - f_{n_k}(x) = \sum_{j=k}^\infty \left(f_{n_{j+1}}(x) - f_{n_j}(x) \right).$$

From the triangle inequality, it follows that

$$\|f - f_{n_k}\|_p \leq \sum_{j=k}^\infty \|f_{n_{j+1}} - f_{n_j}\|_p < \sum_{j=k}^\infty \frac{1}{2^j} = \frac{1}{2^{k-1}}.$$

Given any $\epsilon > 0$, choose k so that $\epsilon > 3/2^k$. For all $n \geq n_k$, we see that

$$\|f - f_n\|_p \leq \|f - f_{n_k}\|_p + \|f_{n_k} - f_n\|_p < \frac{1}{2^{k-1}} + \frac{1}{2^k} = \frac{3}{2^k} < \epsilon.$$

$\square$

The Riesz–Fischer Theorem

Recall that we began our study of L^2 with the observation that we had an inner product. We then used the inner product to define the norm, and from the norm we defined distance. But we do not want to lose that inner product, because inner products enable us to write vectors in terms of basis vectors.

For example, in $\mathbb{R}^3$, we can use $\vec{e}_1 = (1, 0, 0)$, $\vec{e}_2 = (0, 1, 0)$, and $\vec{e}_3 = (0, 0, 1)$ as our basis vectors. Any two of these are orthogonal, and every vector is a unique linear combination of these three vectors. We could use other basis vectors. For example,

$$\vec{v}_1 = (1, -1, 2), \quad \vec{v}_2 = (2, 0, -1), \quad \vec{v}_3 = (1, 5, 2).$$

In Exercise 8.3.1, you are asked to verify that each pair from $\{\vec{v}_1, \vec{v}_2, \vec{v}_3\}$ is orthogonal. Each vector has a unique decomposition into a linear combination of these three vectors, and we can use the dot product to find this decomposition. The component of $\vec{v}$ in the direction of $\vec{v}_k$ is

$$\frac{\vec{v} \cdot \vec{v}_k}{\vec{v}_k \cdot \vec{v}_k} \vec{v}_k.$$

Thus, if $\vec{v} = (1, 2, 3)$, the coefficient of $\vec{v}_1$ is

$$\frac{1 \cdot 1 + 2 \cdot (-1) + 3 \cdot 2}{1 \cdot 1 + (-1) \cdot (-1) + 2 \cdot 2} = \frac{5}{6}.$$

We can write $(1, 2, 3)$ in terms of our basis as

$$(1, 2, 3) = \frac{5}{6} (1, -1, 2) + \frac{-1}{5} (2, 0, -1) + \frac{17}{30} (1, 5, 2).$$

Now think about the Fourier series representation of a function in $L^2[-\pi, \pi]$, say $f(x) = x$:

$$f(x) = 2 \sin(x) - \frac{2}{2} \sin(2x) + \frac{2}{3} \sin(3x) - \frac{2}{4} \sin(4x) + \cdots. \tag{8.20}$$

The trigonometric functions used in the Fourier series are

$$1, \cos x, \cos 2x, \ldots, \sin x, \sin 2x, \ldots$$

These functions are orthogonal using the L^2 inner product! For $n \geq 1$, we note that

$$\int_{-\pi}^{\pi} 1 \cdot \cos(nx) \, dx = \left. \frac{1}{n} \sin(nx) \right|_{-\pi}^{\pi} = 0,$$

$$\int_{-\pi}^{\pi} 1 \cdot \sin(nx) \, dx = \left. \frac{-1}{n} \cos(nx) \right|_{-\pi}^{\pi} = 0,$$

$$\int_{-\pi}^{\pi} \sin(nx) \cdot \cos(nx) \, dx = \left. \frac{-\cos^2(nx)}{2n} \right|_{-\pi}^{\pi} = 0.$$

For $m \neq n$, we observe that

$$\int_{-\pi}^{\pi} \cos(mx) \cdot \cos(nx) \, dx = \left. \frac{\sin((m-n)x)}{2(m-n)} + \frac{\sin((m+n)x)}{2(m+n)} \right|_{-\pi}^{\pi} = 0,$$

$$\int_{-\pi}^{\pi} \sin(mx) \cdot \sin(nx) \, dx = \left. \frac{\sin((m-n)x)}{2(m-n)} - \frac{\sin((m+n)x)}{2(m+n)} \right|_{-\pi}^{\pi} = 0,$$

$$\int_{-\pi}^{\pi} \sin(mx) \cdot \cos(nx) \, dx = \left. \frac{-\cos((m-n)x)}{2(m-n)} - \frac{\cos((m+n)x)}{2(m+n)} \right|_{-\pi}^{\pi} = 0.$$

It appears that these functions form a basis, but one that is infinite. Even so, we should be able to use the inner product to find the coefficients:

$$\frac{\langle x, \cos(nx) \rangle}{\langle \cos(nx), \cos(nx) \rangle} = \frac{\int_{-\pi}^{\pi} x \cos(nx)\, dx}{\int_{-\pi}^{\pi} \cos^2(nx)\, dx} = 0,$$

$$\frac{\langle x, \sin(nx) \rangle}{\langle \sin(nx), \sin(nx) \rangle} = \frac{\int_{-\pi}^{\pi} x \sin(nx)\, dx}{\int_{-\pi}^{\pi} \sin^2(nx)\, dx}$$

$$= \frac{(-1)^{n+1} \pi / n}{\pi} = \frac{(-1)^{n+1}}{n}.$$

The Fourier series expansion of our function f defined by $f(x) = x, -\pi < x < \pi$, is simply the representation of f in terms of this orthogonal basis of sines and cosines.

The amazing result discovered by Fischer and Riesz is that *every* function in L^2 has such a representation and every suitably convergent trigonometric series corresponds to a function in L^2.

Theorem 8.8 (Riesz–Fischer Theorem). *Let $f \in L^2[-\pi, \pi]$, then f has a unique Fourier series representation*

$$f(x) = \frac{a_0}{2} + \sum_{n=1}^{\infty} \left(a_n \cos(nx) + b_n \sin(nx) \right),$$

where

$$a_n = \frac{\langle f(x), \cos(nx) \rangle}{\langle \cos(nx), \cos(nx) \rangle}, \; n \geq 0, \quad b_n = \frac{\langle f(x), \sin(nx) \rangle}{\langle \sin(nx), \sin(nx) \rangle}, \; n \geq 1.$$

The convergence of this series is convergence in the sense of the L^2 norm. Furthermore,

$$\frac{a_0^2}{4} + \sum_{k=1}^{\infty} \left(a_k^2 + b_k^2 \right) < \infty. \tag{8.21}$$

The implication also goes the other way. If $(a_0, a_1, b_1, a_2, b_2, \ldots)$ is any sequence of real numbers for which $\frac{a_0^2}{4} + \sum_{k=1}^{\infty} \left(a_k^2 + b_k^2 \right)$ converges, then

$$\frac{a_0}{2} + \sum_{n=1}^{\infty} \left(a_n \cos(nx) + b_n \sin(nx) \right)$$

is a function in L^2.

We shall prove this theorem in the next (and last) section.

Comparing this to results on Fourier series in Section 8.1, we see that we have strengthened Lebesgue's assumption. Instead of simply being integrable, we insist that the square of the function must be integrable. In exchange, we get a much stronger conclusion. We no longer need the Cesàro limit. Nevertheless, we get convergence only in the L^2 norm. But in 1966, the Swedish mathematician Lennart Carleson (b. 1928) showed that convergence is not just in the L^2 norm, it is pointwise convergence almost everywhere. In 2006, Carleson received the Abel Prize "for his profound and seminal contributions to harmonic analysis and the theory of smooth dynamical systems." In 1970, Richard A. Hunt of Purdue University showed that there is nothing special about L^2. The same is true for functions in any L^p space, provided only that p is strictly greater than 1. All of the problematic functions that require a Cesàro limit live in L^1 but not in L^p, $p > 1$.

Theorem 8.9 (Carleson–Hunt Theorem). *If $f \in L^p$, $p > 1$, then the Fourier series representation of f converges almost everywhere to f.*

The proof of the Carleson–Hunt theorem is beyond the scope of this book. The remainder of this chapter will be devoted to proving the Riesz–Fischer theorem. Because it requires little additional effort to prove a far more general result, I shall present this proof in the context of Hilbert spaces.

Exercises

8.3.1. Verify that each pair of the vectors

$$\vec{v}_1 = (1, -1, 2), \quad \vec{v}_2 = (2, 0, -1), \quad \vec{v}_3 = (1, 5, 2)$$

is orthogonal.

8.3.2. Show that for functions in $L^\infty[a, b]$, each of the following sets has measure 0:

$$A_k = \left\{ x \in [a, b] \,\big|\, |f_k(x)| > \|f_k\|_\infty \right\},$$
$$B_{m,n} = \left\{ x \in [a, b] \,\big|\, |f_m(x) - f_n(x)| > \|f_m - f_n\|_\infty \right\}.$$

8.3.3. Let $C = C[0, 1]$ be the set of all continuous functions on $[0, 1]$. For $f \in C$, define the max norm by $\|f\|_{\max} = \max |f(x)|$. Show that C equipped with the max norm is a Banach space.

8.3.4. Let $C = C[0, 1]$ be the set of all continuous functions on $[0, 1]$. Show that C equipped with the L^2 norm, $\|f\|_2 = \int_0^1 f^2(x) \, dx$, is not a Banach space.

8.3.5. Show that for $f \in L^p[a, b]$, $1 \le p < \infty$ and any $\epsilon > 0$, there exists a continuous function ϕ and a step function ψ such that $\|f - \phi\|_p < \epsilon$ and $\|f - \psi\|_p < \epsilon$.

8.3.6. Let (f_n) be a sequence of function in $L^p[a, b]$, $1 \leq p < \infty$, that converges almost everywhere to $f \in L^p[a, b]$. Show that (f_n) converges to f in the L^p norm if and only if $\|f_n\|_p \to \|f\|_p$.

8.3.7. We say that a sequence (f_n) of integrable functions is **equi-integrable** on $[a, b]$ if for each $\epsilon > 0$ there is a response $\delta > 0$ so that for any $S \subseteq [a, b]$ with $m(S) < \delta$, we have that $\int_S |f_n(x)| \, dx < \epsilon$ for all $n \geq 1$. Show that if (f_n) is an equi-integrable sequence that converges almost everywhere to f, then

$$\int_a^b f(x) \, dx = \lim_{n \to \infty} \int_a^b f_n(x) \, dx.$$

8.3.8. Let (f_n) be a sequence of functions in $L^p[a, b]$, $1 < p < \infty$, that converges almost everywhere to $f \in L^p[a, b]$. Show that if the sequence of norms is bounded, $\|f_n\|_p \leq M$ for all $n \geq 1$, then for all $g \in L^q[a, b]$, $q = p/(p - 1)$, we have that

$$\int_a^b f(x)g(x) \, dx = \lim_{n \to \infty} \int_a^b f_n(x)g(x) \, dx.$$

Identify how $p > 1$ is used in your proof and give an example that shows that this result is not correct for $(f_n) \subseteq L^1[a, b]$, $g \in L^\infty[a, b]$.

8.3.9. Let (f_n) be a sequence of functions that converges in the L^p norm to f, $1 \leq p < \infty$. Let (g_n) be a sequence of measurable functions such that $|g_n| \leq M$, $n \geq 1$, and $g_n \to g$ almost everywhere. Show that $g_n f_n$ converges to gf in the L^p norm.

8.4 Hilbert Spaces

We begin with the definition of a Hilbert space. Note that the inner product on L^2 satisfies all of the criteria.

It is ironic that these Banach spaces equipped with an inner product are today called Hilbert spaces, because Hilbert resisted this formalism. Nevertheless, it was

---

Definition: Hilbert space

A **Hilbert space** is a Banach space equipped with an **inner product**, $\langle x, y \rangle$, that maps pairs of elements to real numbers and satisfies the following properties:

1. $\langle x, y \rangle = \langle y, x \rangle$ (symmetry).
2. for $\alpha, \beta \in \mathbb{R}$, $\langle \alpha x + \beta y, z \rangle = \alpha \langle x, z \rangle + \beta \langle y, z \rangle$ (linearity).
3. $\langle x, x \rangle \geq 0$ (nonegativity).
4. $\langle x, x \rangle = \|x\|^2$ (defines norm).

his work on orthogonal systems of functions, especially in finding solutions of the general integral equation

$$f(s) = \phi(s) + \int_a^b K(s, t)\phi(t)\,dt,$$

where f and K are given and the task is to find an appropriate function ϕ, that would lead others to develop this concept.

David Hilbert (1862–1943) was from Königsberg (modern Kaliningrad in the little piece of Russia between Poland and Lithuania), earned his doctorate under the direction of Ferdinand Lindemann in 1885, and in 1895 became chair of mathematics at the University of Göttingen. He was probably the most influential and highly acclaimed German mathematician of the early twentieth century. In addition to his work on integral equations and the related field of calculus of variations, he is noted for his fundamental contributions to invariant theory, algebraic number theory, mathematical physics, and, especially, geometry.

We first observe a basic inequality and identity for Hilbert spaces. Recall that the dot product in $\mathbb{R}^n$ satisfies

$$\vec{v} \cdot \vec{w} = \|\vec{v}\| \, \|\vec{w}\| \, \cos\theta,$$

where θ is the angle between $\vec{v}$ and $\vec{w}$. It follows that

$$|\vec{v} \cdot \vec{w}| \le \|\vec{v}\| \, \|\vec{w}\|.$$

This is true of any inner product.

Proposition 8.10 (Cauchy–Schwarz–Bunyakovski Inequality). *The inner product of any Hilbert space satisfies the inequality*

$$|\langle x, y \rangle| \le \|x\| \, \|y\|.$$

This result was proved for $\mathbb{R}^\infty$ by Cauchy in 1821 and for L^2 (though they did not call it that) independently by Victor Bunyakovsky in 1859 and Hermann A. Schwarz in 1885.

Proof. By linearity, $\langle x, 0 \rangle = 0$. We now assume $x, y \ne 0$, set $\lambda = \|x\|/\|y\|$, and observe that

$$
\begin{aligned}
0 &\le \langle x - \lambda y, x - \lambda y \rangle \\
&= \langle x, x \rangle + \lambda^2 \langle y, y \rangle - 2\lambda \langle x, y \rangle \\
&= 2\|x\|^2 - 2\frac{\|x\|}{\|y\|} \langle x, y \rangle, \\
0 &\le \|x\| \, \|y\| - \langle x, y \rangle. \qquad \square
\end{aligned}
$$

The parallelogram law says that the sum of the squares of the diagonals of a parallelogram equals the sum of the squares of the sides,

$$\|\vec{v} + \vec{w}\|^2 + \|\vec{v} - \vec{w}\|^2 = 2\left(\|\vec{v}\|^2 + \|\vec{w}\|^2\right).$$

This is also true of any Hilbert space, and the proof is the same.

Proposition 8.11 (Parallelogram Law). *The inner product of any Hilbert space satisfies the equality*

$$\|x + y\|^2 + \|x - y\|^2 = 2\left(\|x\|^2 + \|y\|^2\right). \tag{8.22}$$

This proof is left as Exercise 8.4.2. Equipped with this result, we can quickly verify that L^2 is the only L^p space whose norm arises from an inner product.

Proposition 8.12 (L^2 Alone). *The only L^p space that is also a Hilbert space is L^2.*

Proof. In $L^p[a, b]$, take two subsets, $S, T \subseteq [a, b]$, such that $S \cap T = \emptyset$ and $m(S) = m(T) \neq 0$. Define $\lambda = m(S)^{-1/p}$. In Exercise 8.4.3, you are asked to show that

$$\|\lambda\chi_S\|_p = 1, \quad \|\lambda\chi_T\|_p = 1, \quad \|\lambda\chi_S + \lambda\chi_T\|_p = 2^{1/p}, \quad \|\lambda\chi_S - \lambda\chi_T\|_p = 2^{1/p}. \tag{8.23}$$

The parallelogram law for this example becomes $2^{2/p} + 2^{2/p} = 2(1^2 + 1^2)$, which is true only when $p = 2$. $\qquad\square$

Complete Orthogonal Set

We know that the set $\{1, \cos x, \sin x, \cos 2x, \sin 2x, \ldots\}$ is an orthogonal set in L^2. Is anything missing? Is there any function (other than the constant function 0) that is orthogonal to all of these in L^2? In other words, do we have a **complete** orthogonal set?

Definition: Orthogonality

If x and y are elements of a Hilbert space such that $\langle x, y \rangle = 0$, then we say that x and y are **orthogonal**. A set Ω of elements of a Hilbert space is called an **orthogonal set** if for each pair $x \neq y \in \Omega$, we have that $\langle x, y \rangle = 0$.

Definition: Complete orthogonal set

An orthogonal set Ω is **complete** if $\langle x, y \rangle = 0$ for all $y \in \Omega$ means that $x = 0$.

Theorem 8.13 (L^2 Complete Orthogonal Set). *The set*

$$\{1, \cos x, \sin x, \cos 2x, \sin 2x, \cos 3x, \ldots\}$$

is a complete orthogonal set in $L^2[-\pi, \pi]$.

Our proof will be spread over Lemmas 8.14–8.16 and will proceed by contradiction. We assume that we have a nonzero function $f \in L^2[-\pi, \pi]$ for which

$$\int_{-\pi}^{\pi} f(x) \cos nx \, dx = 0, \; n \geq 0, \quad \text{and} \quad \int_{-\pi}^{\pi} f(x) \sin nx \, dx = 0, \; n \geq 1.$$

The proof breaks up into three pieces. We first establish the existence of finite trigonometric series with certain properties that help us isolate the value of f near specific points. Next, we use the existence of such a finite series with the fact that f is continuous to find a contradiction. Finally, we pull this all together to find a contradiction when all we assume about f is that it is integrable. Notice that we need to prove our result only for $f \in L^2$. What we shall actually prove is stronger than we need.

Lemma 8.14 (Special Trigonometric Polynomial). *Given any $\delta > 0$ and any $\epsilon > 0$, we can find a finite trigonometric series,*

$$T(x) = a_0 + \sum_{n=1}^{N} \left(a_n \cos(nx) + b_n \sin(nx) \right),$$

for which

1. $T(x) \geq 0$ *for all $x \in [-\pi, \pi]$,*
2. $\int_{-\pi}^{\pi} T(x) \, dx = 1$, *and*
3. $T(x) < \epsilon$ *for all $\delta \leq |x| \leq \pi$.*

These functions are all nonnegative, and the area underneath the graph is always 1, but that area is concentrated as close to zero as we wish (see Figure 8.2). The effect of integrating $f \, T$ is to pick out the values of f closest to 0. If f is nonzero at any point, say $f(z) \neq 0$, then integrating $f(x + z)T(x)$ picks out those values of f in an arbitrarily small neighborhood of z.

Proof. Define

$$T_n(x) = (1 + \cos x)^n \left(\int_{-\pi}^{\pi} (1 + \cos x)^n \, dx \right)^{-1}$$

$$= \left(\cos(x/2) \right)^{2n} \left(\int_{-\pi}^{\pi} \left(\cos(x/2) \right)^{2n} \, dx \right)^{-1}.$$

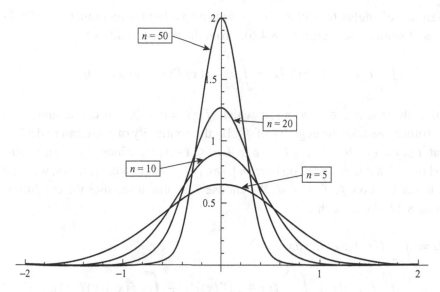

Figure 8.2. Graphs of T_5, T_{10}, T_{20}, and T_{50}.

The first two properties are clearly satisfied by this function. For the third property, we observe that for $\delta \leq |x| \leq \pi$ we have that

$$
\begin{aligned}
T_n(x) &< \left(\cos(\delta/2) \right)^{2n} \left(\int_0^{\delta/2} \left(\cos(x/2) \right)^{2n} dx \right)^{-1} \\
&< \left(\cos(\delta/2) \right)^{2n} \left(\frac{\delta}{2} \left(\cos(\delta/4) \right)^{2n} \right)^{-1} \\
&= \frac{2}{\delta} \left(\frac{\cos(\delta/2)}{\cos(\delta/4)} \right)^{2n}.
\end{aligned}
$$

Since

$$
0 < \frac{\cos(\delta/2)}{\cos(\delta/4)} < 1,
$$

we can find an n for which $T_n(x) < \epsilon$ for all $\delta \leq |x| \leq \pi$. $\qquad \square$

Lemma 8.15 (Continuous f). *If f is continuous on $[-\pi, \pi]$ and f is orthogonal to $\cos(nx)$, $n \geq 0$, and to $\sin(nx)$, $n \geq 1$, then f is the constant function 0 on this interval.*

Proof. We extend the definition of f to all of $\mathbb{R}$ by assuming that $f(x + 2\pi) = f(x)$. This might entail changing the value of f at one of the endpoints of $[-\pi, \pi]$, but that will not change the values of any of the integrals. Let $T(x)$ be any finite trigonometric series in $\cos(nx)$ and $\sin(nx)$, and let y be any real number. Using the difference of angles formulas, $T(x - y)$ is also a finite trigonometric series in $\cos(nx)$ and $\sin(nx)$ (see Exercise 8.4.6). From the orthogonality of f,

$$\int_{-\pi}^{\pi} f(x + y) T(x) \, dx = \int_{y-\pi}^{y+\pi} f(x) T(x - y) \, dx = 0.$$

Assume there is a $z \in (-\pi, \pi)$ at which $f(z) = c \neq 0$. We can assume c is positive (otherwise take the negative of f). By the continuity of f we can find a $\delta > 0$ so that $f(x) > c/2$ for all $x \in (z - \delta, z + \delta) \subseteq (-\pi, \pi)$. Since f is continuous, we also know that it is bounded on $[-\pi, \pi]$, say $|f| \leq M$. In what follows, we shall use the lower limit on f, $f \geq -M$. We choose T so that it satisfies the conditions of Lemma 8.14. We have that

$$0 = \int_{-\pi}^{\pi} f(x + z) T(x) \, dx$$

$$\geq \frac{c}{2} \int_{-\delta}^{\delta} T(x) \, dx + \int_{-\pi}^{-\delta} f(x + z) T(x) \, dx + \int_{\delta}^{\pi} f(x + z) T(x) \, dx$$

$$= \frac{c}{2} \int_{-\pi}^{\pi} T(x) \, dx + \int_{-\pi}^{-\delta} \left(f(x + z) - c/2 \right) T(x) \, dx$$

$$+ \int_{\delta}^{\pi} \left(f(x + z) - c/2 \right) T(x) \, dx$$

$$\geq \frac{c}{2} - \left(M + \frac{c}{2} \right) \int_{-\pi}^{-\delta} T(x) \, dx - \left(M + \frac{c}{2} \right) \int_{\delta}^{\pi} T(x) \, dx$$

$$\geq \frac{c}{2} - 2 \left(M + \frac{c}{2} \right) (\pi - \delta) \epsilon,$$

where we can get any positive value we wish for ϵ by suitable choice of T. This tells us that

$$0 = \int_{-\pi}^{\pi} f(x + z) T(x) \, dx \geq \frac{c}{2} > 0,$$

a contradiction. Therefore, there is no $z \in (-\pi, \pi)$ at which $f(z) \neq 0$. $\square$

Lemma 8.16 (Integrable f). *Let f be integrable on $[-\pi, \pi]$ where f is orthogonal to $\cos(nx)$, $n \geq 0$, and to $\sin(nx)$, $n \geq 1$, then f is the constant function 0 on this interval.*

Proof. We use the fact that the integral of f,

$$F(x) = \int_{-\pi}^{x} f(t) \, dt,$$

is continuous and that we can relate the Fourier coefficients of f and F by means of integration by parts (see Exercise 7.4.11 on p. 239). For $n \geq 1$, we have that

$$\int_{-\pi}^{\pi} F(x) \cos(nx) \, dx = F(x) \frac{\sin(nx)}{n} \Big|_{-\pi}^{\pi} - \frac{1}{n} \int_{-\pi}^{\pi} f(x) \sin(nx) \, dx = 0,$$

$$\int_{-\pi}^{\pi} F(x) \sin(nx) \, dx = -F(x) \frac{\cos(nx)}{n} \Big|_{-\pi}^{\pi} + \frac{1}{n} \int_{-\pi}^{\pi} f(x) \cos(nx) \, dx = 0.$$

The only nonzero Fourier coefficient of F is the constant term, say A_0. All Fourier coefficients of $F - A_0$ are zero, so F is constant. This implies that $f = F' = 0$. $\qquad\square$

Complete Orthonormal Sets

We have seen that our sequence of sines and cosines is complete. Before working with arbitrary complete orthogonal sets, it is useful to consider those sets for which each element has norm 1. This can be done just by dividing each element by its norm.

In $L^2[-\pi, \pi]$, we turn our sequence into an orthonormal set if we replace the constant function function 1 with $1/\sqrt{2\pi}$ and divide each of the other functions by $\sqrt{\pi}$. The set

$$\left\{ \frac{1}{\sqrt{2\pi}}, \frac{\cos x}{\sqrt{\pi}}, \frac{\sin x}{\sqrt{\pi}}, \frac{\cos 2x}{\sqrt{\pi}}, \frac{\sin 2x}{\sqrt{\pi}}, \frac{\cos 3x}{\sqrt{\pi}}, \dots \right\}$$

is a complete orthonormal set in $L^2[-\pi, \pi]$. The advantage of working with this orthonormal set is that the Fourier coefficient of, say, $\cos(nx)/\sqrt{\pi}$ is simply the inner product of $f \in L^2$ with $\cos(nx)/\sqrt{\pi}$.

All of the remaining results needed to establish the Riesz–Fischer theorem will be done in the context of an arbitrary Hilbert space H for which there exits a countable, complete, orthonormal set $\{\phi_1, \phi_2, \dots\}$. For $f \in H$, the real number $\langle f, \phi_k \rangle$ is called the **generalized Fourier coefficient**.

For $f \in H$, we need to show that

$$f = \sum_{k=1}^{\infty} \langle f, \phi_k \rangle \phi_k.$$

Definition: Orthonormal set

An orthogonal set Ω is **orthonormal** if each element has norm 1.

This equality is to be understood in terms of convergence in the norm. If we want to find the finite sum $\sum_{k=1}^{N} c_k \phi_k$ that most closely approximates f using our norm, we see that

$$\left\| f - \sum_{k=1}^{N} c_k \phi_k \right\|^2 = \left\langle f - \sum_{j=1}^{N} c_j \phi_j, \, f - \sum_{k=1}^{N} c_k \phi_k \right\rangle$$

$$= \langle f, f \rangle - \sum_{j=1}^{N} c_j \langle \phi_j, f \rangle - \sum_{k=1}^{N} c_k \langle f, \phi_k \rangle$$

$$+ \sum_{j=1}^{N} \sum_{k=1}^{N} c_j c_k \langle \phi_j, \phi_k \rangle$$

$$= \|f\|^2 - 2 \sum_{k=1}^{N} c_k \langle f, \phi_k \rangle + \sum_{k=1}^{N} c_k^2$$

$$= \|f\|^2 - \sum_{k=1}^{N} \langle f, \phi_k \rangle^2 + \sum_{k=1}^{N} \left(\langle f, \phi_k \rangle - c_k \right)^2. \quad (8.24)$$

As we would hope, the distance between f and $\sum_{k=1}^{N} c_k \phi_k$ is minimized when $c_k = \langle f, \phi_k \rangle$. Equation (8.24) implies the following two results, equation (8.25) and inequality (8.26), known as **Bessel's identity** and **Bessel's inequality**. They are named for Friedrich Wilhelm Bessel (1784–1846), a mathematical astronomer who spent most of his career at the Königsberg observatory (then in Germany, today it is the Russian city Kaliningrad). His most widely known mathematical work is the development of the **Bessel functions**, a complete orthonormal set for L^2 that arose in his study of planetary perturbations.

$$\left\| f - \sum_{k=1}^{N} \langle f, \phi_k \rangle \phi_k \right\|^2 = \|f\|^2 - \sum_{k=1}^{N} \langle f, \phi_k \rangle^2, \quad (8.25)$$

$$\sum_{k=1}^{N} \langle f, \phi_k \rangle^2 \leq \|f\|^2. \quad (8.26)$$

Since inequality (8.26) holds for all N, we have that

$$\sum_{k=1}^{\infty} \langle f, \phi_k \rangle^2 \leq \|f\|^2, \quad (8.27)$$

and therefore if $f \in H$, then the sum of the squares of the generalized Fourier coefficients must converge. This proves inequality (8.21) in the Riesz–Fischer theorem (Theorem 8.8 on p. 269).

Completing the Proof of the Riesz–Fischer Theorem

There are still two pieces of the Riesz–Fischer theorem that must be established. First, that the series $\sum_{k=1}^{\infty} \langle f, \phi_k \rangle \phi_k$ converges in norm to f. Second, that any series $\sum_{k=1}^{\infty} c_k \phi_k$ is in H if $\sum_{k=1}^{\infty} c_k^2$ converges. Equation (8.29), known as **Parseval's equation**, is a pleasant surprise that appears in the course of the proof. It is named for Marc-Antoine Parseval des Chênes (1755–1836).

Theorem 8.17 (Convergence of Fourier Series). *Let (ϕ_k) be a complete orthonormal set in the Hilbert space H. For each $f \in H$, we have that*

$$f = \sum_{k=1}^{\infty} \langle f, \phi_k \rangle \phi_k. \tag{8.28}$$

Furthermore, we have that

$$\sum_{k=1}^{\infty} \langle f, \phi_k \rangle^2 = \| f \|^2. \tag{8.29}$$

Finally, if (c_k) is any sequence of real numbers for which $\sum_{k=1}^{\infty} c_k^2$ converges, then

$$\sum_{k=1}^{\infty} c_k \phi_k$$

converges to an element of H.

Proof. Let

$$f_n = \sum_{k=1}^{n} \langle f, \phi_k \rangle \phi_k.$$

We need to prove that

$$\lim_{n \to \infty} \| f - f_n \| = 0.$$

We have that for $m < n$,

$$f_n - f_m = \sum_{k=m+1}^{n} \langle f, \phi_k \rangle \phi_k,$$

and, therefore,

$$\| f_n - f_m \|^2 = \sum_{k=m+1}^{n} \langle f, \phi_k \rangle^2.$$

Bessel's inequality (8.26) implies that $\sum_{k=1}^{\infty} \langle f, \phi_k \rangle^2$ converges, so we can force the distance between f_n and f_m to be as small as we wish by taking m and n

sufficiently large. That means that our sequence (f_n) is Cauchy, and therefore it must converge. Let g be its limit,

$$\lim_{n\to\infty} f_n = g = \sum_{k=1}^{\infty} \langle f, \phi_k \rangle \phi_k.$$

We fix k. Then for all $n \geq k$ we have that

$$\langle f_n, \phi_k \rangle = \left\langle \sum_{j=1}^{n} \langle f, \phi_j \rangle \phi_j, \phi_k \right\rangle$$

$$= \sum_{j=1}^{n} \langle f, \phi_j \rangle \langle \phi_j, \phi_k \rangle$$

$$= \langle f, \phi_k \rangle.$$

It follows that

$$\langle g, \phi_k \rangle = \lim_{n\to\infty} \langle f_n, \phi_k \rangle = \langle f, \phi_k \rangle,$$

and, therefore,

$$\langle g - f, \phi_k \rangle = 0, \quad \text{for all } k \geq 1.$$

Since our orthonormal set is complete, $g = f$.

We now use Bessel's equality, equation (8.25), to prove Parseval's equality:

$$\|f\|^2 - \sum_{k=1}^{\infty} \langle f, \phi_k \rangle^2 = \lim_{N\to\infty} \left(\|f\|^2 - \sum_{k=1}^{N} \langle f, \phi_k \rangle^2 \right)$$

$$= \lim_{N\to\infty} \left(\left\| f - \sum_{k=1}^{N} \langle f, \phi_k \rangle \phi_k \right\|^2 \right) = 0.$$

Finally, as seen above, if $\sum_{k=1}^{\infty} c_k^2$ converges, then the sequence of partial sums of $\sum_{k=1}^{\infty} c_k \phi_k$ is Cauchy. Since every Hilbert space is a Banach space, H is complete. Therefore this series converges to an element of H. $\square$

Exercises

8.4.1. Show that if x is orthogonal to $\vec{y}_1, \ldots, \vec{y}_n$, then x is orthogonal to any linear combination

$$a_1 \vec{y}_1 + \cdots + a_n \vec{y}_n.$$

8.4.2. Using the definition of the norm in terms of the inner product, prove that

$$\|x + y\|^2 + \|x - y\|^2 = 2 \left(\|x\|^2 + \|y\|^2 \right).$$

8.4.3. Finish the proof of Proposition 8.12 by proving the four identities in equation (8.23).

8.4.4. Use integration by parts (Exercise 7.4.12) to show that if F is a differentiable function over $[-\pi, \pi]$ and if $F' = f$ has the Fourier series representation

$$f(x) = a_0 + \sum_{n=1}^{\infty} \left(a_n \cos(nx) + b_n \sin(nx) \right),$$

then the Fourier series for F is

$$F(x) = A_0 + \sum_{n=1}^{\infty} \left(\frac{-b_n}{n} \cos(nx) + \frac{a_n}{n} \sin(nx) \right).$$

8.4.5. Explore the values of $\int_{-\pi}^{\pi}(1 + \cos x)^n dx = 2^n \int_{-\pi}^{\pi}(\cos(x/2))^n dx$ for $n = 1, 2, 3, \ldots$. Find a general formula for the value of this function of the positive integers n.

8.4.6. Using the difference of angles formula, show that if $T(x)$ is a finite trigonometric series in $\cos(nx)$ and $\sin(nx)$, then $T(x - y)$ is also a finite trigonometric series in $\cos(nx)$ and $\sin(nx)$.

8.4.7. Show that any inner product is continuous: if $x_n \to x$ and $y_n \to y$, then $\langle x_n, y_n \rangle \to \langle x, y \rangle$.

8.4.8. Let B be a Banach space whose norm satisfies the parallelogram law, (equation (8.22)). Show that if we define the inner product by

$$\langle x, y \rangle = \frac{1}{4} \left(\|x + y\|^2 - \|x - y\|^2 \right),$$

this makes B into a Hilbert space.

8.4.9. Let $C[0, 1]$ be the set of continuous functions with the max norm (see Exercise 8.3.3). Does this space satisfy the parallelogram law? Justify your answer.

8.4.10. Let x be orthogonal to each of the elements $y_1, y_2, \ldots$ and let $y = \lim_{n \to \infty} y_n$. Show that x is orthogonal to y.

9

Epilogue

Does anyone believe that the difference between the Lebesgue and Riemann integrals can have physical significance, and that whether say, an airplane would or would not fly could depend on this difference? If such were claimed, I should not care to fly in that plane.

— Richard W. Hamming[1]

Hamming's comment, though cast in a more prosaic style, echoes that of Luzin with which we began the preface. When all is said and done, the Lebesgue integral has moved us so far from the intuitive, practical notion of integration that one can begin to question whether the journey was worth the price.

Before undertaking this study of the development of analysis in the late nineteenth, early twentieth centuries, I had been under the misapprehension that what convinced mathematicians to adopt the Lebesgue integral was the newfound ability to integrate the characteristic function of the rationals. In fact, the evidence is that they were quite content to leave that function unintegrable. The ability to integrate the derivative of Volterra's function was important, but less for the fact that the Lebesgue integral expanded the realm of integrable functions than that, in so doing, this integral simplified the fundamental theorem of calculus. This begins to get to the heart of what made the Lebesgue integral so attractive: It simplifies analysis. I have included Osgood's proof to show how difficult it can be to make progress when chained to the Riemann integral and how easily such a powerful result as the dominated convergence theorem flows from the machinery of Lebesgue measure and integration.

The real significance of the Lebesgue integral was the reappraisal of the notion of function that enabled and, in turn, was promoted by its creation. Following

---

[1] See Hamming (1998) for a full elaboration on this theme.

282

Dirichlet and Riemann, mathematicians had begun to grasp how very significant it would be to take seriously the notion of a function as an arbitrary rule mapping elements of one set to another. Through the second half of the nineteenth century, they came to realize that the study of real-valued functions of a real variable is the study of the structure of $\mathbb{R}$. Set theory and the geometry of $\mathbb{R}$ took on an importance that was totally new.

This insight was solidified in Jordan's *Cours d'analyse*, the textbook of the mid-1890s that would shape the mathematical thinking of Borel, Lebesgue, and their contemporaries. Jordan established the principle that the integral is fundamentally a geometric object whose definition rests on the concept of measure. Jordan got the principle correct, but the details – his choice of a measure based on finite covers – were discovered to be flawed. This was the direct inspiration for the work of Borel and Lebesgue. The great simplification that came out of their work was the recognition that what happens on a set of measure zero can be ignored.

Chapter 8 hints at the fundamental shift that occurred in the early twentieth century when the theory of functions as points in a vector space – the basis for functional analysis – emerged. To appreciate this new field, we must view it in the context of all that was happening in mathematics. This book has followed a single strand from the historical development of mathematics and so ignored much else that was happening, influencing and being influenced by the development of the theory of integration. We saw a hint of this in Borel's work in complex analysis that led him to the Heine–Borel theorem and in occasional references to multidimensional integrals. But we have ignored the entire development of complex analysis, the insights into the calculus of variations, the study of partial differential equations, and the nascent work in probability theory. Most of the nineteenth-century mathematicians working in what today we call real analysis were working broadly on analytical questions, and especially on the practical questions of finding and describing solutions to situations modeled by partial differential equations.

The manuscript that Joseph Fourier deposited at the Institut de France on December 21, 1807, the event that I identitified in *A Radical Approach to Real Analysis* as the beginning of real analysis, showed how to solve Laplace's equation, a partial differential equation,

$$\frac{\partial^2 z}{\partial x^2} + \frac{\partial^2 z}{\partial w^2} = 0.$$

The same trick that was used there, the assumption that

$$z(x, w) = \sum_k a_k \psi_k(x)\phi_k(w),$$

would be shown to work on many other partial differential equations. The difficulty would come in expressing the distribution of values along the boundary in terms of the basis functions, ψ_k,

$$f(x) = z(x, 0) = \sum_k \left(a_k \phi_k(0) \right) \psi_k(x).$$

If the sequence (ψ_k) forms a complete, orthogonal basis with respect to an appropriately defined inner product over some Hilbert space, then we are home free. The Hilbert spaces that we saw in Section 8.4 unify our understanding of many different partial differential equations.

Exact solutions to partial differential equations are rare. But as physicists, astronomers, and mathematicians came to realize, it is often possible, even when an exact solution cannot be found, to say something definite about the existence, uniqueness, and stability of solutions. Often, this is all that really is needed.

The classic example of this is Henri Poincaré's *Les Méthodes Nouvelles de la Mécanique Céleste* (New methods of celestial mechanics). Newton had established the physical laws that govern the motion of the sun and the planets within our solar system, laws that are easily interpreted into the language of partial differential equations. But even a system as simple as three bodies – the sun, the earth, and the moon – does not yield an exact solution. From 1799 until 1825, Laplace published his five-volume *Traité de Mécanique Céleste* (Treatise on celestial mechanics), greatly extending and simplifying the tools needed to study celestial motion. Yet even he was stymied by the three-body problem. The great problem that neither Newton nor Laplace could solve was to determine whether or not our solar system is intrinsically stable. Orbits will vary under the pull of the constantly shifting planets. Do we need to worry that next year the earth will reach a tipping point where it suddenly begins a rapid spiral into the sun?

Poincaré solved this problem. He did not find an exact solution. Rather, he invented entirely new tools for analyzing solutions to partial differential equations, tools that enabled him to conclude that our solar system is indeed stable. On each revolution, planetary orbits will return to within a certain clearly delimited window. Poincaré's work was published in three volumes appearing in 1892, 1893, and 1899. It considered each orbit as a point in space and examined the possible perturbations of these points.

By 1910, there was strong evidence coming from many directions that it was very fruitful to work with functions as single points in an abstract function space. As we have seen, such spaces have many possible definitions for distance, the L^p norms giving some indication of what is possible. Topology begins when we study properties that remained unchanged as we modify our definition of distance, properties that include compactness, completeness, and separability.[2]

An even more fundamental transformation in our understanding of functions was about to take place. At the start of Chapter 3, I described the power of the realization that two very different mathematical patterns exhibit points of similarity. Sets of functions were now beginning to look like Euclidean spaces with their notions of dimension, distance, and orthogonality. This suggests that we might usefully apply the tools that have been used to analyze Euclidean space, tools that include linear transformations with their eigenvalues and eigenvectors and the algebraic structures that have been created to probe and explain symmetries.

This was done with astounding if often complex results. The twentieth century saw new and powerful methods: spectral theory, algebraic topology, harmonic analysis, K-theory, and C^*-algebras. This is the mathematics at the heart of quantum mechanics, string theory, and modern physics. These are the methods that would lead to breakthroughs in number theory. Analysis today looks very little like calculus. Today's advances are more likely to look like geometry and algebra, but acting on strange, twisted spaces with operations that are only vaguely reminiscent of multiplication or addition. The floodgates had been opened.

---

[2] A set is **separable** if it contains a subset that is both countable and dense.

Appendix A

Other Directions

A.1 The Cardinality of the Collection of Borel Sets

The collection[1] of Borel sets, $\mathcal{B}$, was defined on p. 127 as the smallest σ-algebra that contains all closed intervals. This is a top-down definition, starting with all σ-algebras and finding the smallest that contains all closed intervals. Our approach to proving that this collection has cardinality $\mathfrak{c}$ is to seek a bottom-up definition, recursively adding Borel sets until we have them all and showing that at each step the collection we have built has cardinality $\mathfrak{c}$. The bottom-up approach runs into difficulties with transfinite induction. Today, these difficulties are bulldozed by the machinery of transfinite ordinals, machinery that was built from the study of problems such as this. While efficient, it can obscure the real issues. We shall instead proceed naively as did the early investigators and see where we get into trouble.

We start with B_0, the set of all closed intervals. Since each closed interval is determined by its endpoints, B_0 has cardinality $\mathfrak{c}$. Let B_1 be the collection that consists of all countable unions of sets from B_0, all countable intersections of sets from B_0, and all differences of sets in B_0. Since we have defined countable to include finite, B_0 is contained in B_1. The cardinality of the collection of countable unions is at most the cardinality of the collection of sequences of sets taken from B_0. The collection of sequences of sets from B_0 is in one-to-one correspondence with the set of mappings from $\mathbb{N}$ to B_0. (For each positive integer, choose a set to go in that position.) It follows that the collection of unions has cardinality at most $\left| B_0^{\mathbb{N}} \right| = \mathfrak{c}$. The same is true for the cardinality of the collection of countable intersections. And the collection of differences has cardinality bounded by $\mathfrak{c} \cdot \mathfrak{c} = \mathfrak{c}$. Since the cardinality of B_1 is at least the cardinality of B_0, the cardinality of B_1 is exactly $\mathfrak{c}$.

---

[1] To avoid confusion, from here on we shall use "collection" when we are speaking of a set of sets.

We proceed inductively, defining B_n to be the collection of all countable unions of sets from B_{n-1}, all countable intersections of sets from B_{n-1}, and all differences of sets in B_{n-1}. Again, $B_{n-1} \subseteq B_n$, and, using the same argument given before, the cardinality of B_n is $\mathfrak{c}$. It might seem that we get all Borel sets if we simply take the union $\bigcup_{n=0}^{\infty} B_n$, but there are many Borel sets that are not yet accounted for. It can be shown that for each positive integer n, there is a Borel set that is in B_n and not in B_{n-1}. Demonstrating the existence of such a set is not easy. It was proven by Lebesgue in 1905 (Lebesgue 1905c), at the same time that he proved almost everything else in this appendix. For our purposes, we shall simply accept the existence of this Borel set and move on. Since there is a Borel set in $B_n - B_{n-1}$, there is at least one Borel set in $B_n - B_{n-1}$ that is contained within the interval $(n, n + 1)$ (see Exercise A.1.1). Call it E_n. It follows that $E = \bigcup_{n=1}^{\infty} E_n$ is a Borel set, but it is not in any of the B_n, and so it is not in the collection $\bigcup_{n=1}^{\infty} B_n$. We need to define a larger collection of Borel sets, B_{∞}, the collection of all countable unions, countable intersections, and differences of sets in $\bigcup_{n=1}^{\infty} B_n$. We note that the cardinality of B_{∞} is $\mathfrak{c}$.

We still are not done. We can find a countable collection of sets in B_{∞} whose union is not in B_{∞}. We need another collection that consists of all countable unions of sets from B_{∞}, all countable intersections of sets from B_{∞}, and all differences of sets in B_{∞}. We shall call this collection of sets $B_{\infty+1}$, and now we begin to see the problem. We are going to get many transfinite subscripts.

In set theory, this first infinite subscript is usually denoted by ω rather than ∞ to avoid confusion with the transfinite cardinal numbers for which adding 1 results in no change ($\aleph_0 + 1 = \aleph_0$). The subscript ω or ∞ is referred to as the **first transfinite ordinal number**, distinguishing these from the transfinite cardinal numbers. Because we do not really need a new symbol for infinity – we shall use ∞ consistently in this appendix – we shall stick with ∞. We are in good company. Our notation is what Cantor used in his early explorations of transfinite induction when he attempted to classify second species sets by continuing the notion of derived sets into transfinite iterations: If $S^{(n)}$ is the nth derived set of S, then $S^{(\infty)} = \bigcap_{n=1}^{\infty} S^{(n)}$ and $S^{(\infty+1)}$ is the derived set of $S^{(\infty)}$.

We continue, aware that we have moved into the realm of transfinite numbers where great care must be taken. By taking unions, intersections, and differences of sets in $B_{\infty+1}$, we get $B_{\infty+2}$, and so on through $B_{\infty+n}$ for all $n \in \mathbb{N}$. Again, Lebesgue showed that each $B_{\infty+n}$ is a proper subcollection of $B_{\infty+n+1}$, and so $\bigcup_{n=0}^{\infty} B_{\infty+n}$ does not contain all of the Borel sets. The collection of all unions, intersections, and differences of sets in $\bigcup_{n=1}^{\infty} B_{\infty+n}$ can be denoted by $B_{\infty+\infty}$, or, more succinctly, as $B_{\infty \cdot 2}$. This collection also has cardinality $\mathfrak{c}$. Of course, from here we get $B_{\infty \cdot 2+1}$ and so on through $B_{\infty \cdot 2+n}$, $n \in \mathbb{N}$. It should by now be clear how we can build $B_{\infty \cdot n}$ for any $n \in \mathbb{N}$. What about the set built from all unions, intersections, and differences of sets in $\bigcup_{n=1}^{\infty} B_{\infty \cdot n}$? We shall call this B_{∞^2}. Continuing in this way, it is possible to define $B_{p(\infty)}$, where p is any polynomial with nonnegative integer coefficients.

We still are not done. We can define the collection of unions, intersections, and differences of sets in $\bigcup_{n=1}^{\infty} B_{\infty^n}$ to be B_{∞^∞}. The subscript ∞^∞ should not be confused with $\aleph_0^{\aleph_0}$. The former still corresponds to a countable union of countable sets, and so is a countable ordinal (an ordinal number that can be reached in countably many steps); the latter is the cardinality of the set of mappings from $\mathbb{N}$ to $\mathbb{N}$, and this is c. In fact, we can build towers of infinities, ∞^{∞^∞}, $\infty^{\infty^{\infty^\infty}}$, ... Their limit is an infinite tower of infinites,

$$T = \infty^{\infty^{\infty^{\cdots}}}.$$

At each stage, we have taken countable unions, countable intersections, and differences of Borel sets chosen from a collection with cardinality c, and so the cardinality of each collection of Borel sets remains c. The cardinality of the collection B_T is c, and it is still not the last such collection. We can add 1 to this subscript.

We have barely scratched the surface. Every transfinite ordinal we have seen can be reached in countably many steps. Lebesgue was able to show that no matter how large a subscript we consider, if it can be reached in countably many steps, then B with that subscript cannot contain all Borel sets. It should be clear that we never get to the end.

Nevertheless, we can talk about Ω, the smallest set of ordinal numbers that contains 0, is closed under addition by 1, and is closed under limits of monotonically increasing sequences of elements of this set. In fact, Ω is the set of all ordinal numbers that can be reached in countably many steps. This includes both those that can be described using finitely many symbols, such as $T^{T^T} + 1$, and those that cannot. Most famous among the latter is the Church–Kleene ordinal, ω_1^{CK}. This is defined to be the smallest ordinal for which the ordinals that are less than it cannot all be described using a finite number of symbols. (Note that we have just defined the Church–Kleene ordinal using a finite number of symbols, which is why it is incorrect to define it as the smallest ordinal that cannot be described using a finite number of symbols.)

The set Ω is well ordered, and its cardinality is $\aleph_1$, the smallest cardinal strictly larger than $\aleph_0$. The proofs of these statements are not difficult, but do require a formal treatment of ordinals and limit sequences.[2]

By the same inductive arguments we have been using up until this point, the cardinality of B_c is c for every $c \in \Omega$. Furthermore, $\bigcup_{c \in \Omega} B_c \subseteq B$, the collection of Borel sets. It only remains to show that every Borel set is in B_c for some $c \in \Omega$. This

---

[2] See, for example, Devlin (1993).

means that we need to show that $\bigcup_{c \in \Omega} B_c$ is a σ-algebra. If S, T are in $\bigcup_{c \in \Omega} B_c$, then we can find $a, b \in \Omega$ so that $S \in B_a$, $T \in B_b$. Choose which ever subscript is larger, say $a \geq b$, and it follows that $S, T \in B_b$, so $S - T \in B_{b+1}$. Given $\{S_n\}$, a countable collection of sets in $\bigcup_{c \in \mathcal{C}} B_c$, we choose $a_n \in \Omega$ so that $S_n \in B_{a_n}$. Ordering these so that $a_1 \leq a_2 \leq \cdots$ and setting λ as the limit of the a_n, we have that $S_n \in B_\lambda$ for all $n \geq 1$, and therefore both the union and the intersection of the S_n are contained in $B_{\lambda+1}$.

It follows that the collection of Borel sets, $\mathcal{B}$, is the union of $\aleph_1$-many collections of cardinality $\mathfrak{c}$, and therefore $\mathcal{B}$ has cardinality $\mathfrak{c}$. One way to see this is to note that we can assign to each Borel set a unique ordered pair, (c, d), whose first coordinate, c, is the ordinal that identifies the first collection of Borel sets, B_c, in which this set appears, and whose second coordinate identifies which of the Borel sets in B_c we have chosen. The first coordinate is chosen from a set of order $\aleph_1$; the second coordinate is chosen from a set of order $\mathfrak{c}$. The set of all such pairs has cardinality $\aleph_1 \cdot \mathfrak{c} \leq \mathfrak{c} \cdot \mathfrak{c} = \mathfrak{c}$. It follows that there are at most $\mathfrak{c}$ Borel sets.[3]

Connection to Baire

It was this 1905 paper (Lebesgue 1905c) in which Lebesgue solved the question of the existence of functions in class $3, 4, \ldots$ (see definition of class, p. 115). Not only was he able to prove the existence of functions in each finite class, he was able to prove the existence of functions in class ∞, functions that are not in class n for any finite n but are limits of functions taken from the union of all finite classes. Just as with Borel sets, there are functions that are not in class ∞ but are limits of functions in class ∞, that is to say, functions in class $\infty + 1$. The entire process repeats. For every countable ordinal, there are functions in that class.

Exercises

A.1.1. Using an appropriate linear function composed with the arctangent function, construct a one-to-one, onto, and continuous function, f, from $\mathbb{R}$ to $(n, n + 1)$. Show that for each $n \geq 0$, S is a set in B_n if and only if $f(S)$ is a set in B_n. Use this result to show that if E is in $B_n - B_{n-1}$, then $f(E)$ is also in $B_n - B_{n-1}$.

A.1.2. Assume $E_n \in B_n - B_{n-1}$ for each positive integer n. Prove that $\bigcup_{n=1}^{\infty} E_n$ is not an element of $\bigcup_{n=1}^{\infty} B_n$.

---

[3] Dave Renfro has pointed out that there is another totally different approach to proving that $|\mathcal{B}| = \mathfrak{c}$. See Renfro (2007).

A.2 The Generalized Riemann Integral

It might seem that the problems of the fundamental theorem of calculus were put to rest with the discovery of the Lebesgue integral, but there is one nagging loose end. It concerns the integrals of Exercise 6.3.5 on p. 189. If we begin with the function f defined by

$$f(x) = x^2 \sin(x^{-2}), \quad x \neq 0; \quad f(0) = 0,$$

then f is differentiable for all x:

$$f'(x) = 2x \sin(x^{-2}) - 2x^{-1} \cos(x^{-2}), \quad x \neq 0; \quad f'(0) = 0.$$

The derivative is not continuous at $x = 0$. It is not even bounded in any neighborhood of $x = 0$. But even though the Riemann integral of f' does not exist, the improper integral is always finite because for $a > 0$,

$$
\begin{aligned}
\int_0^a f'(x)\,dx &= \lim_{\epsilon \to 0^+} \int_\epsilon^a f'(x)\,dx \\
&= \lim_{\epsilon \to 0^+} \big(f(a) - f(\epsilon)\big) \\
&= f(a) - f(0) = f(a).
\end{aligned}
$$

Surely, this should be an example of the fundamental theorem of calculus (evaluation part) in action.

But it is not. As shown in Exercise 6.3.5, f' is not Lebesgue integrable over $[0, a]$. The problem is that the nonnegative part of this function has an integral that is not bounded.

If we go back to the evaluation part of the fundamental theorem of calculus, Theorem 7.18, we see that the only assumption that this theorem requires of f is to be absolutely continuous. Our f is not. Given any $\delta > 0$, choose K so that

$$\frac{1}{\sqrt{(K - 1/2)\pi}} < \delta.$$

Define

$$a_k = \frac{1}{\sqrt{(k + 1/2)\pi}} \quad \text{and} \quad b_k = \frac{1}{\sqrt{(k - 1/2)\pi}}.$$

For every $N \geq K$, the open intervals $\big\{(a_k, b_k)\big\}_{k=K}^N$ are pairwise disjoint and their union is contained in $(0, \delta)$, but

$$\sum_{k=K}^N \big| f(b_k) - f(a_k) \big| = \sum_{k=K}^N \left(\frac{1}{k + 1/2} + \frac{1}{k - 1/2} \right) \geq \sum_{k=K}^N \frac{2}{k}.$$

The sum of the oscillations is unbounded. This tells us that f cannot be expressed as a Lebesgue integral of any function.

Maybe absolute continuity is too strict. Is it possible to define an integral so that if a function is differentiable on the interval $[a, b]$, then its derivative is always integrable and the evaluation part of the fundamental theorem of calculus always holds?

In fact, this is possible. Several mathematicians, beginning with Arnaud Denjoy in 1912, found ways of extending the Lebesgue integral. Denjoy's original definition was greatly simplified by Luzin. In 1914, Oskar Perron came up with a different formulation. Jaroslav Kurzweil, in 1957, discovered an extension of the Riemann integral that, in the 1960s, Ralph Henstock would rediscover, realizing that it could do the trick. The integrals of Denjoy, Perron, and Kurzweil–Henstock are equivalent, though this would not be established until the 1980s. The description we shall use is that of Kurzweil and Henstock. Because of the many people to whom this integral could be attributed, we shall refer to it simply as the **generalized Riemann integral**.

Recall that a tagged partition of $[a, b]$ is a partition, $(x_0 = a < x_1 < \cdots < x_n = b)$, of $[a, b]$ together with a set of tags, $\{x_1^*, x_2^*, \ldots, x_n^*\}$, one value taken from each of intervals, $x_{j-1} \leq x_j^* \leq x_j$. A function fails to be Riemann integrable when we cannot control the oscillation over an interval simply by restricting the length of the interval. The Kurzweil–Henstock idea is to tie the allowed length of the interval to the choice of tag.

Specifically, given an error-bound $\epsilon > 0$ (how much the approximating sum is allowed to deviate from the desired value), we define a **gauge** δ_ϵ, a mapping from the set of possible tags, $[a, b]$, to the set of positive real numbers, chosen so that problematic points are assigned appropriately small values. We restrict ourselves to those Riemann sums for which $x_j - x_{j-1} \leq \delta_\epsilon(x_j^*)$. If we can find a suitable gauge for every $\epsilon > 0$, then the generalized Riemann integral exists.

Definition: Generalized Riemann integral

The generalized Riemann integral of a function f over the interval $[a, b]$ exists and has the value V if for every error-bound $\epsilon > 0$ there is a gauge function δ_ϵ, $\delta_\epsilon(x) > 0$ for all $x \in [a, b]$, such that for any tagged partition $(x_0 = a < x_1 < \cdots < x_n = b)$, $\{x_1^*, x_2^*, \ldots, x_n^*\}$, $x_{j-1} \leq x_j^* \leq x_j$, with $|x_j - x_{j-1}| < \delta_\epsilon(x_j^*)$ for all j, the corresponding Riemann sum lies within ϵ of the value V:

$$\left| \sum_{j=1}^{n} f(x_j^*)(x_j - x_{j-1}) - V \right| < \epsilon.$$

Dirichlet's Function

To see that this works for Dirichlet's function, the characteristic function of the rationals over $[0, 1]$, we let $(q_n)_{n=1}^{\infty}$ be an ordering of the set of rational numbers in $[0, 1]$. We define

$$\delta_{\epsilon}(x) = \begin{cases} \epsilon/2^n, & x = q_n \in \mathbb{Q}, \\ 1, & x \notin \mathbb{Q}. \end{cases}$$

The sum of the lengths of the intervals for which the tag is rational is strictly less than ϵ, and therefore

$$0 \le \sum_{j=1}^{n} \chi_{\mathbb{Q}}(x_j^*)(x_j - x_{j-1}) < \epsilon.$$

The generalized Riemann integral of $\chi_{\mathbb{Q}}$ over $[0, 1]$ exists and is equal to 0.

The Fundamental Theorem of Calculus

One of the advantages of this characterization of the generalized Riemann integral is that it yields a simple proof of the evaluation part of the fundamental theorem of calculus.

Theorem A.1 (FTC, Evaluation, Generalized Riemann Integrals). *If f is differentiable at every point in $[a, b]$, then the generalized Riemann integral of f' over $[a, b]$ exists and*

$$\int_a^b f'(x)\,dx = f(b) - f(a). \tag{A.1}$$

Proof. We need to show that for each $\epsilon > 0$, we can find a gauge, δ_{ϵ}, so that

$$\left| \sum_{j=1}^{n} f'(x_j^*)(x_j - x_{j-1}) - \big(f(b) - f(a)\big) \right| < \epsilon.$$

We choose our gauge so that for each $x \in [a, b]$, $0 < |z - x| \le \delta_{\epsilon}(x)$ implies that

$$\left| \frac{f(z) - f(x)}{z - x} - f'(x) \right| \le \frac{\epsilon}{b - a}.$$

Differentiability at x guarantees that we can do this. This is equivalent to

$$\left| f(z) - f(x) - f'(x)(z - x) \right| \le \frac{\epsilon}{b - a} |z - x|.$$

We now observe that

$$\left| \sum_{j=1}^{n} f'(x_j^*)(x_j - x_{j-1}) - \left(f(b) - f(a)\right) \right|$$

$$= \left| \sum_{j=1}^{n} f'(x_j^*)(x_j - x_{j-1}) - \sum_{j=1}^{n} \left(f(x_j) - f(x_{j-1})\right) \right|$$

$$= \left| \sum_{j=1}^{n} \left(f'(x_j^*)(x_j - x_j^*) + f'(x_j^*)(x_j^* - x_{j-1})\right) \right.$$

$$\left. - \left(f(x_j) - f(x_j^*) + f(x_j^*) - f(x_{j-1})\right) \right|$$

$$\leq \sum_{j=1}^{n} \left| f'(x_j^*)(x_j - x_j^*) - \left(f(x_j) - f(x_j^*)\right) \right|$$

$$+ \left| f'(x_j^*)(x_j^* - x_{j-1}) - \left(f(x_j^*) - f(x_{j-1})\right) \right|$$

$$\leq \sum_{j=1}^{n} \frac{\epsilon}{b - a} \left((x_j - x_j^*) + (x_j^* - x_{j-1})\right)$$

$$= \epsilon. \qquad \qquad \square$$

Since $f(x) = x^2 \sin(x^{-2})$ is differentiable everywhere, its derivative is integrable using the generalized Riemann integral.

The other part of the fundamental theorem of calculus is also true, though its proof is much subtler and we shall not pursue it here. The interested reader can find this proof in Gordon's *The Integrals of Lebesgue, Denjoy, Perron, and Henstock* (1994, p. 145).

Theorem A.2 (FTC, Antidifferentiation, Generalized Riemann Integrals). *If the generalized Riemann integral of f exists over* $[a, b]$, *then*

$$\frac{d}{dx} \int_a^x f(t)\, dt = f(x), \quad \textit{almost everywhere on } [a, b]. \tag{A.2}$$

The antiderivative is still continuous, though not necessarily absolutely continuous. The fact that the antiderivative is differentiable almost everywhere requires proof.

Comparison with the Lebesgue Integral

We have seen that there is at least one function that is integrable using the generalized Riemann integral that is not integrable using the Lebesgue integral. Are there any Lebesgue integrable functions that cannot be integrated using the generalized Riemann integral? No.

Theorem A.3 (Lebesgue $\Longrightarrow$ Generalized Riemann). *If f is Lebesgue integrable on $[a, b]$, then the generalized Riemann integral of f exists over this interval and equals the Lebesgue integral.*

Proof. Given $\epsilon > 0$, we must show how to construct a gauge function δ so that given any Riemann sum $\sum_{j=1}^{n} f(x_j^*)(x_j - x_{j-1})$ for which $x_j - x_{j-1} < \delta(x_j^*)$, we have that

$$\left| \sum_{j=1}^{n} f(x_j^*)(x_j - x_{j-1}) - \int_a^b f(x)\,dx \right| < \epsilon.$$

We begin by partitioning the range of f and identifying those values of x in $[a, b]$ at which f lies in each interval. Let β be a positive quantity to be determined later. For $-\infty < k < \infty$, define

$$E_k = \left\{ x \in [a, b] \,\middle|\, (k-1)\beta < f(x) \leq k\beta \right\}.$$

Since each E_k is measurable, we can find an open set $G_k \supseteq E_k$ for which $m(G_k - E_k) < \nu(k)$, where ν is a strictly positive function of $\mathbb{Z}$ that will be determined later.

Given $x \in [a, b]$, we define our gauge by finding the unique E_k such that $x \in E_k$ and choosing $\delta(x)$ so that

$$0 < \delta(x) < \inf_{y \in G_k^C} |x - y|. \tag{A.3}$$

The fact that G_k is open guarantees that we can choose $\delta(x) > 0$. Given a Riemann sum that satisfies this gauge, $\sum_{j=1}^{n} f(x_j^*)(x_j - x_{j-1})$, let E_{k_j} be the set that contains x_j^*. Since $|x_j - x_{j-1}| < \delta(x_j^*) < \inf_{x \in G_k^C} |x_k^* - x|$, we know that $G_{k_j} \supseteq [x_{j-1}, x_j]$.

We now take the difference between the Riemann sum and the Lebesgue integral and bound it by pieces that we can bound,

$$\left| \sum_{j=1}^{n} f(x_j^*)(x_j - x_{j-1}) - \int_a^b f(x)\,dx \right| = \left| \sum_{j=1}^{n} \int_{x_{j-1}}^{x_j} \left(f(x_j^*) - f(x) \right) dx \right|$$

$$\leq \sum_{j=1}^{n} \int_{[x_{j-1}, x_j] \cap E_{k_j}} \left| f(x_j^*) - f(x) \right| dx + \sum_{j=1}^{n} \int_{[x_{j-1}, x_j] - E_{k_j}} \left| f(x_j^*) \right| dx$$

$$+ \sum_{j=1}^{n} \int_{[x_{j-1}, x_j] - E_{k_j}} \left| f(x) \right| dx$$

$$\leq \beta \sum_{j=1}^{n}(x_j - x_{j-1}) + \sum_{k=-\infty}^{\infty} \int_{G_k - E_k} (|k| + 1)\beta \, dx + \int_{\bigcup(G_k - E_k)} |f(x)| \, dx$$

$$\leq \beta(b - a) + \sum_{k=-\infty}^{\infty} (|k| + 1)\beta \, \nu(k) + \int_{\bigcup(G_k - E_k)} |f(x)| \, dx.$$

We want to choose β and ν so that each of these pieces is strictly less than $\epsilon/3$. We pick $\beta = \epsilon/(4(b - a))$. Since f is Lebesgue integrable, so is $|f|$. Corollary 6.18 tells us there is an $\eta > 0$, such that $m(\bigcup(G_k - E_k)) < \eta$ implies that

$$\int_{\bigcup(G_k - E_k)} |f(x)| \, dx < \epsilon/3.$$

Since $m(\bigcup(G_k - E_k)) \leq \sum m(G_k - E_k) < \sum \nu(k)$, we need to choose ν so that $\sum_{k=-\infty}^{\infty} \nu(k) < \eta$, say $\nu(k) \leq \eta \cdot 2^{-|k|-2}$. And we need to choose $\nu(k)$ so that

$$\sum_{k=-\infty}^{\infty} 2(|k| + 1)\beta \, \nu(k) = \frac{\epsilon}{2(b - a)} \sum_{k=-\infty}^{\infty} (|k| + 1)\nu(k) < \frac{\epsilon}{3}.$$

This inequality will be satisfied if

$$\nu(k) \leq \frac{b - a}{3(|k| + 1)2^{|k|+1}}.$$

We choose $\nu(k)$ to be whichever of the two bounds is smaller. $\qquad\square$

Final Thoughts

Over the past few decades, several mathematicians have made the argument that the generalized Riemann integral should be introduced in undergraduate real analysis and that at the undergraduate level it should take priority over the Lebesgue integral. Their argument is based on three premises: that these students are already familiar with the Riemann integral, that the Lebesgue integral, resting as it does on the notion of measure theory, is too complicated to introduce at the level of undergraduate analysis, and that the generalized Riemann integral is much more satisfying because it can handle a strictly larger class of functions and provides a clean and simple proof of the most general statement of the evaluation part of the fundamental theorem of calculus.

I disagree. While many calculus texts introduce the Riemann integral, the fact is that Riemann's definition is both subtle and sophisticated. Its only real purpose is to explore how discontinuous a function can be while remaining integrable. For the functions encountered in the first year of calculus, which should be limited to piecewise analytic functions, there is no reason to work with tagged partitions in their full generality. Even in undergraduate analysis, it requires a certain level of mathematical maturity before students are ready to appreciate the Riemann integral.

In countering the second point, the beauty of the Lebesgue integral is that it defines the integral of a nonnegative function as the area of the set of points that lie between the x-axis and the values of f. This is the intuitive understanding of the integral that is lost in Riemann's definition. What is sophisticated about the Lebesgue integral is how this area must be defined and how the principal results are then proven. I fully agree that this is a topic that is seldom appropriate for undergraduate mathematics, but given a choice between spending time introducing measure theory or the generalized Riemann integral, I would always see the former as more useful and interesting.

Finally, there is the issue of the fundamental theorem of calculus. I find it instructive that while mathematicians of the late nineteenth century considered Volterra's example of a function with a bounded derivative that could not be integrated disturbing, most were perfectly content to live with this anomaly. It was the usefulness of measure theory, not its ability to handle integration of exotic functions, that drove the adoption of the Lebesgue integral. Regarding the fundamental theorem of calculus, it is not clear that Theorem A.1 is most useful or satisfying than Theorem 7.18. In the former, we make the assumption that f is differentiable at every point. Lebesgue's form of this theorem is based on the assumption that f is absolutely continuous. These two conditions overlap, but neither implies the other. The two theorems are not directly comparable.

There is no hope for a theorem using the generalized Riemann integral that assumes only that f is differentiable almost everywhere. Once a function is not required to be absolutely continuous, it is possible for the derivative to exist almost everywhere but $\int_a^b f'(x)\,dx \neq f(b) - f(a)$. The devil's staircase is the prime example.

It is possible, however, to weaken the assumption of differentiability at all points of $[a, b]$ in Theorem A.1. It is not too hard to show (see Exercise A.2.5) that the conclusion of the evaluation part of the fundamental theorem of calculus still holds if we assume that f is differentiable at all but a countable set of points. It is possible to push even further, exploring when the fundamental theorem of calculus holds for something called an "approximate derivative." Here the results are neither as clean nor as compelling as Theorems 7.18 and A.1, and they begin to call for ever more complex extensions of the generalized Riemann integral.

My personal conclusion is that there is no single best possible statement of the fundamental theorem of calculus that would hold if only we used the correct definition of the integral. I consider the generalized Riemann integral to be interesting and worthy of attention, but only as an appendix to the main show, the Lebesgue integral.

Exercises

A.2.1. Prove that if f is Riemann integrable over $[a, b]$, then the generalized Riemann integral of f over $[a, b]$ also exists.

A.2.2. Prove that for any gauge over any closed interval, there is at least one tagged partition that corresponds to that gauge.

A.2.3. Explain why we can choose a gauge that satisfies the bounds given in inequality (A.3).

A.2.4. Find an example of a function that is differentiable on $[0, 1]$ but not absolutely continuous on this interval. Find an example of a function that is absolutely continuous on $[0, 1]$ but not differentiable at every point of this interval.

A.2.5. Modify the proof of Theorem A.1 to weaken the assumption on f so that it is differentiable at all but countably many points in $[a, b]$.

Appendix B

Hints to Selected Exercises

Exercises that can also be found in Kaczor and Nowak are listed at the start of each section following the symbol $\boxed{\text{KN}}$. The significance of $3.1.2 = \text{II}:2.1.1$ is that Exercise 3.1.2 in this book can be found in Kaczor and Nowak, volume II, problem 2.1.1.

1.1.4 Use the trigonometric identities,

$$\sin x \ \cos y = \frac{1}{2}\left(\sin(x+y) + \sin(x-y)\right), \qquad (\text{B.1})$$

$$\cos x \ \cos y = \frac{1}{2}\left(\cos(x+y) + \cos(x-y)\right), \qquad (\text{B.2})$$

$$\sin x \ \sin y = \frac{1}{2}\left(\cos(x-y) - \cos(x+y)\right). \qquad (\text{B.3})$$

1.1.6 Write $C = A + E_1(\Delta x)$ and $I = A - E_2(\Delta x)$, where E_1 and E_2 are monotonically decreasing functions that are greater than or equal to 0 for $\Delta x > 0$.

1.1.7 In the ring between distance x and distance $x + \Delta x$, the total population can be approximated by the density, $\rho(x)$, multiplied by the area, $\pi(x + \Delta x)^2 - \pi x^2 = 2\pi x \Delta x + \pi(\Delta x)^2$.

1.1.10 Given any partition of $[-1, 0]$, let $-a$ be the left endpoint of the rightmost interval (the interval whose right endpoint is 0). Show that any Riemann sum using left endpoints for the tags is bounded above by $2 - \sqrt{a}$.

1.1.13 Lagrange's remainder theorem says that

$$F(x + t) = F(x) + t \, F(x) + \frac{t^2}{2} F''(\zeta),$$

for some ζ, $x < \zeta < x + t$. Let $x = a + (j - 1)t$.

299

1.2.3 Try to find a sequence of nested intervals $[a_1, b_1] \supseteq [a_2, b_2] \supseteq [a_3, b_3] \supseteq \cdots$ so that the endpoints are elements of the sequence and a_k equals or precedes a_{k+1}, and b_k equals or precedes b_{k+1} in the sequence.

1.2.6 Let $S_n = \sum_{k \geq n} a_k$. Show that $S_1 \geq S_2 \geq \cdots$. If A is the infimum of this sequence, then it is also the limit. Show that A must satisfy the ϵ-definition of the lim sup. Show that if A satisfies the ϵ-definition of the lim sup, then it is the greatest lower bound of the sequence of S_n.

1.2.17 Since f is differentiable at c, $\lim_{x \to c^+} \left(f(x) - f(c) \right) / \left(x - c \right)$ exists and equals $f'(c)$. Use the fact that for $x > c$, $\left(f(x) - f(c) \right) / \left(x - c \right) = f'(y)$ for some y between c and x. Show that $\lim_{y \to c^+} f'(y) = f'(c)$.

1.2.21 Show that $\sum_{n=1}^{\infty} x^n / n^2$ converges uniformly over $[-1, 1]$.

1.2.23 Use the intermediate value theorem.

2.1.5 The discontinuities of this function will occur when $(2^n x + 1)/2 \in \mathbb{Z}$.

2.1.7 Show that $g(x) = 1 - g(1 - x)$, and therefore $\int_0^1 g(x)\,dx = 1 - \int_0^1 g(x)\,dx$.

2.1.2 The key is to prove that for any two partitions P and Q, $\overline{S}(P; f) \geq \underline{S}(Q; f)$. Consider their common refinement.

2.1.14 Show that m is discontinuous at every rational number but continuous at every irrational number. For the latter, show that if x is irrational, then for every $\epsilon > 0$ there are at most finitely many y such that $\left| m(x) - m(y) \right| \geq \epsilon$.

2.1.18 Integrate from $1/n$ to 1 and then take the limit as $n \to \infty$. Rewrite as

$$\int_{1/n}^1 \left(\left\lfloor \frac{\alpha}{x} \right\rfloor - \alpha \left\lfloor \frac{1}{x} \right\rfloor \right) dx = \int_{\alpha/n}^1 \left\lfloor \frac{\alpha}{x} \right\rfloor dx - \alpha \int_{1/n}^1 \left\lfloor \frac{1}{x} \right\rfloor dx - \int_{\alpha/n}^{1/n} \left\lfloor \frac{\alpha}{x} \right\rfloor dx.$$

Show that the first two integrals cancel and the third integral approaches $\alpha \ln \alpha$ as n approaches infinity.

2.1.19 Using the fact that the oscillation over $[a, b]$ is bounded, show that for any partition P and any $\epsilon > 0$, we can find a refinement of P, $Q \supseteq P$, so that the Riemann sum for Q with tags at the left-hand endpoints is strictly larger than $\overline{S}(P; f) - \epsilon$.

2.2.5 Use the Taylor polynomial expansion with Lagrange remainder term.

2.2.6 One approach is to prove this by induction on N.

2.2.7 Uniform convergence will imply the first inequality.

KN 2.3.3 = III:1.7.7

2.3.6 8. Fix a prime p. Consider the set of rational numbers in $[0, 1]$ with numerator equal to p. What is the greatest distance between any real number in this interval and the nearest rational number with numerator equal to p?

2.3.7 Show that m is continuous at every irrational number and discontinuous at every rational.

2.3.8 Show that h is continuous at $x = 0$ and nowhere else.

2.3.9 Prove that g is continuous at c if $2^n c \notin \mathbb{Z}$ for any integer n. Given $\epsilon > 0$, let M be the smallest integer such that $2^{-M} < \epsilon$. Find $\delta > 0$ so that $2^M y$ is not an integer for any $y \in (c - \delta, c + \delta)$. Show that $\sum_{n=1}^{M} 2^{-(2n-1)} \lfloor (2^n x + 1)/2 \rfloor$ is constant on $(x - \delta, x + \delta)$. Show that $|g(x) - g(c)| < \epsilon$ for $|x - c| < \delta$.

2.3.12 Proceed by induction on k and assume that the derived set of T_{k-1} is $T_{k-2} \cup \{0\}$. Let $x \neq 0$ be an element of $T_k - T_{k-1}$. Explain why x is not in the derived set of T_{k-1}. Explain why there must be an $\epsilon > 0$ so that $T_{k-1} \cap (x - \epsilon, x + \epsilon) = \emptyset$. Show that $(x - \epsilon/2, x + \epsilon/2)$ contains at most finitely many elements of T_k.

2.3.17 Let $[a, b]$ be any closed subinterval of $[0, 1]$, $a < b$. Show that there must be a closed subinterval $[a_1, b_1] \subseteq [a, b]$, $a_1 < b_1$, on which the oscillation stays less than 1. Show that this has a closed subinterval $[a_2, b_2] \subseteq [a_1, b_1]$, $a_2 < b_2$ on which the oscillation stays less than $1/2$. In general, once you have found $[a_{k-1}, b_{k-1}]$, show that it must contain a closed subinterval $[a_k, b_k] \subseteq [a_{k-1}, b_{k-1}]$, $a_k < b_k$, on which the oscillation stays less than $1/k$. Let α be contained in the intersection of all of these intervals. Show that f must be continuous at α. Thus, every closed interval contains a point of continuity.

3.1.6 7. For each $x \in \mathbb{R}$, consider the set of fractions strictly less than x and of the form $\pm 2^k / b$, b odd. Does this set have a largest element?

KN 3.2.12 = I:3.2.27

3.2.3 Order the intervals so that $I_1 = (a_1, b_1)$, $I_2 = (a_2, b_2)$, ..., $I_n = (a_n, b_n)$, where $b_1 \leq b_2 \leq \cdots \leq b_n$. Why can we assume that no two of the b_i are equal? Why can we assume that $a_1 < a_2 < \cdots < a_n$? Why can we assume that $a_1 < 0 < b_1$? Now finish the proof.

3.2.6 Show that $|f(a) - f(s)| < \epsilon$ implies that there is an element larger than s that is in this set. Show that $|f(a) - f(s)| > \epsilon$ implies that there is an element smaller than s that is greater than or equal to every element in this set.

3.2.8 Show that every Cauchy sequence is bounded and that the limit point of this set of points is equal to its limit as a sequence.

3.2.9 Let S be a bounded set and consider the sequence: a_1 chosen from S, a_2 chosen from among the upper bounds of S, $a_3 = (a_1 + a_2)/2$, a_n is the average of the largest element of $\{a_1, a_2, \ldots, a_{n-1}\}$ that is less than or equal to some element of S and the smallest element of $\{a_1, a_2, \ldots, a_{n-1}\}$ that is an upper bound for S. Show that this sequence is Cauchy and therefore converges. Show that the element to which it converges must be the least upper bound of S.

3.2.10 Consider the set of left-hand endpoints of the nested intervals and prove that the least upper bound of this set must lie in all of the intervals.

3.2.12 To prove that $\sum A_n^{1/2}$ converges if such a sequence, (b_n), exists, first show that $A_n^{1/2} = \sqrt{b_n(A_n/b_n)} \leq (1/2)(b_n + A_n/b_n)$.

3.2.14 Show that $1/3$ is the only point of this set that is in the open interval $(5/18, 7/18)$.

3.3.2 Figure 3.1 shows how to get the correspondence between $\mathbb{N}$ and the positive rational numbers.

3.3.3 Pick a countable subset of $\mathbb{R}$ that is disjoint from the rationals, say $\mathbb{Q} + \pi = \{a + \pi \,|\, a \in \mathbb{Q}\}$. Define the correspondence so that if $x \notin \mathbb{Q} \cup (\mathbb{Q} + \pi)$, then x gets mapped to itself, and the union $\mathbb{Q} \cup (\mathbb{Q} + \pi)$ gets mapped to just $\mathbb{Q} + \pi$.

3.3.4 2. First find a one-to-one correspondence between $\mathbb{R}$ and $(0, 1)$. Find a way to use the arctangent function. Now find a one-to-one correspondence between the rational numbers in $(0, 1)$ and the rational numbers in $[0, 1]$. 5. You need to be able to combine every pair of real numbers into a single real number in a way that you can recover the original pair. Think of using the decimal expansions.

3.3.6 See hint to 3.3.3.

3.3.7 First define a one-to-one mapping $\phi : \mathbb{N} \times \mathbb{Z} \to \mathbb{Z}$ and then map $(a, b) \to \phi(a, \lfloor b \rfloor) + b - \lfloor b \rfloor$.

3.3.8 Explain the natural bijection between real numbers in $[0, 1]$ and the set of mappings from $\mathbb{N}$ to $\{0, 1, 2, 3, 4, 5, 6, 7, 8, 9\}$ (which has cardinality $10^{\mathbb{N}}$). Note that each rational number with a denominator that is a power of 10 – other than 0 and 1 – corresponds to two different mappings. We assume that the set of mappings is countable, and assign one such mapping (equivalently, we assign the decimal expansion of one real number) to each positive integer. We let ψ be a one-to-one and onto mapping from $\mathbb{N}$ to $10^{\mathbb{N}}$. We now define a mapping

T (equivalently, a real number) with the property that $T(n)$ (the nth digit of T) is not the image of n in $\psi(n)$. (Equivalently, it is not equal to the nth digit of $\psi(n)$.)

3.3.10 Let ψ be the mapping from A onto B. Select $A' \subseteq A$ so that $\psi : A' \to B$ is one to one. Define $\psi^1 = \psi$ and $\psi^n = \psi \circ \psi^{n-1}$. We define the one-to-one correspondence between A and B as follows: For each $a \in A'$, if we can find an n so that $\psi^n(a) \in B - A'$, then a is mapped to $\psi(a)$. If a is not in A' or there is no such n (if each time we apply ψ we get back an element of A'), then a is mapped to itself. Show that under this mapping, each element of B is the image of exactly one element of A.

3.3.13 These are countably infinite sequences of natural numbers.

3.3.15 There is always one map from the empty set to any other set, the trivial map that takes nothing to nothing.

4.1.3 To prove that nowhere dense implies the existence of subintervals with no points of S, you may find it easier to prove the contrapositive: If every subinterval of (a, b) contains at least one point of S, then every point in (a, b) is an accumulation point of S.

4.1.10 One approach is to modify Cantor's first proof that $\mathbb{R}$ is not countable. Assume $S = (s_1, s_2, \ldots)$ is perfect. Start with s_1, s_2, and s_3. For simplicity, first assume that $s_1 < s_2 < s_3$. Since s_2 is an accumulation point of S, there are infinitely many points of S between s_1 and s_3. Pick the one with smallest subscript larger than 3. If the new point is larger than s_2, discard s_3. If it is smaller, discard s_1. Continue doing this to create a sequence of nested intervals.

4.1.12 Show that if $|x - y| \le 1/3$, then $|DS(x) - DS(y)| \le 1/2$. What if $|x - y| \le 1/9$?

4.2.3 First show that for any partition of $[a, b]$, there is a Riemann sum approximation to the integral of f' whose value is zero.

4.2.4 Note that every neighborhood of any point in SVC(4) contains at least part of one of the removed intervals, and since it is not entirely contained in this interval, it must contain an endpoint of one of the removed intervals.

4.2.5 Show that for every $\epsilon > 0$, there is some $n \in \mathbb{N}$ so that after removing all intervals of length $1/4^n$, the intervals that remain all have length strictly less than ϵ.

4.2.10 Find the open interval centered at 0. Find the open intervals on either side of $\pm 1/3$. For each $M \in \mathbb{N}$, find the open intervals on either side of $\sum_{k=1}^{M} \pm 1/3^k$.

4.2.14 Show that if $x \in S - S'$, then we cannot have points of S' that are arbitrarily close to x.

4.3.3 If we have any point in Γ_ϵ, then we can always find an increasing sequence of integers $(n_1 < n_2 < n_3 < \cdots)$ and a sequence of intervals, $[a_1, b_1], [a_2, b_2],$ $[a_3, b_3] \ldots$, for which $|f_{n_k}|$ is greater than or equal to ϵ at all points in $[a_k, b_k]$. Think about our three examples.

4.3.5 Since F_1 has finite outer content, it must be bounded, say $F_1 \subseteq [a, b]$. Define $G_k = [a, b] - F_k$.

4.3.10 Show that for every n, the oscillation of g_n at $x = \pm 1/m$ is $1/m$.

4.4.2 Consider the examples you know for which the limit of the integrals does not equal the integral of the limit.

4.4.5 The Cantor set consists of numbers that can be represented in base 3 without the use of the digit 1. The set C does include numbers that have a 1 in their base 3 representation. How many 1s?

4.4.10 Show that the series of the derivatives of $n^{-2} f(x - r_n)$ converges uniformly.

4.4.13 Let $R_N + \mathbb{Q} = \{r_k + q \mid 1 \leq k \leq N, \ q \in \mathbb{Q}\}$. Show that the characteristic function of this set is in class 2.

(KN) 5.1.3 = III:1.7.1, 5.1.4 = III:1.7.2, 5.1.1 = III:1.7.3, 5.1.6 = III:1.7.8, 5.1.7 = III:1.7.13, 5.1.8 = III:1.7.14, 5.1.9 = III:1.7.15, 5.1.10 = III:1.7.15

5.1.1 We know that for any $\sigma > 0$, the set of points with oscillation $\geq \sigma$ must have outer content 0. Thus for each $k \geq 1$, we can find a closed interval (not a single point) on which the oscillation is less than $1/k$. Show how to find a sequence of nested closed intervals so that all points in the kth interval have oscillation less than $1/k$. Prove that the point in all of these intervals must be a point of continuity.

5.1.2 Consider the characteristic function of a suitable set.

5.1.4 The interior of S is $S - \partial S$. Show that if S is Jordan measurable, then the inner content of the interior of S is the same as the content of S.

5.1.5 $S \cup T = (S - T) \cup (S \cap T) \cup (T - S)$.

5.1.6 Let S_I denote the interior of S and S_C the closure. $S_I \cup (S \cap \partial S) = S$. $S \cup (S^C \cap \partial S) = S_C$.

5.1.12 Are there any Borel sets that are not in this σ-algebra?

5.1.14 $\left(\bigcap S_k \right)^C = \bigcup \left(S_k^C \right)$.

5.1.16 One possibility is to show that C as defined in Exercise 4.4.5 is such a set.

5.1.20 Start by showing that for each $\sigma > 0$, the set of points with oscillation strictly less than σ is an open set.

5.1.21 Let $S_{N,k}$ be the set of all x such that for all $m, n \geq N$, we have that $\left| f_m(x) - f_n(x) \right| \leq 1/k$. Show that $S_{N,k}^C$ is open, and therefore $S_{N,k}$ is closed. Show that the set of points of convergence is precisely $\bigcap_{k=1}^{\infty} \bigcup_{N=1}^{\infty} S_{N,k}$.

5.1.22 For part 1, show that the the set in question is

$$\bigcap_{k=1}^{\infty} \bigcup_{m=1}^{\infty} \bigcap_{n=m}^{\infty} \left\{ y \,\middle|\, \left| \frac{N(y, d, n)}{n} - \frac{1}{10} \right| \leq \frac{1}{k} \right\}.$$

(KN) 5.2.12 = III:2.1.3, 5.2.13 = III:2.1.4

5.2.4 Since $S \subseteq [\alpha, \beta]$, it follows that $S^C \cap [a, b] = [a, \alpha) \cup \left(S^C \cap [\alpha, \beta] \right) \cup (\beta, b]$, and these three sets are pairwise disjoint.

5.2.6 $m_e(S \cup T) \leq m_e(S) + m_e(T)$.

5.2.11 Use the assumption that the measure of a countable union of pairwise disjoint intervals is the sum of the lengths of the intervals to show that if $K = \bigcup_{j=1}^{\infty} I_j$, where the I_j are pairwise disjoint intervals, then

$$m(K) = \sup_{J \geq 1} m \left(\bigcup_{j=1}^{J} I_j \right).$$

5.2.12 Use Exercise 5.1.5 and the fact that $m(S) = \lim_{N \to \infty} m \left(\bigcup_{n=1}^{N} I_n \right)$.

5.2.13 Given any $\epsilon > 0$, choose covers U of S and V of T so that $m_e(S) > m(U) - \epsilon$ and $m_e(T) > m(V) - \epsilon$. Now use the result proven in Exercise 5.2.12.

5.2.14 Choose $\delta > 0$ that is strictly less than $\inf \left\{ |x - y| \,\middle|\, x \in S, \ y \in T \right\}$. Define $U = \bigcup_{x \in S} N_\delta(x)$ and $V = \bigcup_{y \in T} N_\delta(y)$. We can restrict the collection of covers of $S \cup T$ to those that are contained in $U \cup V$.

5.2.15 By subadditivity, $m_e(T - S) \geq m_e(T) - m(S)$. Use the result of Exercise 5.2.13 to prove the other inequality.

(**KN**) 5.3.7 = III:2.1.5, 5.3.8 = III:2.1.9, 5.3.9 = III:2.1.10, 5.3.10 = III:2.1.11, 5.3.11 = III:2.1.12, 5.3.13 = III:2.1.16, 5.3.14 = III:2.1.17, 5.3.15 = III:2.1.20, 5.3.16 = III:2.1.21, 5.3.17 = III:2.1.23, 5.3.18 = III:2.1.26

5.3.2 $S = \bigcup_{n=-\infty}^{\infty} (S \cap [n, n+1))$.

5.3.4 We already know the middle inequality. The third inequality is fairly easy to establish. The toughest part of this problem is to prove that if $S \subseteq [a, b]$, then

$$c_i(S) = b - a - c_e(S^C \cap [a, b]).$$

5.3.7 Use the fact that U is measurable and thus satisfies Carathéodory's condition.

5.3.9 Show that $m(T) \leq m_i(S)$.

5.3.11 Show that 1 implies 2 implies 3 implies 1. Show that 1 implies 4 implies 5 implies 1.

5.3.12 Every set is a countable union of bounded sets.

5.3.13 One direction, use the Carathéodory condition for both S_1 and T_1. Other direction, show that there are measurable sets S_1 and T_1 that contain S and T, respectively, and for which $m_e(S) = m(S_1)$ and $m_e(T) = m(T_1)$. Show that $m(S_1 \cap T_1) = 0$.

5.3.14 Let S_1, T_1 be the sets whose existence is established in Exercise 5.3.13. Let $C \subseteq S \cup T$ be a measurable set for which $m(C) > m_i(S \cup T) - \epsilon$. Show that

$$\begin{aligned}
m_i(S) + m_i(T) - \epsilon &\leq m_i(S \cup T) - \epsilon \\
&< m(C) \\
&= m(C \cap S_1) + m(C \cap T_1) \\
&\leq m_i(S) + m_i(T).
\end{aligned}$$

5.3.15 Show that

$$m\left(\bigcup_{k=1}^{\infty} \bigcap_{n=k}^{\infty} S_n\right) = \lim_{k \to \infty} m\left(\bigcap_{n=k}^{\infty} S_n\right),$$

$$\sup_{k \geq 1} \inf_{n \geq k} m(S_n) = \lim_{k \to \infty} \inf_{n \geq k} m(S_n).$$

5.4.4 Show that $\inf_{x \in S, \, y \in U^c} |x - y| > 0$.

(KN) 6.1.9 = III:2.2.3, 6.1.10 = III:2.2.4, 6.1.11 = III:2.2.5, 6.1.12 = III:2.2.6, 6.1.13 = III:2.2.7, 6.1.14 = III:2.2.8, 6.1.15 = III:2.2.15, 6.1.16 = III:2.2.16, 6.1.17 = III:2.2.17, 6.1.18 = III:2.2.18, 6.1.19 = III:2.2.19, 6.1.20 = III:2.2.20

6.1.9 Let $\mathcal{N}$ be a nonmeasurable set and define f so that the evaluation of $f(x)$ depends on whether or not x is in $\mathcal{N}$.

6.1.10 Use the result of Exercise 1.2.3, that every infinite sequence contains a monotonic sequence.

6.1.11 Show that $f^{-1}(U)$ is measurable for every open set U if and only if $f^{-1}\big((a, b)\big)$ is measurable for every open interval (a, b).

6.1.13 Use the result of Exercise 6.1.11.

6.1.15 In one direction, use the result of Exercise 5.4.7. In the other direction, use the result of Exercise 5.4.9.

6.1.16 Show that $\big\{y \,\big|\, f(y) > c\big\}$ is an open set. Show that for any open set U in the range of g, $\big\{x \,\big|\, g(x) \in U\big\}$ is measurable.

6.1.17 Consider the function ψ defined in Exercise 5.4.10. Let $\mathcal{M}$ be a nonmeasurable set contained in the image of the Cantor set. Show that if $g = \psi^{-1}$ and $h = \chi_{\psi^{-1}(M)}$, then g is continuous, h is measurable, and $h \circ g$ is not measurable.

6.1.18 Use Exercise 6.1.12.

6.1.19 Start with the function ψ from Exercise 5.4.10.

6.1.20 Write f' as a limit of a sequence of continuous (and hence measurable) functions.

(KN) 6.2.1 = III:2.3.1, 6.2.2 = III:2.3.2, 6.2.3 = III:2.3.3, 6.2.12 = III:2.3.5, 6.2.16 = III:2.3.7, 6.2.18 = III:2.3.8, 6.2.19 = III:2.3.11, 6.2.20 = III:2.3.12, 6.2.21 = III:2.3.13

6.2.1 Remember that for the Lebesgue integral, changing the values of a function on a set of measure zero does not change the value of the integral.

6.2.2 Use the fact that $\sum_{n=1}^{\infty} nx^{n-1} = (1 - x)^{-2}$.

6.2.3 The outer content of the Cantor set SVC(3) is zero.

6.2.5 Show that
$$\big\{x \,\big|\, f(x) < g(x)\big\} = \bigcup_{q \in \mathbb{Q}} \big(\{x \,|\, f(x) < q\} \cap \{x \,|\, q < g(x)\}\big).$$

6.2.7 Explain why if $\sum_{k=1}^{\infty} \left(\int_E f_k(x) \, dx \right)$ converges, then $\int_E \left(\sum_{k=1}^{N} f_k(x) \right) dx$ is bounded for all N.

6.2.8 Note the conditions are not sufficient to guarantee the convergence of the series $\sum f_k(x)$. Even if the series converges, it could happen that the series of integrals does not converge.

6.2.9 Consider the sequences (f_n^+) and (f_n^-).

6.2.10 Find a sequence of functions (f_n) for which $\sum_{n=1}^{\infty} \int_a^b f_n(x) \, dx$ converges conditionally and $\sum f_n(x)$ does not converge for any x in the chosen interval $[a, b]$.

6.2.12 First show that there is a bound B such that $n \cdot m(E_n) \leq B$ for all n. Show that this implies that given any $\epsilon > 0$, there is a response N so that for all $n \geq N$, $\int_{E_n} |f(x)| \, dx < \epsilon$.

6.2.17 Choose E so that its only open subset is the empty set.

6.2.18 The equality is equivalent to

$$\left| \int_E f^+(x) \, dx - \int_E f^-(x) \, dx \right| = \int_E f^+(x) \, dx + \int_E f^-(x) \, dx.$$

6.2.21 First show that if $E = \bigcup_{k=1}^{\infty} E_k$, where the E_k are pairwise disjoint measurable sets, and ϕ is any simple function, then $\int_E \phi(x) \, dx = \sum_{k=1}^{\infty} \int_{E_k} \phi(x) \, dx$. Then extend this result to any integrable function, $\int_E f(x) \, dx = \sum_{k=1}^{\infty} \int_{E_k} f(x) \, dx$.

(KN) 6.3.12 = III:2.3.14, 6.3.13 = III:2.3.15, 6.3.14 = III:2.3.18, 6.3.15 = III:2.3.19

6.3.2 $|f| = f^+ + f^-$.

6.3.4 It is enough to prove this if f is unbounded in all neighborhoods of a single point $c \in (a, b)$. To say that the improper integral of f exists is to say that

$$\lim_{\epsilon_1, \epsilon_2 \to 0} \int_a^{c-\epsilon_1} f(x) \, dx + \int_{c+\epsilon_2}^b f(x) \, dx$$

exists. Define $f_n(x) = f(x) \chi_{[a, c-1/n] \cup [c+1/n, b]}$ and use the monotone convergence theorem.

6.3.5 The integrand is the derivative of $x^2 \sin(x^{-2})$. Show that the integral of the positive part of this function is not bounded.

6.3.8 Show that the assumptions of Lebesgue's dominated convergence theorem are satisfied.

6.3.9 Use Corollary 6.17 to show that it is enough to prove this theorem when f is a simple function. Then use the definition of a simple function to show that it is enough to prove this when f is the characteristic function of a measurable set. Then use Theorem 5.11 to show that it is enough to prove this when f is the characteristic function of an interval.

6.3.11 Use Fatou's lemma.

6.3.12 Apply Fatou's lemma to both $\lim_{n \to \infty} \int_E f_n(x)\,dx$ and $\lim_{n \to \infty} \int_{E^C} f_n(x)\,dx$.

6.3.16 See hint to Exercise 6.3.9.

6.3.17 See hint to Exercise 6.3.9.

(KN) 6.4.2 = III:2.3.10, 6.4.16 = III:2.3.9, 6.4.17 = III:2.3.20, 6.4.18 = III:2.3.21, 6.4.19 = III:2.3.22, 6.4.20 = III:2.3.23, 6.4.21 = III:2.3.24

6.4.1 If $m(S) = 0$, then $[-2, 2] - S$ is dense in $[-2, 2]$.

6.4.6 Show that Baire's sequence, Example 4.5 on p. 100 converges in measure but does not converge in Kronecker's sense.

6.4.7 Consider the characteristic function of a suitable set.

6.4.9 Choose any two numbers $a, b \in \mathcal{N}$. Show that for any $\epsilon > 0$ and any $x \in \mathbb{R}$ there exist $x_1, x_2 \in N_\epsilon(x)$ for which $|f(x_1) - f(x_2)| \geq |b - a|$. Show that this implies that f is discontinuous at x.

6.4.10 Let $E_n = \left\{ x \in E \mid |f_n(x)| > \epsilon \right\}$. Show that

$$\frac{\epsilon}{1 + \epsilon}\, m(E_n) \leq \int_E \frac{|f_n(x)|}{1 + |f_n(x)|}\, dx \leq m(E_n) + \frac{\epsilon}{1 + \epsilon}\, m(E - E_n).$$

To show that the measure of E must be finite, consider the sequence defined by $f_n(x) = 1/(nx)$.

6.4.11 Consider Exercises 6.4.1 and 6.4.2.

6.4.14 If f is continuous relative to $[a, b] - E$, then $\left\{ x \in [a, b - E \mid f(x) > c \right\}$ is an open set, and therefore measurable. Any subset of a set of measure zero is measurable.

6.4.15 Show that

$$E = \bigcap_{k=1}^{\infty} \bigcup_{n=1}^{\infty} \bigcap_{m=1}^{\infty} \left\{ x \in [a, b] \mid |f_n(x) - f_m(x)| < 1/k \right\}.$$

6.4.18 The sequence $\left(\int_E f_n(x)\,dx\right)_{n=1}^{\infty}$ is bounded. If it does not converge to $\int_E f(x)\,dx$, then it has a convergent subsequence that converges to a different value. Show that this leads to a contradiction.

6.4.19 See the hint to Exercise 6.4.18.

6.4.20 By Exercise 6.1.16, $g \circ f_n$ and $g \circ f$ are measurable. Use the fact that g is uniformly continuous on $[-C, C]$ to show that $g \circ f_n$ converges in measure to $g \circ f$.

7.1.1 Start with the function $x \sin(1/x)$. Create a piecewise-defined function that is continuous at 0 but with four different Dini derivatives at $x = 0$. Add an appropriate linear function to make this function strictly increasing.

7.1.3 Assume there exists $b > a$ such that $f(b) + cb < f(a) + ca$. For each $x \in [a, b]$ show that there is a neighborhood $N_\delta(x)$ so that $y \in N_\delta(x)$, $y \neq x$, implies that
$$\frac{f(y) + cy - (f(x) + cx)}{y - x} > \frac{c - |A|}{2}.$$
These neighborhoods provide a cover of $[a, b]$. Use the Heine–Borel theorem to find a contradiction.

7.1.7 If c is a point of discontinuity of a monotonically increasing function, f, then $\lim_{x \to c^-} f(x)$ and $\lim_{x \to c^+} f(x)$ exist, and $\lim_{x \to c^-} f(x) < \lim_{x \to c^+} f(x)$.

7.1.11 Show that for each partition P of $[a, b]$ and each $\epsilon > 0$, there is a response N so that for all $n \geq N$, $|V(P, f) - V(P, f_n)| < \epsilon$. It follows that for each P and for n sufficiently large (how large depends on P), $V(P, f) < V(P, f_n) + \epsilon \leq V_a^b(f_n) + \epsilon$.

7.1.12 Explain why it is that if we can show that f has bounded variation on $[0, a]$ for some $0 < a < 1$, then f has bounded variation on $[0, 1]$. Find $V(P_N, f)$, where P_N is the partition of $[0, \sqrt{2/(3\pi)}]$ with cut points at $\sqrt{2/((4n - 1)\pi)}$ and $\sqrt{2/((4n + 1)\pi)}$, $1 \leq n \leq N$. Explain why $\lim_{N \to \infty} V(P_N, f)$ is less than the total variation on $[0, \sqrt{2/(3\pi)}]$, and it is larger than the total variation on $[0, \sqrt{2/(7\pi)}]$.

7.1.13 See the hint to Exercise 7.1.12.

7.1.16 Use the assumption that for any A, $f(x) + Ax$ is piecewise monotonic. By decreasing, we mean weakly decreasing: $y > x$ implies $f(y) \leq f(x)$.

7.1.21 Find a sequence (x_n) that converges to $1/4$ and for which $\left(DS(x_n) - DS(1/4)\right) / \left(x - 1/4\right)$ does not converge.

7.2.3 First show this result for a function g for which $D^+g \geq \epsilon > 0$. Apply this to $g(x) = f(x) + \epsilon x$.

(KN) 7.3.3 = III:2.4.7, 7.3.14 = III:2.4.3, 7.3.16 = III:2.4.4, 7.3.17 = III:2.4.5, 7.3.19 = III:2.4.6

7.3.2 See Exercise 7.1.12.

7.3.4 For the second question, from Corollary 7.3 we know that $T(x) = V_0^x(f)$ is continuous.

7.3.5 Since the summands are positive, if every finite sum is $< \epsilon$, then the infinite sum is $\leq \epsilon$.

7.3.7 Use the mean value theorem.

7.3.14 See Exercises 7.1.13, 7.3.4, 7.3.12, and 7.3.7.

7.3.17 If g is monotonically increasing and the intervals (a_k, b_k) are pairwise disjoint, then so are the intervals $\big(g(a_k), g(b_k)\big)$.

7.3.18 Show that for any $\epsilon > 0$, there is an open cover of $F(S)$ for which the sum of the lengths of the intervals is less than ϵ.

7.3.19 Let S be a measurable set. Explain why we can assume that S is bounded. From the definition of Lebesgue measure on page 138, if S is measurable, then we can find a sequence of closed sets $F_n \subseteq S$ and a set Z of measure zero so that $S = Z \cup \bigcup_{n=1}^{\infty} F_n$. Use Proposition 3.8 and Exercise 7.3.18 to finish the proof.

(KN) 7.4.7 = III:2.4.11, 7.4.9 = III:2.4.12, 7.4.10 = III:2.4.13, 7.4.11 = III:2.4.14, 7.4.12 = III:2.4.15, 7.4.13 = III:2.4.22, 7.4.14 = III:2.4.23, 7.4.15 = III:2.4.24, 7.4.17 = III:2.4.27

7.4.1 Changing the value of f at one point does not change the value of F.

7.4.2 Use the Heine–Borel theorem.

7.4.3 Show that if f is differentiable with a bounded derivative on $[a, b]$ then it is absolutely continuous on this interval.

7.4.4 Use Theorems 6.21 and 6.25.

7.4.5 Show that if $F(x) = \int_a^x f(t)\, dt$, then Theorem 7.18 implies that $\int_a^x \big(F'(t) - f(t)\big)\, dt = 0$ for all $x \in [a, b]$.

7.4.6 Use Theorem 7.18.

7.4.7 First show that if P is any partition of $[a, b]$, then $V(P, f) \leq \int_a^b |f'(x)| \, dx$. For the other inequality, define $S^+ = \{x \in [a, b] \mid f'(x) > 0\}$, $S^- = \{x \in [a, b] \mid f'(x) < 0\}$, so that

$$\int_a^b |f'(x)| \, dx = \int_{S^+} f'(x) \, dx - \int_{S^-} f'(x) \, dx.$$

Using the fact that every measurable set is almost a finite union of pairwise disjoint open intervals, show that for any $\epsilon > 0$ there is a partition P of $[a, b]$ for which

$$\int_{S^+} f'(x) \, dx - \int_{S^-} f'(x) \, dx < V(P, f) + \epsilon.$$

7.4.8 $\int_a^x f'(t) \, dt$ is absolutely continuous.

7.4.9

1. First prove this equality for the case where $g(S)$ is open. Use this to prove it for the case where $g(S)$ is closed. Then use the fact that given any $\epsilon > 0$ we can find an open set $G \supseteq g(S)$ and a closed set $F \subseteq g(S)$ for which $m(g(S)) - \epsilon < m(F) < m(G) < m(g(S)) + \epsilon$.

2. To avoid the possibility that $g^{-1}(B)$ might not be measurable, define a sequence of open sets $[c, d] \supseteq G_1 \supseteq G_2 \supseteq \cdots \supseteq B$ for which $\lim_{n \to \infty} m(G_n) = 0$. We have $B \subseteq G = \bigcap_{n=1}^\infty G_n$ and $g^{-1}(G) = \bigcap_{n=1}^\infty g^{-1}(G_n)$ is measurable. Use part 1 to show that $m\left(A \cap g^{-1}(G)\right) = 0$.

3. Again define a sequence of open sets $[c, d] \supseteq H_1 \supseteq H_2 \supseteq \cdots \supseteq C$ for which $\lim_{n \to \infty} m(H_n) = m(C)$. Then $C = B \cup \bigcap_{n=1}^\infty H_n$, where $m(B) = 0$.

7.4.10 Use Exercise 7.4.9. First prove it when f is a simple function, then when f is nonnegative, and finally for an arbitrary integrable function.

7.4.11 Let $A = \sup_{x \in [a,b]} |F(x)|$ and $B = \sup_{x \in [a,b]} |G(x)|$. Use the inequality

$$\left| F(t)G(t) - F(s)G(s) \right| \leq A \left| G(t) - G(s) \right| + B \left| F(t) - F(s) \right|$$

to prove that FG is absolutely continuous.

7.4.12 Use the result of Exercise 7.4.11.

7.4.15 For each rational number q, consider the function defined by $|f(x) - q|$. Show that

$$\lim_{h \to 0} \frac{1}{h} \int_c^{c+h} |f(t) - q| \, dt = |f(c) - q|, \quad \text{almost everywhere.} \tag{B.4}$$

Let E_q be the set of $x \in [a, b]$ for which equation (B.4) fails to hold and define

$$E = \left(\bigcup_{q \in \mathbb{Q}} E_q \right) \cup \{ x \in [a, b] \,|\, f(x) = \pm\infty \}.$$

Using the fact that we can find a rational value as close as we wish to $f(c)$, show that every point in $[a, b] - E$ is a Lebesgue point.

7.4.17 Use Luzin's theorem, Theorem 6.26.

(KN) 8.1.10–8.1.15 = III:2.5.36

8.1.1 Observe that

$$\lim_{u \to 0} \frac{\sin[(2n + 1)u]}{\sin u} = 2n + 1.$$

For $\delta \le u \le \pi/2$, we have that

$$\left| \frac{\sin[(2n + 1)u]}{\sin u} \right| \le \frac{1}{\sin \delta}.$$

8.1.8 Work backward starting with the fact that $(0, 1, 0, 1, 0, 1, \ldots)$ is $(C, 1)$. If this is the sequence $(a_1^{(1)}, a_2^{(1)}, a_3^{(1)}, a_4^{(1)}, \ldots)$, what is $(a_1, a_2, a_3, a_4, \ldots)$?

8.1.10 Define f on $(-\pi, \pi)$ by $f(0) = 0$, $f(x) = -\pi/4$ for $x < 0$, and $f(x) = \pi/4$ for $x > 0$. The Fourier series for f is a pure sine series. Restrict it to $(0, \pi)$ and then do a change of variables, replacing x by $x/2$.

8.1.11 $\cos(a + b) - \cos(a - b) = 2 \sin a \sin b$. Use the convergence proven in Exercise 8.1.10 to justify the bound.

8.2.1 Use the fact that x^a is integrable over $[0, 1]$ if and only if $a > -1$.

8.2.9 Use Proposition 8.6.

8.2.11 Use Egorov's theorem, Theorem 6.21.

8.2.13 Show that it converges in the Cesàro sense to 0 at every irrational value of x in $[0, 1]$.

8.2.15 Find a sequence of functions that converges uniformly almost everywhere but fails to converge on a set of measure zero.

8.2.16 Find a sequence (a_n) that converges only in the Cesàro sense, and define $f_n(x) = a_n$.

8.2.17 Find a suitable sequence (a_n) so that for $f_{k,n} = a_n \chi_{(\frac{k-1}{n}, \frac{k}{n})}$, the sequence $f_{1,1}, f_{1,2}, f_{2,2}, f_{1,3}, \ldots$ does not converge in the Cesàro sense at any irrational value of x in $[0, 1]$.

8.2.20 If $x < y$, rewrite the term inside the limit as $y \left(1 + (x/y)^p\right)^{1/p}$.

8.2.23 First show that for positive α and β and $1/p + 1/q = 1$, $\alpha\beta = \alpha^p/p + \beta^q/q$ if and only if $\alpha = \beta^{q-1}$.

8.2.26 Show that for $0 < p < 1$ and positive x, $f(x) = xy - x^p/p$ has its minimum at $x = y^{1/(p-1)}$.

8.2.28 Let A and B be disjoint subsets of $[a, b]$ and set $f = \alpha\chi_A$, $g = \beta\chi_B$. Compute the norms and find values of α and β for which these functions satisfy the inequality.

$\boxed{\text{KN}}$ 8.3.7 = III:2.3.20, 8.3.8 = III:2.3.31, 8.3.9 = III:2.3.32

8.3.3 $C[0, 1]$ with the max norm is a subset of $L^\infty[0, 1]$. Observe that it is closed under vector space operations. Since L^∞ is a Banach space, any sequence of continuous functions that is Cauchy in the max norm converges in the max norm to a function in L^∞. Show that this limit is a continuous function.

8.3.4 $C[0, 1]$ with the L^2 norm is a subset of $L^2[0, 1]$. Observe that it is closed under vector space operations. Since $L^2[0, 1]$ is a Banach space, any sequence of continuous functions that is Cauchy in the L^2 norm converges in the L^2 norm to a function in $L^2[0, 1]$. Find a sequence of continuous functions that converges pointwise to $\chi_{\text{SVC}(4)}$. Show that it also converges to $\chi_{\text{SVC}(4)}$ in the L^2 norm.

8.3.7 Use Egorov's theorem and Exercise 6.3.13.

8.3.8 Use the Hölder–Riesz inequality to show that the sequence (f_n) is equi-integrable, and therefore $\lim_{n\to\infty} \int_a^b |f_n(x)|^p \, dx = \int_a^b |f(x)|^p \, dx$. Use Exercise 8.3.6 to show that $\| f_n - f \|_p \to 0$. Finally use the Hölder–Riesz inequality again to show that $f_n g \to fg$ in the L^1 norm.
The condition $p > 1$ is used in the proof that (f_n) is equi-integrable.

8.3.9 Using Lebesgue's dominated convergence theorem, show that $g_n f$ converges to gf in the L^p norm. To show that $g_n f_n$ converges to gf in the L^p norm, use the fact that

$$\left|f_n g_n - fg\right|^p \le 2^p \left(\left|g_n\right|^p \left|f_n - f\right|^p + \left|fg_n - fg\right|^p\right).$$

8.4.8 Linearity is the only property of inner products that does not follow imme-
diately. To prove that $\langle x + y, z \rangle = \langle x, z \rangle + \langle y, z \rangle$, use the parallelogram law to
show that

$$\|x + y + z\|^2 = 2\|x + y\|^2 + 2\|z\|^2 - \|x + y - z\|^2$$
$$= 2\|x + z\|^2 + 2\|y\|^2 - \|x - y + z\|^2$$
$$= 2\|y + z\|^2 + 2\|x\|^2 - \|x - y - z\|^2.$$

There is a similar set of identities for $\|x + y - z\|^2$. To show that $\langle \alpha x, y \rangle =
\alpha \langle x, y \rangle$, first prove this for integer α, then rational α, and then use the continuity
of the inner product (exercise 8.4.7) to finish the proof. But be careful, we do
not yet know that this is an inner product. What properties were needed to prove
continuity?

A.2.2 Use the Heine–Borel theorem.

A.2.5 Let (t_n) be an ordering of the points at which f is not differentiable. Define
the gauge at t_n to be $\epsilon \cdot 2^{-n-1} / \left(|f(t_n)| + 1 \right)$.

Bibliography

Baire, R. 1898. Sur les fonctions discontinues développables en séries de fonctions continues. *C. R. Acad. Sci. Paris*. **126**: 884–887.

———. 1900. Nouvelle démonstration d'un théorème sur les fonctions discontinues. *Bull. Soc. Math. France*. **28**: 173–179.

Bear, H. S. 1995. *A Primer of Lebesgue Integration*. San Diego, CA: Academic Press.

Bhatia, R. 2005. *Fourier Series*. Washington, DC: The Mathematical Association of America.

Birkhoff, G. 1973. *A Source Book in Classical Analysis*. Cambridge, MA: Harvard University Press.

Borel, É. 1895. Sur quelques points de la théorie des fonctions. *Ann. Sci. École Norm. Sup*. (3). **12**: 9–55.

———. 1905. *Leçons sur les Fonctions de Variables Réelles*. Paris: Gauthier-Villars.

———. 1950. *Leçons sur la Théorie des Fonctions*. 4th ed. Paris: Gauthier-Villars. (Original work published 1898.)

Bressoud, D. M. 2007. *A Radical Approach to Real Analysis*, 2nd ed. Washington, DC: The Mathematical Association of America.

Browder, A. 1996. *Mathematical Analysis: An Introduction*. New York: Springer-Verlag.

Burk, F. 1998. *Lebesgue Measure and Integration: An Introduction*. New York: John Wiley & Sons.

Burkill, J. C. 1971. *The Lebesgue Integral*. Cambridge: Cambridge University Press.

Cauchy, A.-L. 1897. *Cours d'Analyse de l'École Royale Polytechnique*, reprinted in *Œuvres Complètes d'Augustin Cauchy*, series 2, vol. 3. Paris: Gauthier-Villars. (Original work published 1821.)

————. 1899a. *Leçons sur le Calcul Différentiel*. Reprinted in *Œuvres Complètes d'Augustin Cauchy*, series 2, vol. 4. Paris: Gauthier-Villars. (Original work published 1829.)

————. 1899b. *Résumé des Leçons données a l'École Royale Polytechnique sur le Calcul Infinitésimal*, series 2, vol. 4. Paris: Gauthier-Villars. (Original work published 1823.)

Chae, S. B. 1995. *Lebesgue Integration*, 2nd ed. New York: Springer-Verlag.

Darboux, G. 1875. Mémoire sur les fonctions discontinues. *Ann. Sci. École Norm. Sup.* 2e série. **4**: 57–112.

————. 1879. Addition au mémoire sur les fonctions discontinues. *Ann. Sci. École Norm. Sup.* 2e série. **8**: 195–202.

Dauben, J. W. 1979. *Georg Cantor: His Mathematics and Philosophy of the Infinite*. Cambridge, MA: Harvard University Press.

de Freycinet, C. 1860. *De L'Analyse Infinitésimale. Étude sur la Métaphysique du Haut Calcul*. Paris: Mallet-Bachelier.

de la Vallée Poussin, C. 1946. *Cours d'Analyse Infinitésimale*, 2nd ed. New York: Dover.

Devlin, K. 1993. *The Joy of Sets: Fundamentals of Contemporary Set Theory*. New York: Springer-Verlag.

Dieudonné, J. 1981. *History of Functional Analysis. Mathematical Studies* Vol. 49 Amsterdam: North-Holland.

Dirichlet, G. L. 1969. *Werke*, reprint. New York: Chelsea.

du Bois-Reymond, P. 1876. Anhang über den fundamenalsatz der integralrechnung. *Abhandlungen der Mathematisch-Physikalischen Classe der Königlich Bayerischen Akademie der Wissenschaften zu München*. **12**: 161–166.

————. 1880. "Der beweis des fundamentalsatzes der integralrechnung: $\int_a^b F'(x)\,dx = F(b) - F(a)$." *Math. Ann.* **16**: 115–128.

Dugac, P. 1989. Sur la correspondance de Borel et le théoreme de Dirichlet-Heine-Weierstrass-Borel-Schoenflies-Lebesgue. *Arch. Int. Hist. Sci.* **39**: 69–110.

Dunham, W. 1990. *Journey through Genius: The Great Theorems of Mathematics*. New York: John Wiley & Sons.

————. 2005. *The Calculus Gallery: Masterpieces from Newton to Lebesgue*. Princeton: Princeton University Press.

Edgar, G. A. 2004. *Classics on Fractals*. Boulder, CO: Westview Press.

Edwards, C. H., Jr. 1979. *The Historical Development of the Calculus*. New York: Springer-Verlag.

Epple, M. 2003. The end of the science of quantity: Foundations of analysis, 1860–1910. In *A History of Analysis*. Edited by H. N. Jahnke. Providence, RI: American Mathematical Society, pp. 291–325.

Ferreirós, J. 1999. *Labyrinth of Thought: A History of Set Theory and its Role in Modern Mathematics*. Basel: Birkhäuser.

Fichera, G. 1994. Vito Volterra and the birth of functional analysis. In *Development of Mathematics, 1900–1950*. Edited by J. P. Pier. Basel: Birkhäuser, pp. 171–184.

Gauss, C. F. 1876. *Werke*, vol. 3. Göttingen: Königlichen Gesellschaft der Wissenschaften.

Gordon, R. A. 1994. *The Integrals of Lebesgue, Denjoy, Perron, and Henstock*. Graduate Studies in Mathematics, vol. 4. Providence, RI: American Mathematical Society.

Grabiner, J. V. 1981. *The Origins of Cauchy's Rigorous Calculus*. Cambridge, MA: MIT Press.

Grattan-Guinness, I. 1970. *The Development of the Foundations of Mathematical Analysis from Euler to Riemann*. Cambridge, MA: MIT Press.

———. 1972. *Joseph Fourier, 1768–1830*. Cambridge, MA: MIT Press.

———. 1990. *Convolutions in French Mathematics, 1800–1840*, vols. I–III. Basel: Birkhäuser Verlag.

Hamming, R. W. 1998. Mathematics on a distant planet. *Am. Math. Mon.* **105**: 640–650.

Hardy, G. H. 1991. *Divergent Series*, 2nd ed. New York: Chelsea.

Hartman, S. and J. Mikusiński. 1961. *The Theory of Lebesgue Measure and Integration*. Translated by L. F. Boron. New York: Pergamon Press.

Hawkins, T. 1975. *Lebesgue's Theory of Integration: Its Origins and Development*, 2nd ed. New York: Chelsea.

Heine, E. 1870. Ueber trigonometrische Reihen. *J. Reine Angew. Math.* **71**: 353–365.

———. 1872. Die elemente der functionenlehre. *J. Reine Angew. Math.* **74**: 172–188.

Hermite, C. and T. J. Stieltjes. 1903–1905. *Correspondance d'Hermite et de Stieltjes*. Edited by B. Baillaud and H. Bourget. Paris: Gauthier-Villars.

Hobson, E. W. 1950. *The Theory of Functions of a Real Variable and the Theory of Fourier's Series*, 3rd ed. Washington, DC: Harren Press.

Hochkirchen, T. 2003. Theory of Measure and Integration from Riemann to Lebesgue. In *A History of Analysis*. Edited by H. N. Jahnke. Providence, RI: American Mathematical Society, pp. 197–212.

Jordan, C. 1881. Sur la série de Fourier. *C. R. Acad. Sci. Paris*. **92**: 228–230.

———. 1892. Remarqes sur les intégrales définies. *J. Math. Pures Appl.* **4**: 69–99.

———. 1893–1896. *Cours d'analyse de l'École Polytechnique*, 3 vols. Paris: Gauthier-Villars.

Kaczor, W. J. and M. T. Nowak. 2000–2003. *Problems in Mathematical Analysis*, vols. I–III. Student Mathematical Library vols. 4, 12, 21. Providence, RI: American Mathematical Society.

Lacroix, S. F. 1816. *An Elementary Treatise on the Differential and Integral Calculus*. Translated by C. Babbage, G. Peacock, and J. Herschel with appendix and notes. Cambridge: J. Deighton and Sons.

———. 1828. *Traité Élémentaire de Calcul Différentiel et de Calcul Intégral*, 4th ed. Paris: Bachelier.

Lardner, D. 1825. *An Elementary Treatise on the Differential and Integral Calculus*. London: John Taylor.

Laugwitz, D. 1999. *Bernhard Riemann 1826–1866: Turning Points in the Conception of Mathematics*. Translated by A. Shenitzer. Basel: Birkhäuser.

———. 2002. Riemann's Dissertation and Its Effect on the Evolution of Mathematics. In *Mathematical Evolutions*. Edited by A. Shenitzer and J. Stillwell. Washington DC: Mathematical Association of America, pp. 55–62.

Lebesgue, H. 1903. Sur une propriété des fonctions. *C. R. Acad. Sci. Paris*. **137**: 1228–1230.

———. 1904. Une propriété caractéristique des fonctions de classe 1. *Bull. Soc. Math. France*. **32**: 229–242.

———. 1905a. Sur une condition de convergence des séries de Fourier. *C. R. Acad. Sci. Paris*. **140**: 1378–1381.

———. 1905b. Recherches sur la convergence des séries de Fourier. *Math. Ann.* **61**: 251–280.

———. 1905c. Sur les fonctions représentables analytiquement. *J. Math. Pures Appl.* **6**: 139–216.

———. 1966. *Measure and the Integral*. Translated and edited by K. O. May. San Francisco: Holden-Day.

———. 2003. *Leçons sur l'Intégration et la Recherche des fonctions primitives*, 3rd ed. New York: Chelsea. Reprinted by American Mathematical Society. Providence, RI. (Original work published 1904.)

Lützen, J. 2003. The foundations of analysis in the 19th century. In *A History of Analysis*. Edited by H. N. Jahnke. Providence, RI: American Mathematical Society, pp. 155–196.

Luzin, N. 2002a. Function. In *Mathematical Evolutions*. Translated by A. Shenitzer and edited by A. Shenitzer and J. Stillwell. Washington, DC: Mathematical Association of America, pp. 17–34.

———. 2002b. Two letters by N. N. Luzin to M. Ya. Vygodskii. In *Mathematical Evolutions*. Translated by A. Shenitzer and edited by A. Shenitzer and J. Stillwell. Washington, DC: Mathematical Association of America, pp. 35–54.

Marek, V. and J. Mycielski. 2002. Foundations of mathematics in the twentieth century. In *Mathematical Evolutions*. Edited by A. Shenitzer and J. Stillwell. Washington, DC: Mathematical Association of America, pp. 225–246.

Medvedev, Fyodor A. 1991. *Scenes from the History of Real Functions*, transl. Roger Cooke. Basel: Birkhäuser Verlag.

Moore, Gregory H. 1982. *Zermelo's Axiom of Choice: Its Origins, Development, and Influence*. New York: Springer-Verlag.

Mykytiuk, S. and A. Shenitzer. 2002. Four significant axiomatic systmes and some of the issues associated with them. In *Mathematical Evolutions*. Edited by A. Shenitzer and J. Stillwell. Washington, DC: Mathematical Association of America, pp. 219–224.

Newton, I. 1999. The Principia: Mathematical Principles of Natural Philosophy. Translated by I. B. Cohen and A. Whitman. Berkeley, CA: University of California Press. (Originally published 1687.)

Osgood, W. F. 1897. Non-uniform convergence and integration of series term by term. *Am. J. Math.* **19**: 155–190.

Pier, J.-P., ed. 1994. Intégration et mesure 1900–1950. In *Development of Mathematics 1900–1950*. Basel: Birhäuser Verlag, pp. 517–564.

Poisson, S.-D. 1820. Suite du mémoire sur les intégrales définies. *J. de l'École Roy. Poly.* 18e cahier. **11**: 295–335.

Renfro, D. L. 2007. Message from discussion *Borel set*. Google Groups. groups.google.com/group/sci.math/msg/66168cf580929605. Accessed August 21, 2007.

Riemann, B. 1990. *Gesammelte Mathematische Werke*. Reprinted with comments by R. Narasimhan. New York: Springer-Verlag.

Rudin, W. 1976. *Principles of Mathematical Analysis*, 3rd ed. New York: McGraw-Hill.

Saxe, K. 2002. *Beginning Functional Analysis*. New York: Springer-Verlag.

Schappacher, N. and R. Schoof. 1995. *Beppo Levi and the Arithmetic of Elliptic Curves*. http://hal.archives-ouvertes.fr/hal-00129719/fr. Accessed 21 August, 2007.

Serret, J.-A. 1894. *Calcul Différentiel et Intégral*, 4th ed. Paris: Gauthier-Villars.

Shenitzer, A. and J. Steprāns. 2002. The evolution of integration. In *Mathematical Evolutions*. Edited by A. Shenitzer and J. Stillwell. Washington, DC: Mathematical Association of America, pp. 63–70.

Siegmund–Schultze, R. 2003. The origins of functional analysis. In *A History of Analysis*. Edited by H. N. Jahnke. Providence, RI: American Mathematical Society, pp. 385–408.

Struik, D. J. 1986. *A Source Book in Mathematics 1200–1800*. Princeton: Princeton University Press.

Vitali, G. 1905. Una proprietà della funzioni misurabili. *Reale Istituto Lonbardo di Scienze e Lettere*. Rendiconti (2). **38**: 600–603.

Wapner, L. M. 2005. *The Pea and the Sun: A Mathematical Paradox*. Wellesley, MA: A. K. Peters.

Weierstrass, K. T. W. 1894–1927. *Mathematische werke von Karl Weierstrass*, 7 vols. Berlin: Mayer & Muller.

Whittaker, E. T. and G. N. Watson. 1978. *A Course of Modern Analysis*, 4th ed. Cambridge: Cambridge University Press.

Index

Printed in the United States
By Bookmasters